Molecular Structures and Dimensions
Vol. 5
Solid State   Classes 1–86

Molecular Structure and Dimensions
Vol. 2
Bibliography Volume 1-2

Molecular Structures and Dimensions

Vol. 5

Bibliography 1972-73
Organic and Organometallic
Crystal Structures

Edited by
Olga Kennard, David G. Watson and William G. Town
University Chemical Laboratory, Cambridge

Springer-Science+Business Media, B.V.

Copyright © 1974 by Springer Science+Business Media Dordrecht
Originally published by International Union of Crystallography in 1974
Softcover reprint of the hardcover 1st edition 1974

First published 1974 by International Union of Crystallography.

Library of Congress catalogue card number 76–133989

ISBN 978-94-017-2346-6   ISBN 978-94-017-2344-2 (eBook)
DOI 10.1007/978-94-017-2344-2

Contents

Introduction

This volume is the fifth classified bibliography of organic and organometallic crystal structures prepared by the Crystallographic Data Centre, University Chemical Laboratory, Cambridge, and published jointly with the International Union of Crystallography.

The first four volumes covered the years 1935–1972. The present volume provides references principally to compounds whose structures were reported in the literature during 1972–1973. A few structures published prior to 1972 and omitted from the previous volumes are also included. The arrangement of entries in the 86 chemical classes is identical with the previous volumes and the reader is referred to the Introduction in Vol. 1 or Vol. 2 for a description of the practical use of the bibliography.

There are three cumulative indexes in the present volume: formula, transition metal and author indexes. All three cover the period 1935–1973 and give references to entries in Vols. 1–5.

The bibliography and indexes were prepared, checked and printed by computer techniques described in the previous volumes. Magnetic tapes of the five volumes are available and anyone interested should contact the Centre for further details.

In the present volume we have continued the special arrangement for literature search with the Centre National de la Recherche Scientifique, Paris, France. Under this arrangement reprints of papers containing crystallographic data are sent directly to the Crystallographic Data Centre, Cambridge, at the same time as they are sent out to abstractors preparing material for the Bulletin Signalétique. As a result this material is incorporated in our files at an estimated 3–6 months following the publication in the primary journals, and even before the appearance of the abstract in the Bulletin Signalétique.

In addition to the above arrangement, 10 journals, covering approximately 78 % of the crystallographic literature, are scanned directly in Cambridge. The cut-off dates for Volume 5 can be summarised as follows:

Acta Cryst. (B), part 6, page 1364, 1973
J. Chem. Soc. Dalton, part 13, page 1420, 1973
J. Chem. Soc. Perkin II, part 8, page 1200, 1973
J. Chem. Soc. Chem. Comm., part 14, page 508, 1973
J. Amer. Chem. Soc., part 11, page 3824, 1973
Acta Chem. Scand., part 3, page 1113, 1973
Inorganic Chemistry, part 6, page 1464, 1973
Tetrahedron Letters, part 28, page 2660, 1973
J. Cryst. Mol. Struct., part 2, page 123, 1973
Cryst. Struct. Comm., part 2, page 374, 1973
Other Journals: complete for 1971
$\qquad\qquad$ ca. 95% complete for 1972
$\qquad\qquad$ ca. 30% complete for 1973

The following Conference Abstracts were included in Vol. 5:

Conference Proceedings of the American Crystallographic Association, Summer 1972, Winter 1973.

The Stockholm Symposium on the Structure of Biological Molecules, July 1973.

First European Crystallography Meeting, Bordeaux, France, September 1973.

Joint meeting of the Italian and Jugoslav Crystallographic Association, Trieste, Jugoslavia, June 1973.

We would like to draw our readers' attention to the first of the numeric tables which has just been published in this series: Vol. A1 'Interatomic Distances 1960–1965, Organic and Organometallic Crystal Structures'. The new volume is a continuation of the 'Tables of Interatomic Distances and Configuration in Molecules and Ions' (Chemical Society Special Publications No. 11, London 1958; No. 18, London 1965) which covered the literature until the end of 1959. Volume A1 contains numeric data, including bond lengths, bond angles and torsion angles for about 1,300 structures analysed by X-ray and neutron diffraction. Numeric volumes for the later years are also planned.

The work of the Crystallographic Data Centre is supported by the Office for Scientific and Technical Information, Department of Education and Science, as part of the British contribution to international data activities.

We are greatly indebted to readers who have notified us of mistakes and omissions in Vols. 1–4. We have attempted to modify our procedures and are at present considering further changes including changes in the contents of forthcoming volumes. We would be grateful to readers for any suggestions on how these volumes could be further improved.

Cambridge November 1973

Olga Kennard, David G. Watson, William G. Town

Acknowledgements

The production of this bibliography was a collaborative effort by members of the Crystallographic Data Centre: Dr F. H. Allen, Mrs A. Doubleday, Dr W. D. S. Motherwell, Mrs K. Watson, Miss C. P. Way and Mrs S. Weeds.

Mrs Weeds was responsible for literature searches, primary abstracting and problems relating to chemical nomenclature.

Mrs Watson has been in charge of the encoding of information and the registration and checking of new material. In the secretarial work of documentation she has been assisted by Miss C. P. Way.

Drs Allen and Motherwell have contributed to the literature scanning, primary editing and proof reading of the final listings.

The work of the Centre was guided by members of the OSTI Scientific Advisory Committee: Professor D. W. J. Cruickshank, Mr O. S. Mills, Dr P. Owston, Professor M. R. Truter and Professor A. J. C. Wilson, FRS (Chairman).

We are grateful to the Medical Research Council for allowing a member of their External Scientific Staff (O. Kennard) to participate in this work.

We thank the University of Cambridge for the provision of accommodation in the University Chemical Laboratory and for the administration of the OSTI grant.

Our task was greatly facilitated by the excellent organisation of the Centre National de la Recherche Scientifique. We are especially grateful to Madam C. Degen of the CNRS who was responsible for the improved literature searches referred to in the Introduction.

We have used the IBM 370/165 computer of the University of Cambridge and we were greatly helped by both the programming staff and operators. We are grateful to INSPEC (Information Service in Physics, Electrotechnology and Computers & Control) and especially to Mr P. Simmons for the use of their computer typesetting programs, which they specially modified for our purposes.

The bibliography was prepared in parallel with the Organic volume of 'Crystal Data' (National Bureau of Standards, Washington D.C., USA). The third edition was published in the summer of 1972 and both publications were strengthened by this collaboration.

List of Classes

ALIPHATIC CARBOXYLIC ACID DERIVATIVES

1.C   **Oxalic acid - acetamide complex**
$C_2H_2O_4$, C_2H_5NO
For complete entry see 60.1

1.C   **Furamide - oxalic acid complex**
$C_2H_2O_4$, $2C_5H_5NO_2$
For complete entry see 60.3

1.1   **Difluoroacetamide**
$C_2H_3F_2NO$
D.O.Hughes, R.W.H.Small   *Acta Cryst. (B),* **28,** 2520, 1972

1.2   **Acetic acid - phosphoric acid complex**
$C_2H_4O_2$, H_3O_4P
P.-G.Jonsson   *Acta Chem. Scand.,* **26,** 1599, 1972

1.3   **Acetic acid - phosphoric acid complex (neutron study)**
$C_2H_4O_2$, H_3O_4P
P.-G.Jonsson   *Acta Chem. Scand.,* **26,** 1599, 1972

1.C   **Oxalic acid - acetamide complex**
C_2H_5NO , $C_2H_2O_4$
For complete entry see 60.1

1.C   **Potassium tri - hydrogen di - malonate (neutron study)**
$C_3H_3O_4^-$, $C_3H_4O_4$, K^+
For complete entry see 2.14

1.C   **3,5 - Di - iodo - L - thyronine - N - methylacetamide complex (absolute configuration)**
C_3H_7NO , $C_{15}H_{13}INO_4$
For complete entry see 60.28

1.4   **2 - (N - Nitrosomethylamino)acetamide**
$C_3H_7N_3O_2$
L.K.Templeton, D.H.Templeton, A.Zalkin   *Acta Cryst. (B),* **29,** 50, 1973
Also classified in 9, 10

1.C   **Malonic dihydrazide monohydrate**
$C_3H_8N_4O_2$, H_2O
For complete entry see 9.1

1.C **Potassium hydrogen acetylenedicarboxylate (α form)**
$C_4HO_4^-$, K^+
For complete entry see 2.15

1.C **Rubidium hydrogen acetylenedicarboxylate (β form,data set I)**
$C_4HO_4^-$, Rb^+
For complete entry see 2.16

1.C **Rubidium hydrogen acetylenedicarboxylate (β form,data set II)**
$C_4HO_4^-$, Rb^+
For complete entry see 2.17

1.5 **Acetylene dicarboxylic acid (for high - angle refinement see Jungk and Schmidt,Chem.Ber.,105,2607,1972)**
$C_4H_2O_4$
V.Benghiat, L.Leiserowitz, G.M.J.Schmidt *J. C. S. Perkin II,* 1769, 1972

1.6 **But - 3 - ynoic acid**
$C_4H_4O_2$
V.Benghiat, L.Leiserowitz *J. C. S. Perkin II,* 1772, 1972

1.7 **Tetrolic acid (α form)**
$C_4H_4O_2$
V.Benghiat, L.Leiserowitz *J. C. S. Perkin II,* 1763, 1972

1.8 **Tetrolic acid (β form)**
$C_4H_4O_2$
V.Benghiat, L.Leiserowitz *J. C. S. Perkin II,* 1763, 1972

1.9 **Maleic acid**
$C_4H_4O_4$
M.N.G.James, G.J.B.Williams
Amer. Cryst. Assoc., Abstr. Papers (Winter Meeting), 85, 1973

1.C **Potassium hydrogen fumarate**
$C_4H_4O_4$, $2C_4H_3O_4^-$, $2K^+$
For complete entry see 2.19

1.10 **(−) - Chlorosuccinic acid (absolute configuration)**
$C_4H_5ClO_4$
L.Kryger, S.E.Rasmussen, J.Danielsen *Acta Chem. Scand.,* **26,** 2339, 1972

1.11 **Fumaramic acid**
$C_4H_5NO_3$
V.Benghiat, H.W.Kaufman, L.Leiserowitz, G.M.J.Schmidt
J. C. S. Perkin II, 1758, 1972

1.C **Benzamide - succinic acid complex**
$C_4H_6O_4$, $2C_7H_7NO$
For complete entry see 60.9

1.C γ - **Aminocrotonic acid hydrobromide**
$C_4H_8NO_2^+$, Br^-
For complete entry see 48.12

1.C γ - **Aminocrotonic acid hydrobromide**
$C_4H_8NO_2^+$, Br^-
For complete entry see 48.13

1.C γ - **Aminobutyric acid**
$C_4H_9NO_2$
For complete entry see 48.18

1.C γ - **Amino** - β - **hydroxybutyric acid**
$C_4H_9NO_3$
For complete entry see 48.19

1.C γ - **Aminobutyric acid hydrochloride**
$C_4H_{10}NO_2^+$, Cl^-
For complete entry see 48.25

1.12 β - **Chloroglutaric acid anhydride**
$C_5H_5ClO_3$
F.J.Koer, A.J.de Kok, C.Romers
Rec. Trav. Chim. Pays-Bas, **91,** 691, 1972
Also classified in 38

1.13 **Mesaconic acid**
$C_5H_6O_4$
M.P.Gupta, S.R.P.Yadav *Acta Cryst. (B),* **28,** 2682, 1972

1.14 **N - (2,2,2 - Trinitro - isopropyl) - chloroacetamide**
$C_5H_7ClN_4O_7$
B.V.Gidaspov, N.V.Grigor'eva, N.V.Margolis, G.V.Makarenko, E.I.Popov,
V.F.Selivanov *Zh. Strukt. Khim.,* **12,** 1117, 1971

1.15 **DL - Methylsuccinic acid**
$C_5H_8O_4$
Y.Schouwstra *Acta Cryst. (B),* **29,** 1, 1973

1.16 γ - **Guanidinobutyric acid hydrobromide**
$C_5H_{12}N_3O_2^+$, Br^-
K.-I.Tomita *Tetrahedron Letters,* 2587, 1971
Residue 1 also classified in 8

1.C **Potassium dihydrogen trans - aconitate**
$C_6H_5O_6^-$, K^+
For complete entry see 2.28

1.17 **trans,trans - Muconic acid**
$C_6H_6O_4$
J.Bernstein, L.Leiserowitz *Israel J. Chem.,* **10,** 601, 1972

1.18   **N - Methyl - citraconamic acid**
$C_6H_9NO_3$
F.H.Allen, O.Kennard   *Cryst. Struct. Comm.*, **2**, 145, 1973

1.19   **-meso - Tartaric acid dimethyl ester**
$C_6H_{10}O_6$
J.Kroon, J.A.Kanters   *Acta Cryst. (B)*, **29**, 1278, 1973

1.20   **ε - Amino - n - caproic acid**
$C_6H_{13}NO_2$
A.Takenaka, T.Yamamoto, K.-I.Yamasaki, A.Furusaki, I.Nitta
Kwansei Gakuin Univ. Ann. Studies, **18**, 127, 1969
Also classified in 48

1.21   **N - Methyl - dipropylacetamide**
$C_7H_{19}NO$
C.Cohen-Addad, A.Grand, J.Lajzerowicz   *Eur. Cryst. Meeting,* 1973

1.22   **Dimethyl - (dimethylaminomethylene) - malonate**
$C_8H_{13}NO_4$
U.Shmueli, H.Shanan-Atidi, H.Horwitz, Y.Shvo
J. C. S. Perkin II, 657, 1973
Also classified in 3

1.C   **3 - Acetoxypropyl - trimethylammonium bromide**
$C_8H_{18}NO_2^+$, Br^-
For complete entry see 3.17

1.23   **β - Chloro - cis - cinnamic acid**
$C_9H_7ClO_2$
S.E.Filippakis, L.Leiserowitz, D.Rabinovich, G.M.J.Schmidt
J. C. S. Perkin II, 1750, 1972

1.24   **β - Chloro - trans - cinnamic acid**
$C_9H_7ClO_2$
S.E.Filippakis, L.Leiserowitz, D.Rabinovich, G.M.J.Schmidt
J. C. S. Perkin II, 1750, 1972

1.25   **N - Methyl - di - n - propylacetamide**
$C_9H_{19}NO$
A.Grand, C.Cohen-Addad   *Acta Cryst. (B)*, **29**, 1149, 1973

1.26   **β - Methyl - cis - cinnamic acid**
$C_{10}H_{10}O_2$
S.E.Filippakis, L.Leiserowitz, D.Rabinovich, G.M.J.Schmidt
J. C. S. Perkin II, 1750, 1972

1.C   **Ammonium 2,3 - di - isopropyl - maleamate**
$C_{10}H_{16}NO_3^-$, H_4N^+
For complete entry see 2.30

1.27 **Ethylenediaminetetra - acetic acid (β form)**
$C_{10}H_{16}N_2O_8$
M.F.C.Ladd, D.C.Povey *J. Cryst. Mol. Struct.*, **3**, 15, 1973
Also classified in 3

1.28 **9 - Keto - trans - 2 - decenoic acid**
$C_{10}H_{16}O_3$
D.T.Cromer, A.C.Larson *Acta Cryst. (B)*, **28**, 2128, 1972

1.C **N - Acetyl - DL - pseudo - leucyl - dimethylamide**
$C_{10}H_{20}N_2O_2$
For complete entry see 48.62

1.29 **N,N' - Diethyladipamide**
$C_{10}H_{20}N_2O_2$
E.Benedetti, M.R.Ciajolo, P.Corradini *Europ. Polymer J.*, **9**, 101, 1973

1.30 **N - Propyl - dipropylacetamide**
$C_{11}H_{23}NO$
C.Cohen-Addad, A.Grand, J.Lajzerowicz *Eur. Cryst. Meeting*, 1973

1.31 **Tripropylacetamide**
$C_{11}H_{23}NO$
C.Cohen-Addad, A.Grand, J.Lajzerowicz *Eur. Cryst. Meeting*, 1973

1.C **bis(11 - Ammonio - undecanoic acid) tetrachlorocuprate**
$2C_{11}H_{24}NO_2^+ , Cl_4Cu^{2-}$
For complete entry see 3.21

1.32 **N,N - Diethyl - phenylacetamide**
$C_{12}H_{17}NO$
G.Allegra, G.Avitabile, E.Benedetti, M.R.Ciajolo, P.Corradini, P.Ganis,
M.Goodman, A.Immirzi, C.Pedone *Acta Cryst. (A)*, **28**, S13, 1972

1.33 **Ethyl p - chloro - α - cyano - β - methyl - cis - cinnamate**
$C_{13}H_{12}ClNO_2$
T.Higuchi, W.Nagai, K.Nakatsu, T.Miwa, A.Shimada
Israel J. Chem., **10**, 221, 1972
Also classified in 19

1.34 **N - Propyl - tripropylacetamide**
$C_{14}H_{29}NO$
C.Cohen-Addad, A.Grand, J.Lajzerowicz *Eur. Cryst. Meeting*, 1973

1.C **3 - (p - Bromophenyl)propionic acid cyclohexanone - cyanohydrin ester**
$C_{16}H_{18}BrN_2O_2$
For complete entry see 21.10

1.C **Mycophenolic acid**
$C_{17}H_{20}O_6$
For complete entry see 17.19

1.35   **13 - Oxo - isostearic acid**
16 - Methyl - 13 - oxo - heptadecanoic acid
$C_{18}H_{34}O_3$
B.Dahlen   *Acta Cryst. (B),* **28,** 2555, 1972

1.36   **Isostearic acid**
$C_{18}H_{36}O_2$
S.Abrahamsson, B.-M.Lunden   *Acta Cryst. (B),* **28,** 2562, 1972

1.C   **Adiphenine hydrochloride**
2 - Diethylaminoethyl diphenylacetate hydrochloride
$C_{20}H_{26}NO_2^+$, Cl^-
For complete entry see 3.23

1.37   **Di - (2,2,6,6 - tetramethyl - 4 - piperidinyl - 1 - oxyl)suberate**
$C_{26}H_{46}N_2O_6$
A.Capiomont   *Acta Cryst. (B),* **28,** 2298, 1972
Also classified in 33, 12

1.38   **Glycerol 1,2 - di(11 - bromoundecanoate) - 3 - (p - toluenesulfonate)**
$C_{32}H_{52}Br_2O_7S$
P.H.Watts Junior, W.A.Pangborn, A.Hybl   *Science,* **175,** 60, 1972
Also classified in 11

6

ALIPHATIC CARBOXYLIC ACID SALTS
(AMMONIUM, IA, IIA METALS)

2.1 **Lithium formate monohydrate (piezoelectric form)**
CHO_2^- , Li^+ , H_2O
J.K.M.Rao, M.A.Viswamitra *Ferroelectrics,* **2,** 209, 1971

2.2 **Lithium formate monohydrate**
CHO_2^- , Li^+ , H_2O
A.Enders-Beumer, S.Harkema *Acta Cryst. (B),* **29,** 682, 1973

2.C **Thallium(i) cryptate formate monohydrate**
CHO_2^- , $C_{18}H_{36}N_2O_6Tl^+$, H_2O
For complete entry see 68.24

2.3 **Sodium oxalate monohydrate**
$C_2O_4^{2-}$, $2Na^+$, H_2O
E.A.Kuzmin, A.Ganem, V.V.Ilyukhin, N.V.Belov
Kristallografija, **17,** 276, 1972

2.4 **Ammonium oxalate monohydrate**
$C_2O_4^{2-}$, $2H_4N^+$, H_2O
J.C.Taylor, T.M.Sabine *Acta Cryst. (B),* **28,** 3340, 1972

2.5 **Ammonium oxalate monohydrate (neutron study)**
$C_2O_4^{2-}$, $2H_4N^+$, H_2O
J.C.Taylor, T.M.Sabine *Acta Cryst. (B),* **28,** 3340, 1972

2.6 **Ammonium oxalate monohydrate, partially deuterated**
$C_2O_4^{2-}$, $2H_4N^+$, H_2O
J.C.Taylor, T.M.Sabine *Acta Cryst. (B),* **28,** 3340, 1972

2.7 **Ammonium oxalate monohydrate, partially deuterated (neutron study)**
$C_2O_4^{2-}$, $2H_4N^+$, H_2O
J.C.Taylor, T.M.Sabine *Acta Cryst. (B),* **28,** 3340, 1972

2.8 **Ammonium oxalate monohydrate, perdeuterated**
$C_2O_4^{2-}$, $2D_4N^+$, D_2O
J.C.Taylor, T.M.Sabine *Acta Cryst. (B),* **28,** 3340, 1972

2.9 **Ammonium oxalate monohydrate, perdeuterated (neutron study)**
$C_2O_4^{2-}$, $2D_4N^+$, D_2O
J.C.Taylor, T.M.Sabine *Acta Cryst. (B),* **28,** 3340, 1972

2.10   **Potassium hydrogen oxalate**
Potassium binoxalate
$C_2HO_4^-$, K^+
H.Einspahr, R.E.Marsh, J.Donohue   *Acta Cryst. (B)*, **28**, 2194, 1972

2.11   **Lithium hydrogen oxalate monohydrate**
$C_2HO_4^-$, Li^+, H_2O
J.O.Thomas   *Acta Cryst. (B)*, **28**, 2037, 1972

2.12   **Ammonium hydrogen oxalate hemihydrate**
$C_2HO_4^-$, H_4N^+, $0.5H_2O$
H.Kuppers   *Acta Cryst. (B)*, **29**, 318, 1973

2.13   **Deuterated lithium hydrogen oxalate monohydrate**
$C_2DO_4^-$, Li^+, D_2O
J.O.Thomas   *Acta Cryst. (B)*, **28**, 2037, 1972

2.14   **Potassium tri - hydrogen di - malonate (neutron study)**
$C_3H_4O_4$, $C_3H_3O_4^-$, K^+
M.Currie   *J. C. S. Chem. Comm.*, 972, 1972
Residue 2 classified in 1

2.15   **Potassium hydrogen acetylenedicarboxylate (α form)**
$C_4HO_4^-$, K^+
I.Leban, L.Golic, J.C.Speakman   *J. C. S. Perkin II*, 703, 1973
Residue 1 also classified in 1

2.16   **Rubidium hydrogen acetylenedicarboxylate (β form, data set I)**
$C_4HO_4^-$, Rb^+
J.Blain, J.C.Speakman, L.A.Stamp, L.Golic, I.Leban
J. C. S. Perkin II, 706, 1973
Residue 1 also classified in 1

2.17   **Rubidium hydrogen acetylenedicarboxylate (β form, data set II)**
$C_4HO_4^-$, Rb^+
J.Blain, J.C.Speakman, L.A.Stamp, L.Golic, I.Leban
J. C. S. Perkin II, 706, 1973
Residue 1 also classified in 1

2.18   **Disodium maleate monohydrate**
$C_4H_2O_4^{2-}$, $2Na^+$, H_2O
M.N.G.James, G.J.B.Williams
Amer. Cryst. Assoc., Abstr. Papers (Winter Meeting), 85, 1973

2.C   **Chloropheniramine maleate**
$C_4H_3O_4^-$, $C_{16}H_{20}ClN_2^+$
For complete entry see 33.50

2.C   **3 - Methoxy - 10 - (3' - dimethylammonium - propyl) - phenothiazine hydrogen maleate**
$C_4H_3O_4^-$, $C_{18}H_{23}N_2OS^+$
For complete entry see 41.44

2.19   **Potassium hydrogen fumarate**
$2C_4H_3O_4^-$, $C_4H_4O_4$, $2K^+$
M.P.Gupta, N.Prasad *Acta Cryst. (B)*, **28**, 2628, 1972
Residue 2 classified in 1

2.20   **Lithium succinate**
$C_4H_4O_4^{2-}$, $2Li^+$
H.Klapper, H.Kuppers *Acta Cryst. (B)*, **29**, 21, 1973

2.21   **Lithium ammonium tartrate monohydrate**
$C_4H_4O_6^{2-}$, Li^+ , H_4N^+ , H_2O
H.Hinazumi, T.Mitsui *Acta Cryst. (B)*, **28**, 3299, 1972

2.22   **Potassium sodium tartrate tetrahydrate (at 33°C)**
$C_4H_4O_6^{2-}$, Na^+ , K^+ , $4H_2O$
Y.Shiozaki, T.Mitsui *Acta Cryst. (A)*, **28**, S185, 1972

2.23   **Potassium sodium tartrate tetrahydrate (at −29°C)**
$C_4H_4O_6^{2-}$, Na^+ , K^+ , $4H_2O$
Y.Shiozaki, T.Mitsui *Acta Cryst. (A)*, **28**, S185, 1972

2.24   **Ammonium tartrate**
$C_4H_4O_6^{2-}$, $2H_4N^+$
V.S.Yadava, V.M.Padmanabhan *Acta Cryst. (A)*, **28**, S31, 1972

2.C   **Ethylenediamine tartrate**
$C_4H_4O_6^{2-}$, $C_2H_{10}N_2^{2+}$
For complete entry see 3.6

2.C   **Calcium oxydiacetate hexahydrate**
$C_4H_{14}CaO_{10}$, H_2O
For complete entry see 67.3

2.25   **Potassium hydrogen DL - methylsuccinate**
$C_5H_7O_4^-$, K^+
Y.Schouwstra *Acta Cryst. (B)*, **28**, 2217, 1972

2.26   **Calcium(+) - allo - hydroxycitrate lactone tetrahydrate**
Hibiscus acid calcium salt tetrahydrate
$C_6H_4O_7^{2-}$, Ca^{2+} , $4H_2O$
J.P.Glusker, J.A.Minkin. F.B.Soule *Acta Cryst. (B)*, **28**, 2499, 1972
Residue 1 also classified in 38, 59

2.27   **Rubidium ammonium hydrogen fluorocitrate dihydrate**
$C_6H_5FO_7^{2-}$, Rb^+ , H_4N^+ , $2H_2O$
H.L.Carrell, J.P.Glusker *Acta Cryst. (B)*, **29**, 674, 1973

2.28   **Potassium dihydrogen trans - aconitate**

$C_6H_5O_6^-$, K^+

K.Fukuyama, S.Kashino, M.Haisa   *Chem. Letters,* 599, 1972

Residue 1 also classified in 1

2.29   **Potassium hydrogen (+) - isocitrate lactone**

$C_6H_5O_6^-$, K^+

H.M.Berman, H.L.Carrell, J.P.Glusker   *Acta Cryst. (B),* **29,** 1163, 1973

Residue 1 also classified in 38

2.30   **Ammonium 2,3 - di - isopropyl - maleamate**

$C_{10}H_{16}NO_3^-$, H_4N^+

P.J.Roberts, O.Kennard   *Cryst. Struct. Comm.,* **2,** 153, 1973

Residue 1 also classified in 1

ALIPHATIC AMINES

3.1 **bis(Methylammonium) tetrabromoferrate(iii) bromide**
CH_6N^{2+} , Br_4Fe^- , Br^-
G.D.Sproul, G.D.Stucky *Inorg. Chem.*, **11**, 1647, 1972

3.2 **Dimethylammonium chloride (α form)**
$C_2H_8N^+$, Cl^-
J.Lindgren, N.Arran *Acta Chem. Scand.*, **26,** 3043, 1972

3.3 **Ethanolamine hydrochloride**
$C_2H_8NO^+$, Cl^-
C.H.Koo, O.Lee, H.S.Shin *Daehan Hwahak Hwoejee,* **16,** 6, 1972

3.4 **Ethylenediamine (at −60°C)**
$C_2H_8N_2$
S.Jamet-Delcroix *Acta Cryst. (B),* **29,** 977, 1973

3.5 **Ethylenediammonium tetrachlorocuprate(ii)**
$C_2H_{10}N_2^{2+}$, Cl_4Cu^{2-}
G.B.Birrell, B.Zaslow *J. Inorg. Nucl. Chem.*, **34,** 1751, 1972

3.6 **Ethylenediamine tartrate**
$C_2H_{10}N_2^{2+}$, $C_4H_4O_6^{2-}$
S.Perez, J.M.Leger, J.Housty *Cryst. Struct. Comm.*, **2,** 303, 1973
Residue 2 classified in 2

3.7 **bis(n - Propylammonium) tetrachloro - manganese(ii)**
$2C_3H_{10}N^+$, Cl_4Mn^{2-}
E.R.Peterson, R.D.Willett *J. Chem. Phys.*, **56,** 1879, 1972

3.8 **tris(Trimethylammonium) catena - tri - μ - chloro - cuprate tetrachloro - cuprate**
$3C_3H_{10}N^+$, Cl_3Cu^- , Cl_4Cu^{2-}
R.M.Clay, P.Murray-Rust, J.Murray-Rust *J. C. S. Dalton*, 595, 1973

3.C **γ - Aminocrotonic acid hydrobromide**
$C_4H_8NO_2^+$, Br^-
For complete entry see 48.12

3.C **γ - Aminocrotonic acid hydrobromide**
$C_4H_8NO_2^+$, Br^-
For complete entry see 48.13

3.C γ - **Aminobutyric acid**
$C_4H_9NO_2$
For complete entry see 48.18

3.C γ - **Amino** - β - **hydroxybutyric acid**
$C_4H_9NO_3$
For complete entry see 48.19

3.C   **Hydrogen tetrakis(benzoylacetonato) europate diethylamine complex**
$xC_4H_{11}N$, $C_{40}H_{37}EuO_8$
For complete entry see 61.10

3.C   **3,5,3′ - Tri - iodothyroacetic acid - N - diethanolamine**
$C_4H_{11}NO_2$, $C_{14}H_9I_3O_4$
For complete entry see 60.26

3.9   **Tetramethylammonium trans - diaquotetrachlororhodate(iii)**
$C_4H_{12}N^+$, $H_4Cl_4O_2Rh^-$
C.K.Thomas, J.A.Stanko   *J. Coord. Chem.*, **2**, 211, 1973

3.10   **Tetramethylammonium tetradecahydrodecaborate**
$2C_4H_{12}N^+$, $H_{14}B_{10}{}^{2-}$
D.S.Kendall, W.N.Lipscomb   *Inorg. Chem.*, **12,** 546, 1973

3.11   **bis(Tetramethylammonium) di(π - (1) - 2 - thiollyl) iron(ii)**
$2C_4H_{12}N^+$, $H_{20}B_{20}FeS_2{}^{2-}$
B.R.Davis, I.Bernal   *J. Cryst. Mol. Struct.*, **2,** 261, 1972

3.12   **Tetramethylammonium nonabromodiantimonate(iii) dibromine**
$3C_4H_{12}N^+$, $Br_9Sb_2{}^{3-}$, Br_2
C.R.Hubbard, R.A.Jacobson   *Inorg. Chem.*, **11,** 2247, 1972

3.13   **Tetramethylammonium cadmium chloride**
$nC_4H_{12}N^+$, $(CdCl_3{}^-)_n$
B.Morosin   *Acta Cryst. (B)*, **28,** 2303, 1972

3.14   **Putrescinium di(diethylphosphate)**
Tetramethylenediammonium di(diethylphosphate)
$C_4H_{14}N_2{}^{2+}$, $2C_4H_{10}O_4P^-$
S.Furberg, J.Solbakk   *Acta Chem. Scand.*, **26,** 2855, 1972
Residue 2 classified in 46

3.C   **Calcium phosphoryl choline chloride**
$C_5H_{13}NO_4P^-$, Ca^{2+} , Cl^-
For complete entry see 59.1

3.C   **L - α - Glycerylphosphoryl - ethanolamine monohydrate (absolute configuration)**
$C_5H_{14}NO_6P$, H_2O
For complete entry see 46.3

3.C   **Carbamoyl - choline chloride**
$C_6H_{15}N_2O_2^+$, Cl^-
For complete entry see 59.2

3.C   **Carbamoyl - choline iodide**
$C_6H_{15}N_2O_2^+$, I^-
For complete entry see 59.3

3.C   **Triethylammonium tris(o - phenylenedioxy)phosphate**
$C_6H_{16}N^+$, $C_{18}H_{12}O_6P^-$
For complete entry see 46.11

3.15   **Triethylammonium tetrachlorocuprate(ii)**
$2C_6H_{16}N^+$, Cl_4Cu^{2-}
J.Lamotte-Brasseur, L.Dupont, O.Dideberg   *Acta Cryst. (B)*, **29,** 241, 1973

3.C   **Homosulfanilamide - sulfathiazole complex**
$C_7H_{11}N_2O_2S^+$, $C_9H_8N_3O_2S_2^-$
For complete entry see 11.5

3.16   **1,3 - bis(Dimethylamino) - trimethinium perchlorate**
$C_7H_{15}N_2^+$, ClO_4^-
B.W.Matthews, R.E.Stenkamp, P.M.Colman
Acta Cryst. (B), **29,** 449, 1973

3.C   **p - Bromo - benzimino methyl ether**
C_8H_8BrNO
For complete entry see 13.9

3.C   **S - (–) - p - Bromo - α - phenylethylammonium (+) - 1,6 - methano(10)annulene - 2 - carboxylate**
$C_8H_{11}BrN^+$, $C_{12}H_9O_2^-$
For complete entry see 31.10

3.C   **5 - Hydroxydopamine hydrochloride**
3,4,5 - Trihydroxyphenyl - ethylamine hydrochloride
$C_8H_{12}NO_3^+$, Cl^-
For complete entry see 59.4

3.C   **Dimethyl - (dimethylaminomethylene) - malonate**
$C_8H_{13}NO_4$
For complete entry see 1.22

3.17   **3 - Acetoxypropyl - trimethylammonium bromide**
$C_8H_{18}NO_2^+$, Br^-
B.M.Craven, G.Hite   *Acta Cryst. (B)*, **29,** 1132, 1973
Residue 1 also classified in 1

3.C   **Tetraethylammonium 2,2' - cobaltobiscarborane**
$C_8H_{20}N^+$, $C_8H_{40}B_{40}Co^-$
For complete entry see 62.6

3.C   **Tetra - ethylammonium (3,11') - commo - (undecahydro - 1,2 - dicarba - 3 - cobalta - closo - dodecaborato)(decahydro - 9' - pyridyl - 7',8' - dicarba - 11' - nido - undecaborate)**
$C_8H_{20}N^+$, $C_9H_{26}B_{17}CoN^-$
For complete entry see 62.8

3.C   **Tetraethylammonium tetra(benzylmercapto - μ_3 - sulfido - iron)**
$2C_8H_{20}N^+$, $C_{28}H_{28}Fe_4S_8^{2-}$
For complete entry see 85.31

3.C   **p - Nitro - benzylimino methyl ether**
$C_9H_{10}N_2O_3$
For complete entry see 15.4

3.C   **(−) - Adrenaline hydrogen (+) - tartrate (absolute configuration)**
$C_9H_{14}NO_3^+$, $C_4H_5O_6^-$
For complete entry see 17.14

3.C   **Hexa - aquo - magnesium ethylenediaminetetra - acetato - aquomagnesium dihydrate**
$C_{10}H_{14}MgN_2O_9^{2-}$, $H_{12}MgO_6^{2+}$, $2H_2O$
For complete entry see 67.6

3.C   **bis(Benzyl - trimethyl - ammonium) carbido - pentadecacarbonyl - hexarhodate**
$2C_{10}H_{16}N^+$, $C_{16}O_{15}Rh^{2-}$
For complete entry see 71.35

3.18   **(−) - Ephedrine dihydrogen phosphate**
$C_{10}H_{16}NO^+$, $H_2O_4P^-$
R.A.Hearn, C.E.Bugg   *Acta Cryst. (B)*, **28,** 3662, 1972

3.C   **Ethylenediaminetetra - acetic acid (β form)**
$C_{10}H_{16}N_2O_8$
For complete entry see 1.27

3.19   **bis(N,N,N′,N′ - Tetraethyl - ethylenediammonium) hexa - μ - chloro - μ_4 - oxo - tetra(chlorocuprate(ii))**
$2C_{10}H_{26}N_2^{2+}$, $Cl_{10}Cu_4O^{4-}$
R.Belford, D.E.Fenton, M.R.Truter   *J. C. S. Dalton*, 2345, 1972

3.C   **DL - 1 - (3,5 - Dihydroxyphenyl) - 2 - (isopropylamino)ethanol sulfate ethanol solvate**
$2C_{11}H_{17}NO_3^+$, O_4S^{2-}, C_2H_6O
For complete entry see 17.15

3.20   **1 - (2 - Aminoethyl) - 3,4,5 - trimethoxybenzene hydrochloride**
Mescaline hydrochloride
$C_{11}H_{18}NO_3^+$, Cl^-
D.Tsoucaris, C.de Rango, G.Tsoucaris, C.Zelwer, R.Parthasarathy,
F.E.Cole   *Cryst. Struct. Comm.*, **2,** 193, 1973
Residue 1 also classified in 17

3.21   **bis(11 - Ammonio - undecanoic acid) tetrachlorocuprate**
$2C_{11}H_{24}NO_2^+$, Cl_4Cu^{2-}
Y.P.Mascarenhas   *Acta Cryst. (A)*, **28,** S86, 1972
Residue 1 also classified in 1

3.C   **N,N - Dimethyltryptamine (form II)**
$C_{12}H_{16}N_2$
For complete entry see 35.16

3.C   **Bufotenine**
5 - Hydroxy - N,N - dimethyltryptamine
$C_{12}H_{16}N_2O$
For complete entry see 35.17

3.C   **Phenyl β - methylcholine ether bromide**
$C_{12}H_{20}NO^+$, Br^-
For complete entry see 59.7

3.C   **Melatonin**
N - Acetyl - 5 - methoxy - tryptamine
$C_{13}H_{16}N_2O_2$
For complete entry see 35.22

3.C   **o - Methylphenyl - β - methylcholine ether bromide**
$C_{13}H_{22}NO^+$, Br^-
For complete entry see 59.8

3.C   **Compound X**
$C_{13}H_{35}BMnNO_4$
For complete entry see 62.10

3.C   **Lidocaine hydrochloride monohydrate**
$C_{14}H_{23}N_2O^+$, Cl^- , H_2O
For complete entry see 16.22

3.C   **Succinyl - choline picrate**
$C_{14}H_{30}N_2O_4^{2+}$, $2C_6H_2N_3O_7^-$
For complete entry see 59.10

3.22   **Alprenolol hydrochloride**
1 - Isopropylamino - 3 - (2 - allylphenoxy) - propan - 2 - ol hydrochloride
$C_{15}H_{24}NO_2^+$, Cl^-
Y.Barrans, M.Cotrait, J.Dangoumau   *Acta Cryst. (B)*, **29,** 1264, 1973
Residue 1 also classified in 17

3.C   **Bromo - (tetraethyl - ethylenediamine) - cyclopentadienyl - magnesium**
$C_{15}H_{29}BrMgN_2$
For complete entry see 67.11

3.C   **Propranolol hydrochloride**
1 - Isopropylamino - 3(1 - oxynaphthyl) - propen - 2 - ol hydrochloride
$C_{16}H_{22}NO_2^+$, Cl^-
For complete entry see 24.7

3.C   **Tetra - n - butylammonium bis(tetramethylene)trihydrodiborane**
$C_{16}H_{36}N^+$, $C_8H_{19}B_2^-$
For complete entry see 61.4

3.C   **Tetrabutylammonium bis(maleonitriledithiolato) copper(ii)**
$2C_{16}H_{36}N^+$, $C_8CuN_4S_4^{2-}$
For complete entry see 85.3

3.C   **bis(Tetra - n - butylammonium)** μ **- S,S' - (tetrakis(ethane - 1,2 - dithiolate) - di - iron(iii))**
$2C_{16}H_{36}N^+$, $C_8H_{16}Fe_2S_8^{2-}$
For complete entry see 85.7

3.C   **10 - (2' - Dimethylamino - propyl) - phenothiazine hydrobromide**
$C_{17}H_{21}N_2S^+$, Br^-
For complete entry see 41.40

3.C   **3 - Methoxy - 10 - (3' - dimethylammonium - propyl) - phenothiazine hydrogen maleate**
$C_{18}H_{23}N_2OS^+$, $C_4H_3O_4^-$
For complete entry see 41.44

3.C   **10 - (2' - Trimethylammonium - n - propyl) - phenothiazine methyl sulfate**
$C_{18}H_{23}N_2S^+$, $CH_3O_4S^-$
For complete entry see 41.45

3.C   **7 - Chloro - 4 - (4' - diethylamino - 1 - methylbutyl - amino)quinoline**
$C_{18}H_{26}ClN_3$
For complete entry see 35.28

3.C   **2 - Diethylaminoethyl - 1 - phenylcyclopentane - carboxylate hydrochloride**
Parpanit
$C_{18}H_{28}NO_2^+$, Cl^-
For complete entry see 20.17

3.23   **Adiphenine hydrochloride**
2 - Diethylaminoethyl diphenylacetate hydrochloride
$C_{20}H_{26}NO_2^+$, Cl^-
J.J.Guy, T.A.Hamor   *J. C. S. Perkin II*, 942, 1973
Residue 1 also classified in 1

3.C   Bicyclo(1,1,0)butan - 1 - yl - lithium tetramethylethylenediamine dimer
$C_{20}H_{42}Li_2N_4$
For complete entry see 67.17

3.24   Tribenzylamine (at $-70°C$)
$C_{21}H_{21}N$
F.Iwasaki, H.Iwasaki *Acta Cryst. (B)*, **28,** 3370, 1972

3.C   Naphthalene - bis(lithium tetramethylethylenediamine)
$C_{22}H_{40}Li_2N_4$
For complete entry see 67.18

3.C   Diphenylphosphino(diphenylphosphinothioyl) methylamine
$C_{25}H_{23}NP_2S$
For complete entry see 64.36

3.C   Triphenylmethyl - lithium tetramethylethylenediamine
$C_{25}H_{31}LiN_2$
For complete entry see 67.19

ALIPHATIC (N AND S) COMPOUNDS

4.1   2 - **Aminoethanethiosulfuric acid**
$C_2H_7NO_3S_2$
W.E.Keefe, J.M.Stewart   *Acta Cryst. (B),* **28,** 2469, 1972

4.2   3 - **Aminopropane sulfonic acid (α form)**
Homotaurine
$C_3H_9NO_3S$
S.Ueoka, T.Fujiwara, K.-I.Tomita   *Bull. Chem. Soc. Jap.,* **45,** 3634, 1972

4.C   **Rubidium dimercapto - maleonitrile monohydrate**
$C_4N_2S_2^{2-}$, $2Rb^+$, H_2O
For complete entry see 7.4

4.3   **S,S - Dimethyl - N - trichloroacetyl sulfilimine**
$C_4H_6Cl_3NOS$
A.Kalman, K.Sasvari, A.Kucsman   *Acta Cryst. (B),* **29,** 1241, 1973

4.4   **1,5 - bis(p - Chlorophenyl) - 2,4 - diaza - 1,3,5 - trithiapenta - 2,3 - diene**
$C_{12}H_8Cl_2N_2S_3$
F.P.Olsen, J.C.Barrick   *Inorg. Chem.,* **12,** 1353, 1973
Also classified in 19

4.5   **bis(Diphenylmethylene)trisulfur tetranitride**
$C_{26}H_{20}N_4S_3$
E.M.Holt, S.L.Holt   *J. C. S. Chem. Comm.,* 36, 1973

ALIPHATIC MISCELLANEOUS

5.1   **Carbon tetrachloride (high pressure form)**
CCl_4
G.J.Piermarini, A.B.Braun *J. Chem. Phys.*, **58**, 1974, 1973

5.2   **Tetrafluoromethane (α form, at 10°K)**
CF_4
D.N.Bol'shutkin, V.M.Gasan, A.I.Prokhvatilov, A.I.Erenburg
Acta Cryst. (B), **28**, 3542, 1972

5.3   **Dibromomethane**
CH_2Br_2
T.Kawaguchi, T.Watanabe *Acta Cryst. (A)*, **28**, S11, 1972

5.4   **Dichloromethane**
CH_2Cl_2
T.Kawaguchi, T.Watanabe *Acta Cryst. (A)*, **28**, S11, 1972

5.5   **Di - iodomethane**
CH_2I_2
T.Kawaguchi, T.Watanabe *Acta Cryst. (A)*, **28**, S11, 1972

5.6   **Monobromomethane**
CH_3Br
T.Kawaguchi, T.Watanabe *Acta Cryst. (A)*, **28**, S11, 1972

5.7   **Monoiodomethane**
CH_3I
T.Kawaguchi, T.Watanabe *Acta Cryst. (A)*, **28**, S11, 1972

5.8   **Deuteromethane (low temp.form, at 4.2°K)**
CD_4
E.Arzi, E.Sandor *Acta Cryst. (A)*, **28**, S188, 1972

5.C α - **Cyclodextrin - methanol**
CH_4O , $C_{36}H_{60}O_{30}$
For complete entry see 61.8

5.9   **Di - iodoacetylene**
C_2I_2
J.D.Dunitz, H.Gehrer, D.Britton *Acta Cryst. (B)*, **28**, 1989, 1972

5.10   **Methylchloroform (at −60°C)**
1,1,1 - Trichloroethane
$C_2H_3Cl_3$
L.Silver, R.Rudman   *J. Chem. Phys.*, **57,** 210, 1972

5.11   **Methyl chloroform (at −145°C)**
1,1,1 - Trichloroethane
$C_2H_3Cl_3$
L.Silver, R.Rudman   *J. Chem. Phys.*, **57,** 210, 1972

5.12   **Chloral hydrate (neutron study)**
$C_2H_3Cl_3O_2$
G.M.Brown, H.A.Levy   *Cryst. Struct. Comm.*, **2,** 107, 1973

5.C   **Cholic acid - ethanol complex**
C_2H_6O , $C_{24}H_{40}O_5$
For complete entry see 60.42

5.C α - **Cyclodextrin - n - propanol**
C_3H_8O , $C_{36}H_{60}O_{30}$
For complete entry see 61.9

5.C   **Urea - 1,4 - dichlorobutane**
$C_4H_8Cl_2$, CH_4N_2O
For complete entry see 61.1

5.13   **Diethyl ether (at 128°K)**
$C_4H_{10}O$
D.Andre, R.Fourme, K.Zechmeister   *Acta Cryst. (B)*, **28,** 2389, 1972

5.14   **2,5 - Dimethyl - 2,5 - hexanediol tetrahydrate**
$C_8H_{18}O_2$, $4H_2O$
G.A.Jeffrey, M.S.Shen   *J. Chem. Phys.*, **57,** 56, 1972

5.15   **3 - (3′,4′ - Dichlorobenzylidene) - pentane - 2,4 - dione**
$C_{12}H_{10}Cl_2O_2$
K.A.Kerr, J.P.Ashmore   *Stockholm Symposium,* 91, 1973
Also classified in 19

5.16   **3 - (p - Bromobenzylidene) - pentane - 2,4 - dione**
$C_{12}H_{11}BrO_2$
K.A.Kerr, J.P.Ashmore   *Stockholm Symposium,* 91, 1973
Also classified in 19

5.17   **n - Octadecane (re - interpretation of data of Hayashida,J.Phys.Soc.Jap., 17,306,1962)**
$C_{18}H_{38}$
S.C.Nyburg, H.Luth   *Acta Cryst. (B)*, **28,** 2992, 1972

5.18 $\alpha,\alpha,\alpha',\alpha'$ - **Tetra - t - butylacetone**
$C_{19}H_{38}O$
G.Lepicard, J.Berthou, J.Delettre, A.Laurent, J.-P.Mornon
C. R. Acad. Sci., Fr., C, **276,** 575, 1973

ENOLATES
(ALIPHATIC AND AROMATIC)

No entries in this volume.

NITRILES (ALIPHATIC AND AROMATIC)

7.1   **Potassium dinitroacetonitrile**
$C_2N_3O_4{}^-$, K^+
B.Klewe   *Acta Chem. Scand.*, **26**, 1921, 1972
Residue 1 also classified in 12

7.2   **Rubidíum dinitroacetonitrile**
$C_2N_3O_4{}^-$, Rb^+
H.J.Bjornstad, B.Klewe   *Acta Chem. Scand.*, **26**, 1874, 1972
Residue 1 also classified in 10, 12

7.3   **Dichlorofumaronitrile**
$C_4Cl_2N_2$
B.Klewe, C.Romming   *Acta Chem. Scand.*, **26**, 2272, 1972

7.4   **Rubidium dimercapto - maleonitrile monohydrate**
$C_4N_2S_2{}^{2-}$, $2Rb^+$, H_2O
M.Drager, G.Gattow   *Z. Anorg. Allg. Chem.*, **391**, 203, 1972
Residue 1 also classified in 4

7.C   **5,7 - (12) - Paracyclophadiyne - tetracyanoethylene complex**
C_6N_4, $2C_{18}H_{20}$
For complete entry see 60.35

7.C   **1,1,2,2 - Tetracyanocyclopropane**
$C_7H_2N_4$
For complete entry see 20.7

7.5   **2,3,5,6 - Tetrachloro - 4 - (methylthio) - benzonitrile**
$C_8H_3Cl_4NS$
D.R.Carter, F.P.Boer   *J. C. S. Perkin II*, 2104, 1972
Also classified in 11

7.C   **Azo - bis(isobutyronitrile) (triclinic form)**
$C_8H_{12}N_4$
For complete entry see 9.5

7.C   **Azo - bis(isobutyronitrile) (monoclinic form)**
$C_8H_{12}N_4$
For complete entry see 9.6

7.C   **Azo - bis(isobutyronitrile),perdeuterated (triclinic form)**
$C_8D_{12}N_4$
For complete entry see 9.7

7.6   **2,4,5 - Trichloro - 6 - (methylthio)isophthalonitrile**
$C_9H_3Cl_3N_2S$
D.R.Carter, J.W.Turley, F.P.Boer   *Acta Cryst. (B),* **28,** 3430, 1972
Also classified in 11

7.7   **1,2,4,5 - Tetracyanobenzene**
$C_{10}H_2N_4$
C.K.Prout, I.J.Tickle   *J. C. S. Perkin II,* 520, 1973

7.C   **p - Phenylenediamine - 1,2,4,5 - tetracyanobenzene complex**
$C_{10}H_2N_4$, $C_6H_8N_2$
For complete entry see 60.7

7.C   **Pyrene - 1,2,4,5 - tetracyanobenzene complex (at 290°K)**
$C_{10}H_2N_4$, $C_{16}H_{10}$
For complete entry see 60.30

7.C   **Pyrene - 1,2,4,5 - tetracyanobenzene complex (at 178°K)**
$C_{10}H_2N_4$, $C_{16}H_{10}$
For complete entry see 60.31

7.8   **o - Chlorobenzylidenemalononitrile**
$C_{10}H_5ClN_2$
D.A.Williams, D.A.Wright   *J. Cryst. Mol. Struct.,* **3,** 55, 1973

7.9   **Tetracyanoquinodimethane - 1 - methyl - N - ethylbenzimidazolium complex acetonitrile solvate**
$C_{12}H_4N_4$, $C_{10}H_{13}N_2^+$, $C_{12}H_4N_4^-$, C_2H_3N
D.Chasseau, J.Gaultier, C.Hauw, J.Jaud
C. R. Acad. Sci., Fr., C, **276,** 661, 1973
Residue 2 classified in 35

7.10   **Tetracyanoquinodimethane - 1,2 - dimethyl - N - ethylbenzimidazolium complex acetonitrile solvate**
$C_{12}H_4N_4$, $C_{11}H_{15}N_2^+$, $C_{12}H_4N_4^-$, C_2H_3N
D.Chasseau, J.Gaultier, C.Hauw, J.Jaud
C. R. Acad. Sci., Fr., C, **276,** 751, 1973
Residue 2 classified in 35

7.C   **7,7,8,8 - Tetracyanoquinodimethane - 1,10 - phenanthroline complex**
$C_{12}H_4N_4$, $C_{12}H_8N_2$
For complete entry see 60.14

7.C   **7,7,8,8 - Tetracyanoquinodimethane - phenazine**
$C_{12}H_4N_4$, $C_{12}H_8N_2$
For complete entry see 60.15

7.C   **7,7,8,8 - Tetracyanoquinodimethane - dibenzo - p - dioxin**
$C_{12}H_4N_4$, $C_{12}H_8O_2$
For complete entry see 60.16

7.C   **Acenaphthene - 7,7,8,8 - tetracyanoquinodimethane complex**
$C_{12}H_4N_4$, $C_{12}H_{10}$
For complete entry see 60.20

7.C   **Benzidine - 7,7,8,8 - tetracyano - p - quinodimethane complex**
$C_{12}H_4N_4$, $C_{12}H_{12}N_2$
For complete entry see 60.21

7.C   **Benzidine - 7,7,8,8 - tetracyano - p - quinodimethane complex dichloromethane solvate**
$C_{12}H_4N_4$, $C_{12}H_{12}N_2$, $1.8CH_2Cl_2$
For complete entry see 60.22

7.C   **Benzidine - 7,7,8,8 - tetracyano - p - quinodimethane complex benzene solvate**
$C_{12}H_4N_4$, $C_{12}H_{12}N_2$, $1.35C_6H_6$
For complete entry see 60.23

7.C   **7,7,8,8 - Tetracyanoquinodimethane - N,N' - dimethyldihydrophenazine (photographic data)**
$C_{12}H_4N_4$, $C_{14}H_{14}N_2$
For complete entry see 60.17

7.C   **7,7,8,8 - Tetracyanoquinodimethane - N,N' - dimethyldihydrophenazine (diffractometer data)**
$C_{12}H_4N_4$, $C_{14}H_{14}N_2$
For complete entry see 60.18

7.C   **Pyrene - 7,7,8,8 - tetracyanoquinodimethane complex (data of Prout and Tickle)**
$C_{12}H_4N_4$, $C_{16}H_{10}$
For complete entry see 60.32

7.C   **Pyrene - 7,7,8,8 - tetracyanoquinodimethane complex (data of Wright)**
$C_{12}H_4N_4$, $C_{16}H_{10}$
For complete entry see 60.33

7.C   **Perylene - 7,7,8,8 - tetracyanoquinodimethane complex**
$C_{12}H_4N_4$, $C_{20}H_{12}$
For complete entry see 60.38

7.C   **Morpholinium 7,7,8,8 - tetracyanoquinodimethane**
$C_{12}H_4N_4^-$, $C_4H_{10}NO^+$
For complete entry see 40.5

7.C   **2,2' - Bi - 1,3 - dithiole 7,7,8,8 - tetracyanoquinodimethanide**
$C_{12}H_4N_4^-$, $C_6H_4S_4^+$
For complete entry see 60.6

7.C   **N - Ethyl - 2 - methyl - thiazoline - bis(7,7,8,8 - tetracyanoquinodimethane)**
$C_{12}H_4N_4^-$, $C_6H_{12}NS^+$, $C_{12}H_4N_4$
For complete entry see 60.8

7.C   **1,2,3 - Trimethyl - benzimidazolium tetracyanoquinodimethanide**
$C_{12}H_4N_4^-$, $C_{10}H_{13}N_2^+$
For complete entry see 35.10

7.C   **Dimethyl - thiacyanine bis(7,7,8,8 - tetracyanoquinodimethane)**
$C_{12}H_4N_4^-$, $C_{17}H_{15}N_2S_2^+$, $C_{12}H_4N_4$
For complete entry see 60.34

7.C   **Dimethyl - thiacarbocyanine bis(7,7,8,8 - tetracyanoquinodimethane)**
$C_{12}H_4N_4^-$, $C_{19}H_{17}N_2S_2^+$, $C_{12}H_4N_4$
For complete entry see 60.36

7.C   **Diethyl - thiacyanine bis(7,7,8,8 - tetracyanoquinodimethane)**
$C_{12}H_4N_4^-$, $C_{19}H_{19}N_2S_2^+$, $C_{12}H_4N_4$
For complete entry see 60.37

7.C   **Diethyl - thiacarbocyanine bis(7,7,8,8 - tetracyanoquinodimethane) (triclinic form)**
$C_{12}H_4N_4^-$, $C_{21}H_{21}N_2S_2^+$, $C_{12}H_4N_4$
For complete entry see 60.39

7.C   **1,1' - Ethylene - 2,2' - bipyridylium 7,7,8,8 - tetracyanoquinodimethane**
$2C_{12}H_4N_4^-$, $C_{12}H_{12}N_2^{2+}$
For complete entry see 36.10

7.C   **N,N' - Dibenzyl - 4,4' - bipyridylium tetra(7,7,8,8 - tetracyanoquinodimethane)**
$2C_{12}H_4N_4^-$, $C_{24}H_{22}N_2^{2+}$, $2C_{12}H_4N_4$
For complete entry see 33.55

7.C   **hexakis(Methylcyanide) magnesium(ii) tetrachloroaluminate(iii)**
$C_{12}H_{18}MgN_6^{2+}$, $2AlCl_4^-$
For complete entry see 67.9

7.C   **Azobis - 3 - cyano - 3 - pentane**
$C_{12}H_{20}N_4$
For complete entry see 9.12

UREA COMPOUNDS
(ALIPHATIC AND AROMATIC)

8.1   Urea - phosphoric acid
Carbamide phosphate
CH_4N_2O , H_3O_4P
D.Mootz, K.-R.Albrand   *Acta Cryst. (B)*, **28**, 2459, 1972

8.2   Urea - phosphoric acid (neutron study)
Carbamide phosphate
CH_4N_2O , H_3O_4P
E.C.Kostansek, W.R.Busing   *Acta Cryst. (B)*, **28**, 2454, 1972

8.C   Urea - 1,4 - dichlorobutane
CH_4N_2O , $C_4H_8Cl_2$
For complete entry see 61.1

8.3   Thiourea - mercury(ii) cyanide complex
$2CH_4N_2S$, C_2HgN_2
E.Moreno, L.Castro   *An. R. Soc. Esp. Fis. Quim., A*, **67**, 371, 1971

8.C   Guanidinium iron(iii) ethylenediaminetetra - acetate dihydrate
$CH_6N_3^+$, $C_{10}H_{14}FeN_2O_9^-$, $2H_2O$
For complete entry see 76.34

8.C   Guanidinium ethylenediaminetetra - acetato - neodymium trihydrate
$CH_6N_3^+$, $C_{10}H_{18}N_2NdO_{11}^-$
For complete entry see 76.37

8.4   Carbohydrazide
CH_6N_4O
P.Domiano, M.A.Pellinghelli, A.Tiripicchio
Acta Cryst. (B), **28**, 2495, 1972
Also classified in 9

8.5   Thiocarbonohydrazide hemihydrochloride
CH_6N_4S , $CH_7N_4S^+$, Cl^-
A.Braibanti, A.Tiripicchio, M.T.Camellini   *J. C. S. Perkin II*, 2116, 1972

8.6   Thiocarbonohydrazide sulfate
$CH_8N_4S^{2+}$, O_4S^{2-}
F.Bigoli, A.Braibanti, A.M.M.Lanfredi, A.Tiripicchio
J. C. S. Perkin II, 2121, 1972

8.C   **Urea - calcium nitrate trihydrate**
$CH_{10}CaN_6O_{11}$
For complete entry see 67.1

8.7   **Dithiobiuret**
$C_2H_5N_3S_2$
W.A.Spofford III, E.L.Amma   *J. Cryst. Mol. Struct.*, **2,** 151, 1972

8.8   **Methylguanidinium dihydrogen phosphate**
$C_2H_8N_3^+$, $H_2O_4P^-$
F.A.Cotton, V.W.Day, E.E.Hazen Junior, S.Larsen, S.T.K.Wong
Amer. Cryst. Assoc., Abstr. Papers (Winter Meeting), 82, 1973

8.9   **bis(Methylguanidinium) monohydrogen phosphate**
$2C_2H_8N_3^+$, HO_4P^{2-}
F.A.Cotton, V.W.Day, E.E.Hazen Junior, S.Larsen, S.T.K.Wong
Amer. Cryst. Assoc., Abstr. Papers (Winter Meeting), 82, 1973

8.10   **N - Ammonium - phosphoryl - taurocyamine**
$C_3H_8N_3O_6PS^{2-}$, $2H_4N^+$
A.Laurent   *Eur. Cryst. Meeting,* 1973
Residue 1 also classified in 11, 64

8.11   **Taurocyamine**
$C_3H_9N_3O_3S$
A.Laurent   *Eur. Cryst. Meeting,* 1973
Also classified in 11

8.C   **Disodium N - phosphorylcreatine hydrate (further refinement)**
$C_4H_8N_3O_5P^{2-}$, $2Na^+$, xH_2O
For complete entry see 48.16

8.C   **Creatine monohydrate (further refinement)**
$C_4H_9N_3O_2$, H_2O
For complete entry see 48.24

8.12 γ - **Guanidino** - β - **hydroxy - propane - sulfonic acid**
$C_4H_{11}N_3O_4S$
Y.B.Kim, A.Wakahara, T.Fujiwara, K.-I.Tomita   *Chem. Letters,* 891, 1972
Also classified in 11

8.13   **Propylguanidinium diethylphosphate**
$C_4H_{12}N_3^+$, $C_4H_{10}O_4P^-$
S.Furberg, J.Solbakk   *Acta Chem. Scand.*, **26,** 3699, 1972
Residue 2 classified in 46

8.C γ - **Guanidinobutyric acid hydrobromide**
$C_5H_{12}N_3O_2^+$, Br^-
For complete entry see 1.16

8.14 γ - **Guanidinobutyric acid hydrochloride**
$C_5H_{12}N_3O_2^+$, Cl^-
T.Maeda, T.Fujiwara, K.-I.Tomita   *Bull. Chem. Soc. Jap.*, **45,** 3628, 1972

8.15   **o - Nitrobenzene - selenyl thiourea thiocyanate**
$C_7H_8N_3O_2SSe^+$, CNS^-
R.Eriksen, S.Hauge   *Acta Chem. Scand.*, **26,** 3153, 1972
Residue 1 also classified in 11, 15

8.C   **2 - Formylpyridine selenosemicarbazide**
$C_7H_8N_4Se$
For complete entry see 33.32

8.16   **4 - Phenyl - thiosemicarbazide**
$C_7H_9N_3S$
A.Kalman, G.Argay, M.Czugler   *Cryst. Struct. Comm.*, **1,** 375, 1972

8.C   **Morpholine biguanide hydrobromide**
$C_8H_{14}N_5O^+$, Br^-
For complete entry see 40.13

8.17   **S - Methyl - dithizone**
$C_{14}H_{14}N_4S$
J.Preuss, A.Gieren   *Eur. Cryst. Meeting,* 1973
Also classified in 9

8.18   **N,N' - Diethyl - N,N' - diphenyl - urea**
$C_{17}H_{20}N_2O$
G.Allegra, G.Avitabile, E.Benedetti, M.R.Ciajolo, P.Corradini, P.Ganis,
M.Goodman, A.Immirzi, C.Pedone   *Acta Cryst. (A),* **28,** S13, 1972

8.19   **N - (4 - (β - 2 - Methoxy - 5 - bromobenzamidoethyl)benzensulfonyl) - N' -**
4 - methyl - cyclohexyl - urea
$C_{24}H_{30}BrN_3O_5S$
D.Kobelt, E.F.Paulus   *Acta Cryst. (B),* **28,** 3452, 1972
Also classified in 11, 13, 17, 21

NITROGEN-NITROGEN COMPOUNDS
(ALIPHATIC AND AROMATIC)

9.C   **Carbohydrazide**
CH_6N_4O
For complete entry see 8.4

9.C   **2 - (N - Nitrosomethylamino)acetamide**
$C_3H_7N_3O_2$
For complete entry see 1.4

9.1   **Malonic dihydrazide monohydrate**
$C_3H_8N_4O_2$, H_2O
C.Miravitlles, J.L.Brianso, M.Font-Altaba, J.P.Declerq, G.Germain
Cryst. Struct. Comm., **2,** 315, 1973
Residue 1 also classified in 1

9.2   **1,4 - bis(Diazo) - 2,3 - butanedione**
$C_4H_2N_4O_2$
H.Hope, K.T.Black   *Acta Cryst. (B),* **28,** 3632, 1972

9.3   **Acetone - semicarbazone**
$C_4H_9N_3O$
D.V.Naik, G.J.Palenik
Amer. Cryst. Assoc., Abstr. Papers (Winter Meeting), 93, 1973

9.C   **tris(2,2 - Dimethylhydrazino)borane**
$C_6H_{21}BN_6$
For complete entry see 62.2

9.C   **2 - Formylpyridine selenosemicarbazide**
$C_7H_8N_4Se$
For complete entry see 33.32

9.4   **Benzaldehyde - semicarbazone**
$C_8H_9N_3O$
D.V.Naik, G.J.Palenik
Amer. Cryst. Assoc., Abstr. Papers (Winter Meeting), 93, 1973

9.5   **Azo - bis(isobutyronitrile) (triclinic form)**
$C_8H_{12}N_4$
A.B.Jaffe, D.S.Malament, E.P.Slisz, J.M.McBride
J. Amer. Chem. Soc., **94**, 8515, 1972
Also classified in 7

9.6   **Azo - bis(isobutyronitrile) (monoclinic form)**
$C_8H_{12}N_4$
A.B.Jaffe, D.S.Malament, E.P.Slisz, J.M.McBride
J. Amer. Chem. Soc., **94**, 8515, 1972
Also classified in 7

9.7   **Azo - bis(isobutyronitrile),perdeuterated (triclinic form)**
$C_8D_{12}N_4$
A.B.Jaffe, D.S.Malament, E.P.Slisz, J.M.McBride
J. Amer. Chem. Soc., **94**, 8515, 1972
Also classified in 7

9.8   **Acetone p - nitrophenylhydrazone**
$C_9H_{11}N_3O_2$
G.Menczel, G.Samay, K.Simon
Acta Chim. Acad. Sci. Hungar., **72**, 441, 1972

9.C   **Pentafluoronitrosobenzene - cis - azo(pentafluorobenzene) - dioxide**
$2C_{12}F_{10}N_2O_2 , C_6F_5NO$
For complete entry see 10.1

9.C   **Hexanitro - azobenzene (β form)**
$C_{12}H_4N_8O_{12}$
For complete entry see 15.5

9.C   **Hexanitro - azobenzene (γ form)**
$C_{12}H_4N_8O_{12}$
For complete entry see 15.6

9.C   **Hexanitro - azobenzene (α form)**
$C_{12}H_4N_8O_{12}$
For complete entry see 15.7

9.9   **p - Bromodiazoaminobenzene (α form)**
$C_{12}H_{10}BrN_3$
Yu.A.Omel'chenko, Yu.D.Kondrashev   *Kristallografija*, **17**, 947, 1972

9.10   **p - Bromodiazoaminobenzene (β form)**
$C_{12}H_{10}BrN_3$
Yu.A.Omel'chenko, Yu.D.Kondrashev   *Kristallografija*, **17**, 947, 1972

9.11   **Diazoaminobenzene (β form)**
$C_{12}H_{11}N_3$
V.F.Gladkova, Yu.D.Kondrashev   *Kristallografija*, **17**, 33, 1972

9.12   **Azobis - 3 - cyano - 3 - pentane**
$C_{12}H_{20}N_4$
A.B.Jaffe, D.S.Malament, E.P.Slisz, J.M.McBride
J. Amer. Chem. Soc., **94**, 8515, 1972
Also classified in 7

9.13   **Sodium 4' - dimethylaminoazobenzene - 4 - sulfonate monohydrate ethanol solvate**
Methyl orange monohydrate ethanol solvate
$C_{14}H_{14}N_3O_3S^-$, Na^+ , H_2O , C_2H_6O
A.W.Hanson *Acta Cryst. (B)*, **29**, 454, 1973
Residue 1 also classified in 16

9.C   **S - Methyl - dithizone**
$C_{14}H_{14}N_4S$
For complete entry see 8.17

9.14   **1 - p - Tolyl - 3 - (α - cyanobenzylidene)triazene**
$C_{15}H_{12}N_4$
J.W.Schilling, C.E.Nordman *Acta Cryst. (B)*, **28**, 2177, 1972

9.15   **4 - Bromobenzophenone 2,4 - dinitrophenylhydrazone (E form)**
$C_{19}H_{13}BrN_4O_4$
M.Tabata, Y.Takada, A.Suzuki, A.Furusaki *Chem. Letters*, 1019, 1972

9.C   **2 - (Diphenylamino - aminylio) - 3,5 - dioxo - 4 - phenyl - 1,2,4 - triazolidine - 1 - ide**
$C_{20}H_{15}N_5O_2$
For complete entry see 32.23

9.16   **β,β' - Dimethyl - cinnamaldazine**
$C_{20}H_{20}N_2$
G.Lepicard, J.Berthou, J.Delettre, A.Laurent, J.-P.Mornon
Acta Cryst. (B), **29**, 1154, 1973

9.17   **(+) - N - Phthalimido - 2 - bromo - 2(S) - octyl - p - tolyl - (R) - sulfoxime (absolute configuration)**
$C_{23}H_{27}BrN_2O_3S$
G.D.Andreetti, G.Bocelli, P.Sgarabotto *Cryst. Struct. Comm.*, **2**, 171, 1973
Also classified in 11, 13

9.C   **Phenyl(triphenylsilyl)diazomethane**
$C_{25}H_{20}N_2Si$
For complete entry see 63.11

9.C   **1,1' - Azo - 2 - phenylimidazo(1,2 - a)pyridinium dibromide**
$C_{26}H_{20}N_6^{2+}$, $2Br^-$
For complete entry see 35.37

9.18   **Diphenyl triketone sym - N - benzoyl - p - bromophenylhydrazone**
$C_{28}H_{19}BrN_2O_3$
D.B.Pendergrass Junior, I.C.Paul, D.Y.Curtin
J. Amer. Chem. Soc., **94,** 8730, 1972

9.19 α **- p - Bromophenylazo -** β **- benzoyloxobenzalacetophenone**
$C_{28}H_{19}BrN_2O_3$
D.B.Pendergrass Junior, D.Y.Curtin, I.C.Paul
J. Amer. Chem. Soc., **94,** 8722, 1972

9.20   **Diphenyl triketone sym - N - benzoylphenylhydrazone**
$C_{28}H_{20}N_2O_3$
D.B.Pendergrass Junior, I.C.Paul, D.Y.Curtin
J. Amer. Chem. Soc., **94,** 8730, 1972

9.21   **Phenylazotribenzoylmethane**
$C_{28}H_{20}N_2O_3$
D.B.Pendergrass Junior, D.Y.Curtin, I.C.Paul
J. Amer. Chem. Soc., **94,** 8722, 1972

NITROGEN-OXYGEN COMPOUNDS
(ALIPHATIC AND AROMATIC)

10.C   **Rubidium dinitroacetonitrile**
$C_2N_3O_4^-$, Rb^+
For complete entry see 7.2

10.C   **2 - (N - Nitrosomethylamino)acetamide**
$C_3H_7N_3O_2$
For complete entry see 1.4

10.C   **4 - Nitropyridine - N - oxide - hydroquinone complex**
$C_5H_4N_2O_3$, $C_6H_6O_2$
For complete entry see 60.2

10.1   **Pentafluoronitrosobenzene - cis - azo(pentafluorobenzene) - dioxide**
C_6F_5NO , $2C_{12}F_{10}N_2O_2$
C.K.Prout, A.Coda, R.A.Forder, B.Kamenar
Abstr. Ital.-Yug. Congr, 113, 1973
Residue 2 classified in 9, 10

10.2   **syn - p - Nitrobenzaldoxime**
$C_7H_6N_2O_3$
L.Brehm, K.J.Watson   *Acta Cryst. (B),* **28,** 3646, 1972
Also classified in 15

10.C   **2 - Chloro - isonitrosacetanilide**
$C_8H_7ClN_2O_2$
For complete entry see 16.7

10.3   **anti - α - Bromoacetophenone oxime**
C_8H_8BrNO
J.B.Wetherington, J.W.Moncrief
Amer. Cryst. Assoc., Abstr. Papers (Winter Meeting), 80, 1973

10.C   **Isonitroso - acetanilide**
$C_8H_8N_2O_2$
For complete entry see 16.8

10.4   **4 - Methoxy - isonitrosoacetanilide**
$C_9H_{10}N_2O_3$
M.Font-Altaba, J.L.Brianso, C.Miravitlles, F.Plana
Eur. Cryst. Meeting, 1973
Also classified in 16

10.C   **p - Dimethylaminobenzaldoxime (α form)**
$C_9H_{12}N_2O$
For complete entry see 16.11

10.C   **2 - Ethoxy - isonitrosoacetanilide**
$C_{10}H_{12}N_2O_3$
For complete entry see 16.14

10.C   **3 - Methyl - N - ethyl - isonitrosoacetanilide**
$C_{11}H_{14}N_2O_2$
For complete entry see 16.16

10.C   **Pentafluoronitrosobenzene - cis - azo(pentafluorobenzene) - dioxide**
$2C_{12}F_{10}N_2O_2$, C_6F_5NO
For complete entry see 10.1

10.5   **α,α,N - Triphenylnitrone**
$C_{19}H_{15}NO$
J.N.Brown, L.M.Trefonas   *Acta Cryst. (B),* **29,** 237, 1973

SULPHUR AND SELENIUM COMPOUNDS

11.C   **3 - Methyl - cytidine methosulfate**
$CH_3O_4S^-$, $C_{10}H_{16}N_3O_5^+$
For complete entry see 47.25

11.C   **10 - (2′ - Trimethylammonium - n - propyl) - phenothiazine methyl sulfate**
$CH_3O_4S^-$, $C_{18}H_{23}N_2S^+$
For complete entry see 41.45

11.C   **N - Ammonium - phosphoryl - taurocyamine**
$C_3H_8N_3O_6PS^{2-}$, $2H_4N^+$
For complete entry see 8.10

11.C   **Taurocyamine**
$C_3H_9N_3O_3S$
For complete entry see 8.11

11.1   **Selenium di(methylxanthate)**
$C_4H_6O_2S_4Se$
N.J.Brondmo, S.Esperas, H.Graver, S.Husebye
Acta Chem. Scand., **27,** 713, 1973

11.2   **Sulfur di(methylxanthate)**
$C_4H_6O_2S_5$
N.J.Brondmo, S.Esperas, H.Graver, S.Husebye
Acta Chem. Scand., **27,** 713, 1973

11.C   **γ - Guanidino - β - hydroxy - propane - sulfonic acid**
$C_4H_{11}N_3O_4S$
For complete entry see 8.12

11.C   **o - Sulfanilamide**
$C_6H_8N_2O_2S$
For complete entry see 16.3

11.3   **p - Toluene sulfonic acid monohydrate (orthorhombic form)**
$C_7H_7O_3S^-$, H_3O^+
D.D.Dexter   *Z. Kristallogr.*, **134,** 350, 1971

11.4   **p - Toluene sulfonic acid monohydrate (neutron study)**
$C_7H_7O_3S^-$, H_3O^+
J.-O.Lundgren, J.M.Williams   *J. Chem. Phys.*, **58,** 788, 1973

11.C **o - Nitrobenzene - selenyl thiourea thiocyanate**
$C_7H_8N_3O_2SSe^+$, CNS^-
For complete entry see 8.15

11.5 **Homosulfanilamide - sulfathiazole complex**
$C_7H_{11}N_2O_2S^+$, $C_9H_8N_3O_2S_2^-$
T.F.Brennan, E.Shefter, P.Sackman   *Chem. Pharm. Bull.*, **19,** 1919, 1971
Residue 1 also classified in 3; residue 2 classified in 41, 11, 16

11.C **2,3,5,6 - Tetrachloro - 4 - (methylthio) - benzonitrile**
$C_8H_3Cl_4NS$
For complete entry see 7.5

11.6 **o - Carboxyphenyl methyl sulfoxide**
$C_8H_8O_3S$
B.Dahlen   *Acta Cryst. (B)*, **29,** 595, 1973
Also classified in 13

11.7 **o - Carboxyphenyl methyl selenium oxide**
$C_8H_8O_3Se$
B.Dahlen   *Acta Cryst. (B)*, **29,** 595, 1973
Also classified in 13

11.C **2,4,5 - Trichloro - 6 - (methylthio)isophthalonitrile**
$C_9H_3Cl_3N_2S$
For complete entry see 7.6

11.C **Homosulfanilamide - sulfathiazole complex**
$C_9H_8N_3O_2S_2^-$, $C_7H_{11}N_2O_2S^+$
For complete entry see 11.5

11.8 **Dipotassium mesitylene - disulfonate dihydrate**
$C_9H_{10}O_6S_2^{2-}$, $2K^+$, $2H_2O$
M.A.M.Meester, H.Schenk   *Rec. Trav. Chim. Pays-Bas,* **91,**213, 1972

11.9 **Propenyl p - tolyl sulfone**
$C_{10}H_{12}O_2S$
A.H.Klazinga, A.Vos   *Rec. Trav. Chim. Pays-Bas,* **92,** 360, 1973

11.C **bis(4 - Morpholinethiocarbonyl) - trisulfide**
$C_{10}H_{16}N_2O_2S_5$
For complete entry see 40.16

11.C **bis(N,N - Cyclopentamethylene - thiocarbonyl)disulfide**
Piperidylthiuram disulfide
$C_{12}H_{20}N_2S_4$
For complete entry see 33.46

11.10 **bis(4 - Bromophenylsulfonyl)methane**
$C_{13}H_{10}Br_2O_4S_2$
J.Berthou, G.Jeminet, A.Laurent   *Acta Cryst. (B)*, **28,** 2480, 1972

11.C **(Ethylsulfinyl) - L - tryptophan (absolute configuration)**
$C_{13}H_{16}N_2O_3S$
For complete entry see 48.69

11.11 **p - Methoxybenzenesulfone - p - anisidide**
$C_{14}H_{15}NO_4S$
S.Pokrywiecki, C.M.Weeks, W.L.Duax   *Cryst. Struct. Comm.*, **2**, 63, 1973
Also classified in 16, 17

11.12 **4 - Dimethylaminodiphenyl sulfide**
$C_{14}H_{15}NS$
C.Panattoni, G.Bandoli, D.A.Clemente, A.Dondoni, A.Mangini
J. Cryst. Mol. Struct., **3**, 65, 1973
Also classified in 16

11.13 **S - (2 - Methoxyphenyl) - N - (2,6 - dimethylphenyl) - dithiourethane S -
oxide S - (2 - methoxyphenyl) - N - (2,6 - dimethylphenyl) - dithiourethane
sulphenic acid solid solution**
$C_{16}H_{17}NO_2S_2$, $C_{16}H_{17}NO_2S_2$
K.Kato   *Acta Cryst. (B)*, **28**, 2653, 1972

11.14 **Methyl 2 - phenyl - (3',4' - dimethyl - 2 - phenyl)vinyl sulfoxide**
$C_{17}H_{18}OS$
D.Tranqui, H.Fillion   *Acta Cryst. (B)*, **28**, 3306, 1972

11.15 **p - Methoxybenzenesulfone - N - isopropyl - p - anisidide**
$C_{17}H_{21}NO_4S$
S.Pokrywiecki, C.M.Weeks, W.L.Duax   *Cryst. Struct. Comm.*, **2**, 67, 1973
Also classified in 16, 17

11.C **(+) - N - Phthalimido - 2 - bromo - 2(S) - octyl - p - tolyl - (R) - sulfoxime
(absolute configuration)**
$C_{23}H_{27}BrN_2O_3S$
For complete entry see 9.17

11.C **N - (4 - (β - 2 - Methoxy - 5 - bromobenzamidoethyl)benzensulfonyl) - N' -
4 - methyl - cyclohexyl - urea**
$C_{24}H_{30}BrN_3O_5S$
For complete entry see 8.19

11.16 **Diphenyl - di((bis(trifluoromethyl) - phenyl)methoxy) - sulfurane**
$C_{30}H_{20}F_{12}O_2S$
I.C.Paul, J.C.Martin, E.F.Perozzi   *J. Amer. Chem. Soc.*, **94**, 5010, 1972

11.C **Glycerol 1,2 - di(11 - bromoundecanoate) - 3 - (p - toluenesulfonate)**
$C_{32}H_{52}Br_2O_7S$
For complete entry see 1.38

CARBONIUM IONS, CARBANIONS, RADICALS

12.1   **Potassium trinitromethanide**
$CN_3O_6^-$, K^+
N.I.Golovina, L.O.Atovmyan *Zh. Strukt. Khim.*, **8,** 307, 1967

12.2   **Triazidocarbonium hexachloroantimonate**
CN_9^+ , Cl_6Sb^-
U.Muller, H.Barnighausen *Acta Cryst. (B)*, **26,** 1671, 1970

12.C   **Potassium dinitroacetonitrile**
$C_2N_3O_4^-$, K^+
For complete entry see 7.1

12.C   **Rubidium dinitroacetonitrile**
$C_2N_3O_4^-$, Rb^+
For complete entry see 7.2

12.3   **2 - Methylphenyl - oxocarbonium hexachloroantimonate**
$C_8H_7O^+$, Cl_6Sb^-
B.Chevrier, J.-M.Le Carpentier, R.Weiss *Acta Cryst. (B)*, **28,** 2673, 1972

12.4   **4 - Methylphenyl - oxocarbonium hexachloroantimonate**
$C_8H_7O^+$, Cl_6Sb^-
B.Chevrier, J.-M.Le Carpentier, R.Weiss
J. Amer. Chem. Soc., **94,** 5718, 1972

12.C   **1,2,3 - Trimethyl - benzimidazolium tetracyanoquinodimethanide**
$C_{12}H_4N_4^-$, $C_{10}H_{13}N_2^+$
For complete entry see 35.10

12.5   **(Cycloheptatrienylidene) - (cycloheptatrienyl) - acetonitrile perchlorate**
$C_{16}H_{12}N^+$, ClO_4^-
C.Kabuto, M.Oda, Y.Kitahara *Tetrahedron Letters,* 4851, 1972
Residue 1 also classified in 22

12.6   **Triphenylcarbenium pentachloroplatinate(iv) dichloromethane solvate**
$C_{19}H_{15}^+$, Cl_5Pt^- , CH_2Cl_2
P.M.Cook, L.F.Dahl, D.W.Dickerhoof
J. Amer. Chem. Soc., **94,** 5511, 1972

12.7   **bis(Triphenylcarbenium) decachlorodiplatinate(iv)**
$2C_{19}H_{15}{}^+$, $Cl_{10}Pt_2{}^{2-}$
P.M.Cook, L.F.Dahl, D.W.Dickerhoof
J. Amer. Chem. Soc., **94**, 5511, 1972

12.8   **bis(Triphenylcarbenium) dodecachlorotriplatinate**
$2C_{19}H_{15}{}^+$, $Cl_{12}Pt_3{}^{2-}$
P.M.Cook, L.F.Dahl, D.W.Dickerhoof
J. Amer. Chem. Soc., **94**, 5511, 1972

12.9   **bis(Triphenylcarbenium) dodecachlorotriplatinate tetrachloroethane solvate**
$2C_{19}H_{15}{}^+$, $Cl_{12}Pt_3{}^{2-}$, $2C_2H_2Cl_4$
P.M.Cook, L.F.Dahl, D.W.Dickerhoof
J. Amer. Chem. Soc., **94**, 5511, 1972

12.10   **bis(Triphenylcarbenium)di - μ - chloro - bis((di - μ - chloro) - (μ - o - phenylene)bis(dichloroplatinum)) dichloromethane solvate**
$2C_{19}H_{15}{}^+$, $C_{12}H_8Cl_{14}Pt_4{}^{2-}$, $2CH_2Cl_2$
P.M.Cook, L.F.Dahl, D.W.Dickerhoof
J. Amer. Chem. Soc., **94**, 5511, 1972
Residue 2 classified in 71

12.11   **Tetramethyl - p - phenylenediamine free radical bis(maleonitriledithiolato) nickel(ii)**
$2C_{20}H_{16}N_2{}^+$, $C_8N_4NiS_4{}^{2-}$
M.J.Hove, B.M.Hofman, J.A.Ibers   *J. Chem. Phys.*, **56**, 3490, 1972
Residue 1 also classified in 16; residue 2 classified in 85

12.12   **2,4,6 - Triphenylverdazyl**
$C_{20}H_{17}N_4$
D.E.Williams   *Acta Cryst. (B),* **29,** 96, 1973
Also classified in 33

12.13   **(Cycloheptatrienylidene) - (2,3 - diphenylcyclopropen - 1 - yl) - acetonitrile tetrafluoroborate**
$C_{24}H_{18}N^+$, $BF_4{}^-$
C.Kabuto, M.Oda, Y.Kitahara   *Tetrahedron Letters,* 4851, 1972
Residue 1 also classified in 20, 22

12.C   **Di - (2,2,6,6 - tetramethyl - 4 - piperidinyl - 1 - oxyl)suberate**
$C_{26}H_{46}N_2O_6$
For complete entry see 1.37

BENZOIC ACID DERIVATIVES

13.1 **Pentafluorobenzoic acid**
$C_7HF_5O_2$
V.Benghiat, L.Leiserowitz *J. C. S. Perkin II,* 1778, 1972

13.2 **p - Bromobenzoic acid**
$C_7H_5BrO_2$
K.Ohkura, S.Kashino, M.Haisa *Bull. Chem. Soc. Jap.,* **45,** 2651, 1972

13.C **Potassium hydrogen bis - m - chlorobenzoate**
$C_7H_5ClO_2$, $C_7H_4ClO_2^-$, K^+
For complete entry see 14.1

13.3 **o - Chlorobenzamide (α form)**
C_7H_6ClNO
K.Sakurai, Y.Takaki, Y.Kato *Acta Cryst. (A),* **28,** S120, 1972

13.4 **o - Chlorobenzamide (β form)**
C_7H_6ClNO
K.Sakurai, Y.Takaki, Y.Kato *Acta Cryst. (A),* **28,** S120, 1972

13.5 **p - Iodobenzamide**
C_7H_6INO
K.Nakata, Y.Kato, Y.Takaki, K.Sakurai
Mem. Osaka Kyoiku Univ., **20,** 93, 1971

13.6 **o - Nitrobenzamide**
$C_7H_6N_2O_3$
K.Fujimori, T.Tsukihara, Y.Katsube, J.Yamamoto
Bull. Chem. Soc. Jap., **45,** 1564, 1972
Also classified in 15

13.7 **Salicylic acid (neutron study)**
$C_7H_6O_3$
G.E.Bacon, R.J.Jude *Acta Cryst. (A),* **28,** S193, 1972
Also classified in 17

13.8 **Benzamide**
C_7H_7NO
C.C.F.Blake, R.W.H.Small *Acta Cryst. (B),* **28,** 2201, 1972

13.C   **Benzamide - succinic acid complex**
$2C_7H_7NO$, $C_4H_6O_4$
For complete entry see 60.9

13.C   **Amobarbital - salicylamide complex**
$C_7H_7NO_2$, $C_{11}H_{18}N_2O_3$
For complete entry see 60.13

13.C   **Ammonium hydrogen terephthalate**
$C_8H_5O_4^-$, H_4N^+
For complete entry see 14.2

13.9   **p - Bromo - benzimino methyl ether**
C_8H_8BrNO
B.Kolakowski   *Stockholm Symposium*, 89, 1973
Also classified in 3

13.10   **Terephthalamide**
$C_8H_8N_2O_2$
R.E.Cobbledick, R.W.H.Small   *Acta Cryst. (B)*, **28,** 2894, 1972

13.C   **o - Carboxyphenyl methyl sulfoxide**
$C_8H_8O_3S$
For complete entry see 11.6

13.C   **o - Carboxyphenyl methyl selenium oxide**
$C_8H_8O_3Se$
For complete entry see 11.7

13.C   **Rubidium hydrogen bis(homophthalate)**
$C_9H_7O_4^-$, $C_9H_8O_4$, Rb^+
For complete entry see 14.4

13.11   **Acetyl benzoyl peroxide (at $-30°C$)**
$C_9H_8O_4$
N.J.Karch, J.M.McBride   *J. Amer. Chem. Soc.*, **94,** 5092, 1972

13.C   **Potassium hydrogen bis(homophthalate)**
$C_9H_8O_4$, $C_9H_7O_4^-$, K^+
For complete entry see 14.3

13.12   **o - Ethoxybenzoic acid (at $-10\pm4°C$)**
$C_9H_{10}O_3$
E.M.Gopalakrishna, L.Cartz   *Acta Cryst. (B)*, **28,** 2917, 1972
Also classified in 17

13.C   **Pyromellitic dithioanhydride**
$C_{10}H_2O_4S_2$
For complete entry see 39.19

13.13 **Mesitoic acid**
$C_{10}H_{12}O_2$
V.Benghiat, L.Leiserowitz *J. C. S. Perkin II,* 1778, 1972

13.14 **Mellite**
Aluminium mellitate hydrate
$C_{12}O_{12}^{6-}$, $H_{30}Al_2O_{15}^{6+}$, H_2O
C.Giacovazzo, S.Menchetti, F.Scordari *Acta Cryst. (B),* **29,** 26, 1973

13.15 **3' - Iodobiphenyl - 4 - carboxylic acid**
$C_{13}H_9IO_2$
H.H.Sutherland, M.J.Mottram *Acta Cryst. (B),* **28,** 2212, 1972

13.C **2 - (2' - Carbomethoxy - 4' - nitrophenoxy) - 1,3,5 - trichlorobenzene**
$C_{14}H_8Cl_3NO_5$
For complete entry see 15.10

13.C **trans - 4 - t - Butylcyclohexanol p - bromobenzoate**
$C_{17}H_{23}BrO_2$
For complete entry see 21.12

13.C **(+) - N - Phthalimido - 2 - bromo - 2(S) - octyl - p - tolyl - (R) - sulfoxime (absolute configuration)**
$C_{23}H_{27}BrN_2O_3S$
For complete entry see 9.17

13.16 **N - t - Butyl - N - benzyl - biphenyl - carboxylic acid amide**
$C_{24}H_{25}NO$
G.Allegra, G.Avitabile, E.Benedetti, M.R.Ciajolo, P.Corradini, P.Ganis, M.Goodman, A.Immirzi, C.Pedone *Acta Cryst. (A),* **28,** S13, 1972

13.C **N - (4 - (β - 2 - Methoxy - 5 - bromobenzamidoethyl)benzensulfonyl) - N' - 4 - methyl - cyclohexyl - urea**
$C_{24}H_{30}BrN_3O_5S$
For complete entry see 8.19

BENZOIC ACID SALTS
(AMMONIUM, IA, IIA METALS)

14.1   **Potassium hydrogen bis - m - chlorobenzoate**

$C_7H_4ClO_2^-$, $C_7H_5ClO_2$, K^+

A.L.Macdonald, J.C.Speakman   *J. C. S. Perkin II,* 1564, 1972

Residue 2 classified in 13

14.2   **Ammonium hydrogen terephthalate**

$C_8H_5O_4^-$, H_4N^+

R.E.Cobbledick, R.W.H.Small   *Acta Cryst. (B),* **28,** 2924, 1972

Residue 1 also classified in 13

14.C   **Triaquo - calcium terephthalate**

$(C_8H_{10}CaO_7)_n$

For complete entry see 67.5

14.3   **Potassium hydrogen bis(homophthalate)**

$C_9H_7O_4^-$, $C_9H_8O_4$, K^+

M.P.Gupta, D.S.Dubey   *Acta Cryst. (B),* **28,** 2677, 1972

Residue 2 classified in 13

14.4   **Rubidium hydrogen bis(homophthalate)**

$C_9H_8O_4$, $C_9H_7O_4^-$, Rb^+

M.P.Gupta, D.S.Dubey   *Z. Kristallogr.,* **135,** 273, 1972

Residue 2 classified in 13

14.C   **Calcium mellitate hydrate**

$C_{12}H_2Ca_2O_{12}$, $9H_2O$

For complete entry see 67.7

BENZENE NITRO COMPOUNDS

15.1 **Lithium picrate monohydrate**
$C_6H_2N_3O_7{}^-$, Li^+ , H_2O
J.F.Griffin, P.Coppens
Amer. Cryst. Assoc., Abstr. Papers (Winter Meeting), 30, 1973
Residue 1 also classified in 17

15.C **Guanine picrate monohydrate**
$C_6H_2N_3O_7{}^-$, $C_5H_6N_5O^+$, H_2O
For complete entry see 44.6

15.C **Thioguanine picrate monohydrate**
$C_6H_2N_3O_7{}^-$, $C_5H_6N_5S^+$, H_2O
For complete entry see 44.7

15.C **3,5,7 - Triphenyl - 4H - 1,2 - diazepine picrate**
$C_6H_2N_3O_7{}^-$, $C_{23}H_{19}N_2{}^+$
For complete entry see 34.7

15.C **Succinyl - choline picrate**
$2C_6H_2N_3O_7{}^-$, $C_{14}H_{30}N_2O_4{}^{2+}$
For complete entry see 59.10

15.C **Trinitrobenzene - benzothiophene complex**
$C_6H_3N_3O_6$, C_8H_6S
For complete entry see 60.5

15.C **Pyrene - 1,3,5 - trinitrobenzene complex**
$C_6H_3N_3O_6$, $C_{16}H_{10}$
For complete entry see 60.29

15.C **2,4,6 - Trinitroaniline**
$C_6H_4N_4O_6$
For complete entry see 16.1

15.C **Mono - (p - nitrophenyl) phosphate**
$C_6H_6NO_6P$
For complete entry see 46.4

15.C **m - Nitroaniline**
$C_6H_6N_2O_2$
For complete entry see 16.2

15.C syn - p - **Nitrobenzaldoxime**
$C_7H_6N_2O_3$
For complete entry see 10.2

15.C o - **Nitrobenzamide**
$C_7H_6N_2O_3$
For complete entry see 13.6

15.C o - **Nitrobenzene - selenyl thiourea thiocyanate**
$C_7H_8N_3O_2SSe^+$, CNS^-
For complete entry see 8.15

15.2 p - **Nitroacetophenone**
$C_8H_7NO_3$
J.K.S.Kim, E.R.Boyko, G.B.Carpenter   *Acta Cryst. (B)*, **29**, 1141, 1973

15.3 p - **Nitrophenyl - acetate**
$C_8H_7NO_4$
R.J.Guttormson, B.E.Robertson   *Acta Cryst. (B)*, **28**, 2702, 1972

15.C N - **Methyl - 2,4,6 - trinitroacetanilide**
$C_9H_8N_4O_7$
For complete entry see 16.9

15.4 p - **Nitro - benzylimino methyl ether**
$C_9H_{10}N_2O_3$
B.Kolakowski   *Stockholm Symposium*, 89, 1973
Also classified in 3

15.C E - 2 - p - **Nitrophenyl - cyclopropyl methyl ketone**
$C_{11}H_{11}NO_3$
For complete entry see 20.13

15.5 **Hexanitro - azobenzene (β form)**
$C_{12}H_4N_8O_{12}$
E.J.Graeber, B.Morosin   *Acta Cryst. (A)*, **28**, S22, 1972
Also classified in 9

15.6 **Hexanitro - azobenzene (γ form)**
$C_{12}H_4N_8O_{12}$
E.J.Graeber, B.Morosin   *Acta Cryst. (A)*, **28**, S22, 1972
Also classified in 9

15.7 **Hexanitro - azobenzene (α form)**
$C_{12}H_4N_8O_{12}$
E.J.Graeber, B.Morosin   *Acta Cryst. (A)*, **28**, S22, 1972
Also classified in 9

15.8 p - **Nitrobiphenyl**
$C_{12}H_9NO_2$
G.Casalone, A.Gavezzotti, M.Simonetta   *J. C. S. Perkin II*, 342, 1973

15.C   **Di - (p - nitrophenyl) phosphate**
$C_{12}H_9N_2O_8P$
For complete entry see 46.8

15.9   **bis - p - Nitrophenyl - carbodi - imide**
$C_{13}H_8N_4O_4$
A.T.Vincent, P.J.Wheatley   *J. C. S. Perkin II,* 1567, 1972

15.10   **2 - (2' - Carbomethoxy - 4' - nitrophenoxy) - 1,3,5 - trichlorobenzene**
$C_{14}H_8Cl_3NO_5$
E.A.H.Griffith, W.D.Chandler, B.E.Robertson
Canad. J. Chem., **50,** 2979, 1972
Also classified in 17, 13

15.C   **4 - Nitro - 4' - methoxy - N - benzylidene - aniline**
$C_{14}H_{12}N_2O_3$
For complete entry see 16.20

15.11   **2 - (3' - Methyl - 4' - nitrophenoxy) - 1,3 - di - isopropylbenzene**
$C_{19}H_{23}NO_3$
E.A.H.Griffith, W.D.Chandler, B.E.Robertson
Canad. J. Chem., **50,** 2963, 1972
Also classified in 17

15.12   **2 - (2',4' - Dinitrophenoxy) - 1,3,5 - tri - t - butylbenzene**
$C_{24}H_{32}N_2O_5$
E.A.H.Griffith, W.D.Chandler, B.E.Robertson
Canad. J. Chem., **50,** 2972, 1972
Also classified in 17

ANILINES

16.1   **2,4,6 - Trinitroaniline**
$C_6H_4N_4O_6$
J.R.Holden, C.Dickinson, C.M.Bock   *J. Phys. Chem.*, **76,** 3597, 1972
Also classified in 15

16.C   **Dichlorophosphinyl(dichlorophosphinothioyl)aniline**
$C_6H_5Cl_4NOP_2S$
For complete entry see 64.3

16.2   **m - Nitroaniline**
$C_6H_6N_2O_2$
A.C.Skapski, J.L.Stevenson   *J. C. S. Perkin II,* 1197, 1973
Also classified in 15

16.C   **p - Phenylenediamine - 1,2,4,5 - tetracyanobenzene complex**
$C_6H_8N_2$, $C_{10}H_2N_4$
For complete entry see 60.7

16.3   **o - Sulfanilamide**
$C_6H_8N_2O_2S$
M.le Bars, M.Alleaume   *C. R. Acad. Sci., Fr., C,* **275,** 187, 1972
Also classified in 11

16.4   **o - Phenylenediamine hydrochloride**
$C_6H_9N_2{}^+$, Cl^-
C.Stalhandske   *Acta Chem. Scand.*, **26,** 2962, 1972

16.5   **o - Phenylenediamine dihydrobromide**
$C_6H_{10}N_2{}^{2+}$, $2Br^-$
C.Stalhandske   *Acta Chem. Scand.*, **26,** 3029, 1972

16.6   **2 - Amino - 5 - bromotoluene**
C_7H_8BrN
H.van der Meer   *Acta Cryst. (B),* **28,** 3098, 1972

16.7   **2 - Chloro - isonitrosacetanilide**
$C_8H_7ClN_2O_2$
M.Font-Altaba, J.L.Brianso, C.Miravitlles, F.Plana
Eur. Cryst. Meeting, 1973
Also classified in 10

16.8   Isonitroso - acetanilide
$C_8H_8N_2O_2$
J.L.Brianso, C.Miravitlles, M.Font-Altaba, J.P.Declerq, G.Germain
Cryst. Struct. Comm., **2,** 319, 1973
Also classified in 10

16.C   Homosulfanilamide - sulfathiazole complex
$C_9H_8N_3O_2S_2^-$, $C_7H_{11}N_2O_2S^+$
For complete entry see 11.5

16.9   N - Methyl - 2,4,6 - trinitroacetanilide
$C_9H_8N_4O_7$
G.G.Christoph, E.B.Fleischer   *Acta Cryst. (B),* **29,** 121, 1973
Also classified in 15

16.C   4 - Methoxy - isonitrosoacetanilide
$C_9H_{10}N_2O_3$
For complete entry see 10.4

16.10   p - Dimethylamino - benzaldehyde hydrobromide
$C_9H_{12}NO^+$, Br^-
J.K.Dattagupta, N.N.Saha   *Acta Cryst. (B),* **29,** 1228, 1973

16.11   p - Dimethylaminobenzaldoxime (α form)
$C_9H_{12}N_2O$
F.Bachechi, L.Zambonelli   *Acta Cryst. (B),* **28,** 2489, 1972
Also classified in 10

16.12   Trimethylphenylammonium nonachlorodirhodate(iii)
Trimethylanilinium nonachlorodirhodate(iii)
$3C_9H_{14}N^+$, $Cl_9Rh_2^{3-}$
F.A.Cotton, D.A.Ucko   *Inorg. Chim. Acta,* **6,** 161, 1972

16.13   4 - Ethoxy - isonitroso - acetanilide
p - Glyoxyl - phenetidine oxime
$C_{10}H_{12}N_2O_3$
M.Font-Altaba, F.Plana, J.L.Brianso   *Acta Cryst. (A),* **28,** S21, 1972
Also classified in 17

16.14   2 - Ethoxy - isonitrosoacetanilide
$C_{10}H_{12}N_2O_3$
M.Font-Altaba, J.L.Brianso, C.Miravitlles, F.Plana
Eur. Cryst. Meeting, 1973
Also classified in 10

16.15   Di(phenacetin) hydrogen penta - iodide
$C_{10}H_{13}NO_2$, $C_{10}H_{14}NO_2^+$, I_3^- , I_2
F.H.Herbstein, M.Kapon   *Nature Phys. Sci.,* **239,** 153, 1972
Residue 1 also classified in 17

16.16   3 - Methyl - N - ethyl - isonitrosoacetanilide
$C_{11}H_{14}N_2O_2$
M.Font-Altaba, J.L.Brianso, C.Miravitlles, F.Plana
Eur. Cryst. Meeting, 1973
Also classified in 10

16.C   Benzidine - 7,7,8,8 - tetracyano - p - quinodimethane complex
$C_{12}H_{12}N_2$, $C_{12}H_4N_4$
For complete entry see 60.21

16.C   Benzidine - 7,7,8,8 - tetracyano - p - quinodimethane complex
dichloromethane solvate
$C_{12}H_{12}N_2$, $C_{12}H_4N_4$, $1.8CH_2Cl_2$
For complete entry see 60.22

16.C   Benzidine - 7,7,8,8 - tetracyano - p - quinodimethane complex benzene
solvate
$C_{12}H_{12}N_2$, $C_{12}H_4N_4$, $1.35C_6H_6$
For complete entry see 60.23

16.17   Benzidine dihydrochloride
$C_{12}H_{14}N_2^{2+}$, $2Cl^-$
C.H.Koo, H.S.Kim, H.S.Shin   *Daehan Hwahak Hwoejee,* **16,** 18, 1972

16.18   tris(4,4' - Diaminodiphenylmethane) - sodium chloride
$3C_{13}H_{14}N_2$, Na^+ , Cl^-
J.W.Swardstrom, L.A.Duvall, D.P.Miller   *Acta Cryst. (B),* **28,** 2510, 1972

16.19   2 - Chloro - 5 - trifluoromethyl - diethylacetanilide
$C_{13}H_{15}ClF_3NO$
C.Cohen-Addad   *Acta Cryst. (B),* **29,** 157, 1973

16.20   4 - Nitro - 4' - methoxy - N - benzylidene - aniline
$C_{14}H_{12}N_2O_3$
J.Meunier-Piret, P.Piret, G.Germain, M.van Meerssche
Bull. Soc. Chim. Belges, **81,** 533, 1972
Also classified in 15, 17

16.C   Sodium 4' - dimethylaminoazobenzene - 4 - sulfonate monohydrate ethanol
solvate
Methyl orange monohydrate ethanol solvate
$C_{14}H_{14}N_3O_3S^-$, Na^+ , H_2O , C_2H_6O
For complete entry see 9.13

16.C   p - Methoxybenzenesulfone - p - anisidide
$C_{14}H_{15}NO_4S$
For complete entry see 11.11

16.C   4 - Dimethylaminodiphenyl sulfide
$C_{14}H_{15}NS$
For complete entry see 11.12

16.21   **4 - Chloro - di - n - propylacetanilide**
$C_{14}H_{20}ClNO$
C.Cohen-Addad   *Acta Cryst. (B)*, **29**, 157, 1973

16.22   **Lidocaine hydrochloride monohydrate**
$C_{14}H_{23}N_2O^+$, Cl^- , H_2O
A.W.Hanson, M.Rohrl   *Acta Cryst. (B)*, **28**, 3567, 1972
Residue 1 also classified in 3

16.23   **2 - Bromo - 5 - trifluoromethyl - dipropylacetanilide**
$C_{15}H_{19}BrF_3NO$
C.Cohen-Addad, A.Grand, J.Lajzerowicz   *Eur. Cryst. Meeting*, 1973

16.C   **p - Methoxybenzenesulfone - N - isopropyl - p - anisidide**
$C_{17}H_{21}NO_4S$
For complete entry see 11.15

16.C   **2 - (p - Dimethylanilino) - 4 - phenyl - 6a - thiathiophthene**
$C_{19}H_{17}NS_3$
For complete entry see 39.40

16.C   **Tetramethyl - p - phenylenediamine free radical bis(maleonitriledithiolato) nickel(ii)**
$2C_{20}H_{16}N_2^+$, $C_8N_4NiS_4^{2-}$
For complete entry see 12.11

16.C   **(p - Dimethylamino - phenyl)diphenylphosphine**
$C_{20}H_{20}NP$
For complete entry see 64.31

16.24   **1,1 - bis(N - Phenylcarbamyl) - 2 - (p - chlorophenyl) - ethylene**
$C_{22}H_{17}ClN_2O_2$
K.A.Kerr, J.P.Ashmore   *Stockholm Symposium*, 91, 1973

PHENOLS AND ETHERS

17.C **bis(Cytosine)resorcinic acid monohydrate**
$C_4H_5N_3O$, $C_7H_5O_4^-$, $C_4H_6N_3O^+$, H_2O
For complete entry see 44.3

17.C **Lithium picrate monohydrate**
$C_6H_2N_3O_7^-$, Li^+ , H_2O
For complete entry see 15.1

17.C **Guanine picrate monohydrate**
$C_6H_2N_3O_7^-$, $C_5H_6N_5O^+$, H_2O
For complete entry see 44.6

17.C **Thioguanine picrate monohydrate**
$C_6H_2N_3O_7^-$, $C_5H_6N_5S^+$, H_2O
For complete entry see 44.7

17.C **3,5,7 - Triphenyl - 4H - 1,2 - diazepine picrate**
$C_6H_2N_3O_7^-$, $C_{23}H_{19}N_2^+$
For complete entry see 34.7

17.C **Succinyl - choline picrate**
$2C_6H_2N_3O_7^-$, $C_{14}H_{30}N_2O_4^{2+}$
For complete entry see 59.10

17.1 **2,5 - Dichlorophenol**
$C_6H_4Cl_2O$
C.Bavoux, M.Perrin   *Acta Cryst. (B)*, **29**, 666, 1973

17.2 **p - Chlorophenol (α form)**
C_6H_5ClO
M.Perrin, P.Michel   *Acta Cryst. (B)*, **29**, 253, 1973

17.3 **p - Chlorophenol (α form)**
C_6H_5ClO
R.Shiono, V.Y.Wu   *Acta Cryst. (A)*, **28**, S17, 1972

17.4 **p - Chlorophenol (β form, at low temp.)**
C_6H_5ClO
M.Perrin, P.Michel   *Acta Cryst. (B)*, **29**, 258, 1973

17.5   **Resorcinol (α form,neutron study)**
$C_6H_6O_2$
G.E.Bacon, R.J.Jude   *Acta Cryst. (A)*, **28**, S193, 1972

17.C   **4 - Nitropyridine - N - oxide - hydroquinone complex**
$C_6H_6O_2$, $C_5H_4N_2O_3$
For complete entry see 60.2

17.C   **Lumiflavinium chloride - hydroquinone complex**
$C_6H_6O_2$, $C_{13}H_{13}N_4O_2{}^+$, Cl^-
For complete entry see 36.14

17.C   **Salicylic acid (neutron study)**
$C_7H_6O_3$
For complete entry see 13.7

17.6   **2 - Methyl - 3 - bromophenol**
C_7H_7BrO
M.Maze-Baudet   *Acta Cryst. (B)*, **29**, 602, 1973

17.C   **Amobarbital - salicylamide complex**
$C_7H_7NO_2$, $C_{11}H_{18}N_2O_3$
For complete entry see 60.13

17.7   **m - Cresol (at $-100°$C)**
C_7H_8O
C.Bois   *Acta Cryst. (B)*, **29**, 1011, 1973

17.8   **2,3 - Dimethylphenol**
$C_8H_{10}O$
M.Maze-Baudet   *Acta Cryst. (B)*, **29**, 602, 1973

17.9   **2,3 - Dimethylphenol (at $-150°$C)**
$C_8H_{10}O$
A.Neuman, H.Gillier-Pandraud   *Acta Cryst. (B)*, **29**, 1017, 1973

17.10   **2,6 - Dimethylphenol**
$C_8H_{10}O$
H.Gillier-Pandraud, P.Becker, F.Longchambon, D.Antona
C. R. Acad. Sci., Fr., C, **275**, 1495, 1972

17.11   **3,5 - Dimethylphenol**
$C_8H_{10}O$
H.Gillier-Pandraud, P.Becker, F.Longchambon, D.Antona
C. R. Acad. Sci., Fr., C, **275**, 1495, 1972

17.12   **2,5 - Dimethylphenol (at $-150°$C)**
$C_8H_{10}O$
A.Neuman, H.Gillier-Pandraud   *Acta Cryst. (B)*, **29**, 1017, 1973

17.13   **3,4 - Dimethylphenol (at −130°C)**
$C_8H_{10}O$
H.Gillier-Pandraud, M.P.Becker, M.-T.Vandenborre, C.Bois
C. R. Acad. Sci., Fr., C, **276,** 411, 1973

17.C   **5 - Hydroxydopamine hydrochloride**
3,4,5 - Trihydroxyphenyl - ethylamine hydrochloride
$C_8H_{12}NO_3^+$, Cl^-
For complete entry see 59.4

17.C   **o - Ethoxybenzoic acid (at −10±4°C)**
$C_9H_{10}O_3$
For complete entry see 13.12

17.14   **(−) - Adrenaline hydrogen (+) - tartrate (absolute configuration)**
$C_9H_{14}NO_3^+$, $C_4H_5O_6^-$
D.Carlstrom   *Acta Cryst. (B),* **29,** 161, 1973
Residue 1 also classified in 3

17.C   **4 - Ethoxy - isonitroso - acetanilide**
p - Glyoxyl - phenetidine oxime
$C_{10}H_{12}N_2O_3$
For complete entry see 16.13

17.C   **Di(phenacetin) hydrogen penta - iodide**
$C_{10}H_{13}NO_2$, $C_{10}H_{14}NO_2^+$, I_3^- , I_2
For complete entry see 16.15

17.C   **2,5 - Dihydroxyphenyl - pyridinium bromide monohydrate**
$C_{11}H_{10}NO_2^+$, Br^- , H_2O
For complete entry see 33.45

17.15   **DL - 1 - (3,5 - Dihydroxyphenyl) - 2 - (isopropylamino)ethanol sulfate ethanol solvate**
$2C_{11}H_{17}NO_3^+$, O_4S^{2-} , C_2H_6O
J.P.Beale   *Cryst. Struct. Comm.,* **1,** 297, 1972
Residue 1 also classified in 3

17.C   **1 - (2 - Aminoethyl) - 3,4,5 - trimethoxybenzene hydrochloride**
Mescaline hydrochloride
$C_{11}H_{18}NO_3^+$, Cl^-
For complete entry see 3.20

17.C   **Phenyl β - methylcholine ether bromide**
$C_{12}H_{20}NO^+$, Br^-
For complete entry see 59.7

17.C   **o - Methylphenyl - β - methylcholine ether bromide**
$C_{13}H_{22}NO^+$, Br^-
For complete entry see 59.8

17.C   2 - (2' - **Carbomethoxy** - **4'** - **nitrophenoxy**) - **1,3,5** - **trichlorobenzene**
$C_{14}H_8Cl_3NO_5$
For complete entry see 15.10

17.C   **3,5,3' - Tri - iodothyroacetic acid** - **N - diethanolamine**
$C_{14}H_9I_3O_4$, $C_4H_{11}NO_2$
For complete entry see 60.26

17.C   **4 - Nitro - 4' - methoxy - N - benzylidene - aniline**
$C_{14}H_{12}N_2O_3$
For complete entry see 16.20

17.C   **p - Methoxybenzenesulfone - p - anisidide**
$C_{14}H_{15}NO_4S$
For complete entry see 11.11

17.16   **4 - Bromo - 2,6 - di - t - butyl - phenol**
$C_{14}H_{21}BrO$
G.Filippini, C.M.Gramaccioli, A.Mugnoli, T.Pilati
Cryst. Struct. Comm., **1,** 305, 1972

17.C   **2 - (5 - Phenyl - 1,2 - dithiole - 3 - ylio)phenolate**
$C_{15}H_{10}OS_2$
For complete entry see 39.27

17.C   **Butein monohydrate**
$C_{15}H_{12}O_5$, H_2O
For complete entry see 59.11

17.C   **Alprenolol hydrochloride**
1 - Isopropylamino - 3 - (2 - allylphenoxy) - propan - 2 - ol hydrochloride
$C_{15}H_{24}NO_2^+$, Cl^-
For complete entry see 3.22

17.17   **2,6 - Di - t - butyl - 4 - methylphenol**
$C_{15}H_{24}O$
M.Maze-Baudet   *Acta Cryst. (B),* **29,** 602, 1973

17.18   **1,4 - bis(p - Bromophenoxy)butane**
$C_{16}H_{16}Br_2O_2$
T.Ishikawa, S.Karino, S.Nagai, N.Yasuoka, N.Kasai, M.Kakudo
Bull. Chem. Soc. Jap., **44,** 2954, 1971

17.C   **1,3 - bis(4 - (2' - Bromoethoxy)phenyl) - 5 - tetrazolone**
$C_{17}H_{16}Br_2N_4O_3$
For complete entry see 32.21

17.19   **Mycophenolic acid**
$C_{17}H_{20}O_6$
W.Harrison, H.M.M.Shearer, J.Trotter   *J. C. S. Perkin II,* 1542, 1972
Also classified in 1

17.C   p - Methoxybenzenesulfone - N - isopropyl - p - anisidide
        $C_{17}H_{21}NO_4S$
        For complete entry see 11.15

17.C   2 - (3' - Methyl - 4' - nitrophenoxy) - 1,3 - di - isopropylbenzene
        $C_{19}H_{23}NO_3$
        For complete entry see 15.11

17.20   bis(o - Ethoxyphenyl)butadiyne
        $C_{20}H_{18}O_2$
        T.Taga, N.Masaki, K.Osaki, T.Watanabe
        Bull. Chem. Soc. Jap., **44,** 2981, 1971

17.C   2,2 - Di - (p - ethoxyphenyl) - 3,3 - dimethyl - oxetan
        $C_{21}H_{26}O_3$
        For complete entry see 38.28

17.21   1,1 - bis - (p - Ethoxyphenyl) - 2,2 - dimethylpropane
        $C_{21}H_{28}O_2$
        T.P.DeLacy, C.H.L.Kennard   *J. C. S. Perkin II,* 2141, 1972

17.C   2' - Bromopodophyllotoxin ethyl acetate solvate (absolute configuration)
        $C_{22}H_{21}BrO_8$, $0.5C_4H_8O_2$
        For complete entry see 59.24

17.C   bis(p - Acetoxyphenyl) - cyclohexylidene - methane
        $C_{23}H_{24}O_4$
        For complete entry see 21.16

17.C   4,5,6 - Tri - (p - methoxyphenyl) - 1,2,3 - triazine
        $C_{24}H_{21}N_3O_3$
        For complete entry see 33.54

17.C   Leuco - thelephoric acid hexamethyl ether
        (2,3,6,8,9,12 - Hexamethoxybenzo - bis(1,2 - b.4,5 - b')benzofuran)
        $C_{24}H_{22}O_8$
        For complete entry see 59.28

17.C   N - (4 - (β - 2 - Methoxy - 5 - bromobenzamidoethyl)benzensulfonyl) - N' -
        4 - methyl - cyclohexyl - urea
        $C_{24}H_{30}BrN_3O_5S$
        For complete entry see 8.19

17.C   2 - (2',4' - Dinitrophenoxy) - 1,3,5 - tri - t - butylbenzene
        $C_{24}H_{32}N_2O_5$
        For complete entry see 15.12

17.22   1,1,2,2 - tetrakis(2 - Methoxophenyl)ethane
        $C_{30}H_{30}O_4$
        J.J.Daly, F.Sanz, R.P.A.Sneeden, H.H.Zeiss   *J. C. S. Perkin II,* 1614, 1972

BENZOQUINONES

18.C   **Acenaphthene - tetrachloro - p - benzoquinone complex**
$C_6Cl_4O_2$, $C_{12}H_{10}$
For complete entry see 60.19

18.C   **9 - Methylanthracene - tetrachloro - p - benzoquinone complex**
$C_6Cl_4O_2$, $C_{15}H_{12}$
For complete entry see 60.27

18.1   **o - Benzoquinone**
$C_6H_4O_2$
A.L.Macdonald, J.Trotter   *J. C. S. Perkin II*, 476, 1973

18.C   **2,5 - bis(Ethyleneimino) - 1,4 - benzoquinone (at 300±5°K)**
$C_{10}H_{10}N_2O_2$
For complete entry see 32.13

18.C   **2,5 - bis(Ethyleneimino) - 1,4 - benzoquinone (at 240±10°K)**
$C_{10}H_{10}N_2O_2$
For complete entry see 32.14

18.C   **2,5 - bis(Ethyleneimino) - 1,4 - benzoquinone (at 110±20°K)**
$C_{10}H_{10}N_2O_2$
For complete entry see 32.15

BENZENE MISCELLANEOUS

19.C   **Hexamethylbenzene - hexafluorobenzene complex (trigonal form, at 5°C)**
C_6F_6, $C_{12}H_{18}$
For complete entry see 60.24

19.C   **Hexamethylbenzene - hexafluorobenzene complex (triclinic form, at
−40°C)**
C_6F_6, $C_{12}H_{18}$
For complete entry see 60.25

19.C   **Deoxycholic acid - p - di - iodobenzene**
$C_6H_4I_2$, $2C_{24}H_{40}O_4$
For complete entry see 60.40

19.C   **Cadmium diammine tetracyanomercury(ii) benzene**
$2C_6H_6$, $C_4H_6CdHgN_6$
For complete entry see 61.3

19.C   **catena - μ - Ethylenediamine - cadmium(ii) tetracyanonickelate(ii) dibenzene**
$2nC_6H_6$, $(C_4H_8CdN_4Ni)_n$
For complete entry see 61.2

19.1   **Pentachlorotoluene**
$C_7H_3Cl_5$
T.L.Khotsyanova, T.A.Babushkina, S.I.Kuznetsov, G.K.Semin
Kristallografija, **17,** 552, 1972

19.2   **p - Iodotoluene**
C_7H_7I
C.-T.Ahn, S.Soled, G.B.Carpenter *Acta Cryst. (B)*, **28,** 2152, 1972

19.3   **2H,5H - Octachloro - p - xylene**
$C_8H_2Cl_8$
J.Silverman, A.P.Krukonis, N.F.Yannoni
Cryst. Struct. Comm., **2,** 37, 1973

19.4   **Tetrachloro - m - xylene**
$C_8H_6Cl_4$
T.L.Khotsyanova, T.A.Babushkina, S.I.Kuznetsov, G.K.Semin
Kristallografija, **17,** 552, 1972

19.5   Tetrachloro - p - xylene
$C_8H_6Cl_4$
T.L.Khotsyanova, T.A.Babushkina, S.I.Kuznetsov, G.K.Semin
Kristallografija, **17,** 552, 1972

19.6   Terephthaldehyde - antimony trichloride complex
$C_8H_6O_2$, Cl_3Sb
W.A.Baker, D.E.Williams
Amer. Cryst. Assoc., Abstr. Papers (Winter Meeting), 43, 1973

19.7   1,2,3 - Trichloro - 4,5,6 - trimethylbenzene (monoclinic form, at 298°K)
$C_9H_9Cl_3$
R.Fourme, M.Renaud, D.Andre   *Molec. Cryst. Liq. Cryst.,* **17,** 209, 1972

19.8   1,2,3 - Trichloro - 4,5,6 - trimethylbenzene (triclinic form, at 173°K)
$C_9H_9Cl_3$
R.Fourme, M.Renaud   *Molec. Cryst. Liq. Cryst.,* **17,** 223, 1972

19.9   p - Diethynyl - benzene
$C_{10}H_6$
N.A.Ahmed, A.I.Kitaigorodskij, M.I.Sirota
Acta Cryst. (B), **28,** 2875, 1972

19.10   p - Diacetylbenzene - antimony trichloride complex
$C_{10}H_{10}O_2$, Cl_3Sb
W.A.Baker, D.E.Williams
Amer. Cryst. Assoc., Abstr. Papers (Winter Meeting), 43, 1973

19.11   Durene
1,2,4,5 - Tetramethylbenzene
$C_{10}H_{14}$
C H.Stam   *Acta Cryst. (B),* **28,** 2630, 1972

19.12   Durene (neutron study, unconstrained refinement)
1,2,4,5 - Tetramethylbenzene
$C_{10}H_{14}$
E.Prince, L.W.Schroeder, J.J.Rush   *Acta Cryst. (B),* **29,** 184, 1973

19.13   Durene (neutron study, constrained refinement)
1,2,4,5 - Tetramethylbenzene.
$C_{10}H_{14}$
E.Prince, L.W.Schroeder, J.J.Rush   *Acta Cryst. (B),* **29,** 184, 1973

19.C   1,5 - bis(p - Chlorophenyl) - 2,4 - diaza - 1,3,5 - trithiapenta - 2,3 - diene
$C_{12}H_8Cl_2N_2S_3$
For complete entry see 4.4

19.C   3 - (3',4' - Dichlorobenzylidene) - pentane - 2,4 - dione
$C_{12}H_{10}Cl_2O_2$
For complete entry see 5.15

19.C   **3 - (p - Bromobenzylidene) - pentane - 2,4 - dione**
$C_{12}H_{11}BrO_2$
For complete entry see 5.16

19.14   **Pentamethyl - trichloromethyl - benzene (at 143°K)**
Pentamethylbenzotrichloride
$C_{12}H_{15}Cl_3$
N.C.Baenziger, R.J.Schultz *Acta Cryst. (B),* **29,** 337, 1973

19.C   **Hexamethylbenzene - hexafluorobenzene complex (trigonal form, at 5°C)**
$C_{12}H_{18}$, C_6F_6
For complete entry see 60.24

19.C   **Hexamethylbenzene - hexafluorobenzene complex (triclinic form, at −40°C)**
$C_{12}H_{18}$, C_6F_6
For complete entry see 60.25

19.15   **3,3′ - Dibromobenzophenone**
$C_{13}H_8Br_2O$
V.Pattabhi, K.Venkatesan *J. Cryst. Mol. Struct.,* **3,** 25, 1973

19.C   **Ethyl p - chloro - α - cyano - β - methyl - cis - cinnamate**
$C_{13}H_{12}ClNO_2$
For complete entry see 1.33

19.16   **1,1 - bis(p - Chlorophenyl) - 2,2,2 - trichloroethane**
DDT
$C_{14}H_9Cl_5$
T.P.DeLacy, C.H.L.Kennard *J. C. S. Perkin II,* 2148, 1972

19.17   **1 - (o - Chlorophenyl) - 1 - (p - chlorophenyl) - 2,2,2 - trichloroethane**
$C_{14}H_9Cl_5$
T.P.DeLacy, C.H.L.Kennard *J. C. S. Perkin II,* 2148, 1972

19.18   **p′ - Bromochalcone**
$C_{15}H_{11}BrO$
D.Rabinovich, G.M.J.Schmidt, Z.Shaked *J. C. S. Perkin II,* 33, 1973

19.C   **1,4 - Diphenylbutadiene iron tricarbonyl - 1,4 - diphenylbutadiene**
$0.5C_{16}H_{12}$, $C_{19}H_{12}FeO_3$
For complete entry see 72.25

19.C   **1,4 - bis(β - Pyridyl - 2 - vinyl) - benzene**
$C_{20}H_{16}N_2$
For complete entry see 33.51

19.19   **1,2,4,5 - Tetra - t - butylbenzene (monoclinic form)**
$C_{20}H_{38}$
C.H.Stam *Acta Cryst. (B),* **28,** 2715, 1972

19.C   **Triphenylmethyl - lithium tetramethylethylenediamine**
$C_{25}H_{31}LiN_2$
For complete entry see 67.19

19.20   **Tetramesityl - ethylene**
$C_{38}H_{44}$
J.F.Blount, K.Mislow, J.Jacobus   *Acta Cryst. (A)*, **28,** S12, 1972

MONOCYCLIC HYDROCARBONS
(3, 4, 5-MEMBERED RINGS)

20.1   **1,2 - Dichloro - cyclobutene - dione**
Squaric acid dichloride
$C_4Cl_2O_2$
R.Mattes, S.Schroebler   *Chem. Ber.*, **105,** 3761, 1972

20.2   **Diketocyclobutenediol**
3,4 - Dihydroxy - 3 - cyclobuten - 1,2 - dione
$C_4H_2O_4$
D.Semmingsen   *Tetrahedron Letters,* 807, 1973

20.3   **anti - cis,cis - 2,2' - Dibromo - bicyclopropyl**
$C_6H_8Br_2$
G.Schrumpf, P.Susse   *Chem. Ber.*, **105,** 3041, 1972

20.4   **cis - 1,2 - Cyclobutanedicarboxylic acid**
$C_6H_8O_4$
D.van der Helm, I.-N.Hsu, J.M.Sims   *Acta Cryst. (B)*, **28,** 3109, 1972

20.5   **Methyl 1 - carbamoyl - cyclopropane - 1 - carboxylate**
$C_6H_9NO_3$
J.G.H.de Jong, H.Schenk   *Cryst. Struct. Comm.*, **2,** 25, 1973

20.6   **trans - 2,trans - 3 - Dimethylcyclopropane - carboxylic acid**
$C_6H_{10}O_2$
P.A.Luhan, A.T.McPhail   *J. C. S. Perkin II,* 2372, 1972

20.7   **1,1,2,2 - Tetracyanocyclopropane**
$C_7H_2N_4$
Y.Wang, G.D.Stucky   *Acta Cryst. (B)*, **29,** 1255, 1973
Also classified in 7

20.8   **Cyclopentane - 1,1 - dicarboxylic acid**
$C_7H_{10}O_4$
T.N.Margulis, L.R.Dalton, A.L.Kwiram   *Nature Phys. Sci.*, **242,** 82, 1973

20.9   **1 - Amino - 3 - methyl - cyclopentane carboxylic acid hemihydrate**
$C_7H_{13}NO_2$, $0.5H_2O$
H.L.Carrell, J.P.Glusker
Amer. Cryst. Assoc., Abstr. Papers (Winter Meeting), 94, 1973
Residue 1 also classified in 48

20.10   **N - Methyl - cyclopentane - 1,2 - dicarboxylic acid monoamide**
$C_8H_{11}NO_3$
F.H.Allen, O.Kennard   *Cryst. Struct. Comm.*, **2**, 149, 1973

20.11   **2,2,4,4 - Tetramethyl - cyclobutane - 1,3 - dione**
Dimethylketen dimer
$C_8H_{12}O_2$
C.Riche   *C. R. Acad. Sci., Fr., C*, **275**, 543, 1972

20.12   **1,1 - Cyclopentane diacetic acid**
$C_9H_{14}O_4$
E.Benedetti, R.Claverini, C.Pedone   *Cryst. Struct. Comm.*, **2**, 141, 1973

20.13   **E - 2 - p - Nitrophenyl - cyclopropyl methyl ketone**
$C_{11}H_{11}NO_3$
J.Bordner, L.A.Jones, R.L.Johnson   *Cryst. Struct. Comm.*, **1**, 389, 1972
Also classified in 15

20.C   **2 - (2 - (3 - Thiacyclohexylidene)ethyl) - 2 - methyl - cyclopentane - 1,3 - dione**
$C_{13}H_{18}O_2S$
For complete entry see 39.23

20.14   **bis(p - Chlorophenyl) - cyclopropenone**
$C_{15}H_8Cl_2O$
K.Peters, H.G.von Schnering   *Chem. Ber.*, **106**, 935, 1973

20.15   **1,1 - bis - (p - Chlorophenyl) - 2,2 - dichlorocyclopropane**
$C_{15}H_{10}Cl_4$
T.P.DeLacy, C.H.L.Kennard   *J. C. S. Perkin II*, 2141, 1972

20.16   **Diphenyl - cyclopropenone monohydrate**
$C_{15}H_{10}O$, H_2O
H.Tsukada, M.Shimanouchi, Y.Sasada   *Tetrahedron Letters*, 2455, 1973

20.C   **Bromo - (tetraethyl - ethylenediamine) - cyclopentadienyl - magnesium**
$C_{15}H_{29}BrMgN_2$
For complete entry see 67.11

20.17   **2 - Diethylaminoethyl - 1 - phenylcyclopentane - carboxylate hydrochloride**
Parpanit
$C_{18}H_{28}NO_2^+$, Cl^-
E.A.H.Griffith, B.E.Robertson   *Acta Cryst. (B)*, **28**, 3377, 1972
Residue 1 also classified in 3

20.18   **1,2 - Dibenzoyl - 4 - nitrocyclopentadiene**
$C_{19}H_{13}NO_4$
G.Ferguson, W.C.Marsh, R.J.Restivo
Amer. Cryst. Assoc., Abstr. Papers (Winter Meeting), 84, 1973

20.C **(Cycloheptatrienylidene) - (2,3 - diphenylcyclopropen - 1 - yl) - acetonitrile tetrafluoroborate**

$C_{24}H_{18}N^+$, BF_4^-

For complete entry see 12.13

20.C **(+) - Allethronyl (+) - trans - chrysanthemate 6 - bromo - 2,4 - dinitrophenyl - hydrazone (absolute configuration)**

$C_{25}H_{29}BrN_4O_6$

For complete entry see 59.29

MONOCYCLIC HYDROCARBONS
(6-MEMBERED RINGS)

21.1 **2,4,6 - Trichloro - cyclohexanone**
$C_6H_7Cl_3O$
A.Lectard, F.Metras, J.Gaultier *Cryst. Struct. Comm.*, **1**, 415, 1972

21.2 **2e,4a,6e - Trichloro - cyclohexanone**
$C_6H_7Cl_3O$
A.Lichanot, J.Petrissans *Eur. Cryst. Meeting*, 1973

21.3 **cis - 1,3,5 - Trichloro - cyclohexane**
$C_6H_9Cl_3$
K.Huml, J.Hasek *Acta Cryst. (B)*, **28**, 2852, 1972

21.C **Di - sodium trans - 2 - hydroxycyclohexyl phosphate trihydrate**
$C_6H_{11}O_5P^{2-}$, $2Na^+$, $3H_2O$
For complete entry see 46.6

21.4 **Cyclohexane (form ii, at 115°K)**
C_6H_{12}
R.Kahn, R.Fourme, D.Andre, M.Renaud *Acta Cryst. (B)*, **29**, 131, 1973

21.5 **Cyclohexane (form i, at 195° K)**
C_6H_{12}
R.Kahn, R.Fourme, D.Andre, M.Renaud *Acta Cryst. (B)*, **29**, 131, 1973

21.C **Iron 1,8 - bis(fluoroboro) - 2,7,9,14,15,20 - hexaoxa - 3,6,10,13,16,19 - hexa - aza - 4,5,11,12,17,18 - hexamethylbicyclo(6.6.6)eicosa - 3,5,10,12,16,18 - hexaene cyclohexane**
$0.5C_6H_{12}$, $C_{12}H_{18}B_2F_2FeN_6O_6$
For complete entry see 61.5

21.C **Validamine hydrobromide (absolute configuration)**
1S - 1 - Amino - 5 - hydroxymethyl - 2,3,4 - cyclohexanetriol hydrobromide
$C_7H_{15}NO_4^+$, Br^-
For complete entry see 50.2

21.C **(+) - Aeroplysinin - I (absolute configuration)**
(+) - 3,5 - Dibromo - 1,6 - dihydroxy - 4 - methoxy - cyclohexa - 1,4 - diene acetonitrile
$C_9H_9Br_2NO_2$
For complete entry see 50.4

21.6   **2,2 - Dichloro - 4 - t - butyl - cyclohexanone**
$C_{10}H_{16}Cl_2O$
A.Lectard, J.Petrissans, C.Hauw   *Cryst. Struct. Comm.*, **2**, 1, 1973

21.7   **4α - t - Butylcyclohexane - 1β,2β - diol**
$C_{10}H_{20}O_2$
M.D.Brice, B.R.Penfold, W.T.Robinson   *Aust. J. Chem.*, **25**, 2117, 1972

21.C   **6 - Cyclohexyladenine - iodine complex**
$C_{11}H_{15}N_5$, I_2
For complete entry see 44.18

21.8   **1e,2e,3e,4e,5e,6e - Hexachloro - 1a - phenyl - cyclohexane**
$C_{12}H_5Cl_6$
C.A.de Maye, A.J.de Kok, J.Lugtenburg, C.Romers
Rec. Trav. Chim. Pays-Bas, **91,** 383, 1972

21.9   **(1S,2S) - 2 - ((S) - Chlorofluoroacetoxy) - p - phenylcyclohexanol (absolute configuration)**
$C_{14}H_{16}ClFO_3$
M.Colapietro, R.Spagna, L.Zambonelli   *J. C. S. Perkin II*, 295, 1973

21.10   **3 - (p - Bromophenyl)propionic acid cyclohexanone - cyanohydrin ester**
$C_{16}H_{18}BrN_2O_2$
L.G.Vorontsova   *Zh. Strukt. Khim.*, **12,** 1032, 1971
Also classified in 1

21.11   **trans - 1,2 - bis(2 - Carboxymethyl - 2 - propyl)cyclohexane**
$C_{16}H_{28}O_4$
H.van Koningsveld   *Acta Cryst. (B),* **29,** 1214, 1973

21.12   **trans - 4 - t - Butylcyclohexanol p - bromobenzoate**
$C_{17}H_{23}BrO_2$
J.Ohrt, R.Parthasarathy   *J. Cryst. Mol. Struct.*, **2,** 213, 1972
Also classified in 13

21.13   **trans - 3 - t - Butylcyclohexyl - p - toluene sulfonate**
$C_{17}H_{26}O_3S$
V.J.James   *Cryst. Struct. Comm.*, **2,** 205, 1973

21.14   **trans - 4 - t - Butylcyclohexyl p - toluene sulfonate (data of James and Moore)**
$C_{17}H_{26}O_3S$
P.L.Johnson, C.J.Cheer, J.P.Schaefer, V.J.James, F.H.Moore
Tetrahedron, **28,** 2893, 1972

21.15   **trans - 4 - t - Butylcyclohexyl p - toluene sulfonate (data of Johnson, Cheer and Schaefer)**
$C_{17}H_{26}O_3S$
P.L.Johnson, C.J.Cheer, J.P.Schaefer, V.J.James, F.H.Moore
Tetrahedron, **28,** 2893, 1972

21.C   **Trihexyphenidyl**
syn - α - Cyclohexyl - α - phenyl - 1 - piperidine - propanol
$C_{20}H_{31}NO$
For complete entry see 33.52

21.16   **bis(p - Acetoxyphenyl) - cyclohexylidene - methane**
$C_{23}H_{24}O_4$
G.Precigoux, B.Busetta, C.Courseille, M.Hospital
Cryst. Struct. Comm., **1,** 341, 1972
Also classified in 17

21.C   **N - (4 - (β - 2 - Methoxy - 5 - bromobenzamidoethyl)benzensulfonyl) - N' - 4 - methyl - cyclohexyl - urea**
$C_{24}H_{30}BrN_3O_5S$
For complete entry see 8.19

21.17   **1,2,5,6 - Di - O - isopropylidene - 3,4 - di - O - tosyl - L - chiro - inositol**
$C_{26}H_{32}O_{10}S_2$
J.F.McConnell, S.J.Angyal, J.D.Stevens   *J. C. S. Perkin II,* 2039, 1972
Also classified in 38

MONOCYCLIC HYDROCARBONS
(7, 8-MEMBERED RINGS)

22.1 **Tropone (at −60°C)**
C_7H_6O
M.J.Barrow, O.S.Mills, G.Filippini *J. C. S. Chem. Comm.*, 66, 1973

22.2 **Tropolone**
$C_7H_6O_2$
H.Shimanouchi, Y.Sasada *Acta Cryst. (B)*, **29**, 81, 1973

22.3 **syn - 3,7, - Dibromo - cis,cis - cyclo - octa - 1,5 - diene**
$C_8H_{10}Br_2$
R.K.Mackenzie, D.D.MacNicol, H.H.Mills, R.A.Raphael, F.B.Wilson,
J.A.Zabkiewicz *J. C. S. Perkin II*, 1632, 1972

22.C **3 - Bromo - 4′a,10′a - dihydrospiro(2,5 - cyclohexadiene - 1,3′ - cyclo - octa - as - trioxin) - 4 - one**
$C_{14}H_{17}BrO_4$
For complete entry see 38.17

22.C **(Cycloheptatrienylidene) - (cycloheptatrienyl) - acetonitrile perchlorate**
$C_{16}H_{12}N^+$, ClO_4^-
For complete entry see 12.5

22.C **(Cycloheptatrienylidene) - (2,3 - diphenylcyclopropen - 1 - yl) - acetonitrile tetrafluoroborate**
$C_{24}H_{18}N^+$, BF_4^-
For complete entry see 12.13

MONOCYCLIC HYDROCARBONS
(9- AND HIGHER-MEMBERED RINGS)

23.1   **(14)Annulene**
$C_{14}H_{14}$
C.C.Chiang, I.C.Paul   *J. Amer. Chem. Soc.*, **94,** 4741, 1972

NAPHTHALENE COMPOUNDS

24.1   **1,5 - Difluoronaphthalene**
$C_{10}H_6F_2$
N.B.Chanh, J.Housty, F.Leroy   *Cryst. Struct. Comm.*, **2**, 291, 1973

24.2 β - **Fluoronaphthalene**
$C_{10}H_7F$
N.B.Chanh, Y.Haget-Bouillaud   *Acta Cryst. (B)*, **28**, 3400, 1972

24.3   **1,5 - Dimethylnaphthalene (neutron study)**
$C_{12}H_{12}$
G.Ferraris, D.W.Jones, J.Yerkess, K.D.Bartle
J. C. S. Perkin II, 1628, 1972

24.4   **trans - 4a - Acetoxy - 8a - chloro - 1,4,4a,5,8,8a - hexahydronaphthalene**
$C_{12}H_{15}ClO_2$
R.Boggs, J.Donohue
Amer. Cryst. Assoc., Abstr. Papers (Winter Meeting), 46, 1973
Also classified in 27

24.C   **1 - (2 - Thiazolylazo) - 6 - bromo - 2 - naphthol**
$C_{13}H_8BrN_3OS$
For complete entry see 41.23

24.5   **1 - Acetyl - 2 - ethoxynaphthalene**
$C_{14}H_{14}O_2$
M.P.Gupta, M.Sahu   *Z. Kristallogr.*, **135**, 262, 1972

24.6   **1,8 - Di(prop - 1 - ynyl)naphthalene**
$C_{16}H_{12}$
A.E.Jungk   *Chem. Ber.*, **105**, 1595, 1972

24.C   **1 - Cyano - 2 - hydroxy - 3,4 - epithiobutane - α - naphthylurethane**
$C_{16}H_{14}N_2O_2S$
For complete entry see 39.31

24.7   **Propranolol hydrochloride**
1 - Isopropylamino - 3(1 - oxynaphthyl) - propen - 2 - ol hydrochloride
$C_{16}H_{22}NO_2{}^+$, Cl^-
Y.Barrans, M.Cotrait, J.Dangoumau   *Acta Cryst. (B)*, **29**, 1264, 1973
Residue 1 also classified in 3

24.8   **3 - (6' - Methoxy - 2' - naphthyl) - 2,2 - dimethyl - butyric acid**
$C_{17}H_{20}O_3$
B.Rerat, C.Stora, C.Rerat, A.Horeau, J.Jacques
C. R. Acad. Sci., Fr., C, **275,** 179, 1972

24.C   **5 - (2' - Bromo - 3',4' - dihydronaphthyl) - acenaphthene**
$C_{22}H_{17}Br$
For complete entry see 28.16

24.C   **Naphthalene - bis(lithium tetramethylethylenediamine)**
$C_{22}H_{40}Li_2N_4$
For complete entry see 67.18

24.C   **cis - 1 - (Phenanthr - 3 - yl) - 2 - (naphth - 1 - yl) - ethylene**
$C_{26}H_{18}$
For complete entry see 28.18

24.C   **trans - 1 - (Phenanthr - 3 - yl) - 2 - (naphth - 1 - yl) - ethylene**
$C_{28}H_{18}$
For complete entry see 28.20

NAPHTHOQUINONES

25.1   **2,3 - Difluoro - 1,4 - naphthoquinone**
$C_{10}H_4F_2O_2$
J.Gaultier, C.Hauw, J.Housty, M.Schvoerer
C. R. Acad. Sci., Fr., C, **275,** 1403, 1972

25.2   **5 - Chloro - 1,4 - naphthoquinone**
$C_{10}H_5ClO_2$
C.Scheringer   *Acta Cryst. (B),* **29,** 618, 1973

ANTHRACENE COMPOUNDS

26.1   **1,8 - Dichloroanthraquinone**
$C_{14}H_6Cl_2O_2$
O.A.Mikhno, Z.I.Ezhkova, B.N.Kolokolov   *Kristallografija*, **17**, 667, 1972

26.C   **Anthraquinoneoxadiazole**
Anthra(1,2 - c)(1,2,5)oxadiazole - 6,11 - dione
$C_{14}H_6N_2O_3$
For complete entry see 40.27

26.2   **Perdeuterio - anthracene (neutron study)**
$C_{14}D_{10}$
M.S.Lehmann, G.S.Pawley   *Acta Chem. Scand.*, **26**, 1996, 1972

26.3   **Anthrone**
$C_{14}H_{10}O$
S.N.Srivastava   *Indian J. Phys.*, **46**, 56, 1972

26.C   **9 - Methylanthracene - tetrachloro - p - benzoquinone complex**
$C_{15}H_{12}$, $C_6Cl_4O_2$
For complete entry see 60.27

26.4   **cis - 9 - Methyl - 10 - ethyl - 9,10 - dihydroanthracene**
$C_{17}H_{18}$
J.Bordner, R.H.Stanford Junior, H.E.Zieger   *Acta Cryst. (B)*, **29**, 313, 1973

26.5   **10 - Chloromethyl - 2,3,9 - trimethylanthracene**
$C_{18}H_{17}Cl$
A.Chomyn, J.P.Glusker, H.M.Berman, H.L.Carrell
Acta Cryst. (B), **28**, 3512, 1972

HYDROCARBONS (2 FUSED RINGS)

27.1   **1,3 - Dicyanobicyclo(1.1.0)butane**
         $C_6H_4N_2$
         P.L.Johnson, J.P.Schaefer   *J. Org. Chem.*, **37,** 2762, 1972

27.2   **Octachloro - 2,4 - dihydropentalene**
         C_8Cl_8
         G.Mandel, J.Donohue   *Acta Cryst. (B),* **29,** 710, 1973

27.3   **cis,exo - Semibullvalene - 1,4 - dibromide**
         $C_8H_6Br_2$
         L.A.Paquette, G.H.Birnberg, J.Clardy, B.Parkinson
         J. C. S. Chem. Comm., 129, 1973

27.4   **1,2,3,3 - Tetrachloro - 4,5 - dimethylspiro(2.3)hexa - 1,4 - diene (at −65°C)**
         $C_8H_6Cl_4$
         R.J.Guttormson, B.E.Robertson   *Acta Cryst. (B),* **29,** 173, 1973

27.5   **(+) - R - Spiro(3.3)heptane - 2,6 - dicarboxylic acid (at −160°C, absolute configuration)**
         d - Fecht acid
         $C_9H_{12}O_4$
         L.A.Hulshof, A.Vos, H.Wynberg   *J. Org. Chem.*, **37,** 1767, 1972

27.6   **Potassium trans - bicyclo(5.1.0)octane - 4 - carboxylate monohydrate**
         $C_9H_{13}O_2^-$, K^+ , H_2O
         R.A.Kershaw, P.W.R.Corfield
         Amer. Cryst. Assoc., Abstr. Papers (Winter Meeting), 48, 1973

27.7   **trans - Bicyclo(5.1.0)octane - 4 - carboxylic acid**
         $C_9H_{14}O_2$
         R.A.Kershaw, P.W.R.Corfield
         Amer. Cryst. Assoc., Abstr. Papers (Winter Meeting), 48, 1973

27.8   **9 - Methyl - $\Delta^{5,10}$ - decalin - 1,6 - dione**
         $C_{11}H_{14}O_2$
         C.R.Jones, D.R.Kearns, R.M.Wing   *J. Chem. Phys.*, **58,** 1370, 1973

27.C   **trans - 4a - Acetoxy - 8a - chloro - 1,4,4a,5,8,8a - hexahydronaphthalene**
         $C_{12}H_{15}ClO_2$
         For complete entry see 24.4

27.9   (+) - 2(S) - N,N - Dimethylamino - 3(S) - hydroxy - 9(S),10(S) - decalin
$C_{12}H_{23}NO$
C.Cabestaing, D.J.Watkin   *Cryst. Struct. Comm.*, **2,** 295, 1973

27.10   2,7 - Dimethyl - 4,5 - benzotropone
$C_{13}H_{12}O$
H.Shimanouchi, K.Ibata, Y.Sasada   *Acta Cryst. (A)*, **28,** S27, 1972

27.11   1,3,4 - Trimethyl - bicyclo(4.4.0)dec - 3,6,8 - triene - 2,5 - dione
$C_{13}H_{14}O_2$
G.D.Andreetti, G.Bocelli, P.Sgarabotto   *Cryst. Struct. Comm.*, **2,** 343, 1973

27.12   2 - (1α,8β - 10β - Bicyclo(6.3.0)undec - 4 - en - 11 - one) - propionic acid
$C_{14}H_{20}O_3$
G.D.Andreetti, G.Bocelli, P.Sgarabotto   *Cryst. Struct. Comm.*, **2,** 371, 1973

27.13   1 - Amino - 2 - p - chlorophenyl - 3 - indenone
$C_{15}H_{10}ClNO$
A.E.Shvets, Ya.Ya.Bleidelis, R.P.Shibaeva, L.O.Atovmyan
Zh. Strukt. Khim., **13,** 745, 1972

27.14   cis - 1,1 - bis(Methoxycarbonyl) - 4a - methyl - decahydronaphthalene
$C_{15}H_{24}O_4$
A.T.McPhail, P.-S.Wong, D.M.S.Wheeler   *J. C. S. Perkin II*, 2369, 1972

27.15   trans,trans - Decalyl - 2 - tosylate
$C_{17}H_{24}O_3S$
V.J.James   *Cryst. Struct. Comm.*, **2,** 307, 1973

27.16   5 - Hydroxy - 1,2,3,4,5,6 - hexamethyl - bicyclo(2.2.0)hex - 2 - ene p - bromobenzoate
$C_{19}H_{23}BrO_2$
L.A.Paquette, S.A.Lang Junior, M.R.Short, B.Parkinson, J.Clardy
Tetrahedron Letters, 3141, 1972

27.17   2 - Methylene - trans - bicyclo(5.4.0)undecan - 4β - ol o - chlorophenylurethane
$C_{19}H_{24}ClNO_2$
N.H.Andersen, H.-S.Uh, S.E.Smith, P.G.M.Wuts
J. C. S. Chem. Comm., 956, 1972

27.C   Bicyclo(1,1,0)butan - 1 - yl - lithium tetramethylethylenediamine dimer
$C_{20}H_{42}Li_2N_4$
For complete entry see 67.17

27.18   Tetraethyl 4 - bromo - 8 - methoxy - decahydronaphthalene - 2,2,6,6 - tetracarboxylate
$C_{23}H_{35}BrO_9$
H.W.Guin, S.H.Simonsen   *Cryst. Struct. Comm.*, **1,** 397, 1972

HYDROCARBONS (3 FUSED RINGS)

28.C   **anti - 1 - Bromo - tricyclo(5.1.0.0^{3,5})octan - 2,6 - dione**
$C_8H_7BrO_2$
For complete entry see 31.1

28.1   **Naphtho(b)cyclobutene**
$C_{12}H_{10}$
J.L.Crawford, R.E.Marsh   *Acta Cryst. (B),* **29,** 1238, 1973

28.C   **Acenaphthene - tetrachloro - p - benzoquinone complex**
$C_{12}H_{10}$, $C_6Cl_4O_2$
For complete entry see 60.19

28.C   **Acenaphthene - 7,7,8,8 - tetracyanoquinodimethane complex**
$C_{12}H_{10}$, $C_{12}H_4N_4$
For complete entry see 60.20

28.2   **2,4,7 - Trinitrofluoren - 9 - one**
$C_{13}H_5N_3O_7$
D.L.Dorset, A.Hybl, H.L.Ammon   *Acta Cryst. (B),* **28,** 3122, 1972

28.3   **Acenaphthylene - 1 - carboxylic acid**
$C_{13}H_8O_2$
H.Bouas-Laurent, A.Castellan, J.-P.Desvergne, G.Dumartin, C.Courseille,
J.Gaultier, C.Hauw   *J. C. S. Chem. Comm.,* 1267, 1972

28.C   **Deoxycholic acid - phenanthrene**
$C_{14}H_{10}$, $3C_{24}H_{40}O_4$
For complete entry see 60.41

28.4   **6,6,12,12 - Tetrachloro - 3,3,9,9 - tetramethoxy -**
tricyclo(9,1,0,0^{5,7})dodecane
$C_{16}H_{14}Cl_4O_4$
R.W.Baker, P.J.Pauling   *J. C. S. Perkin II,* 1451, 1972

28.C   **2 - Hydroxy - (4,2,2)propellane p - nitrobenzoate**
$C_{17}H_{19}NO_4$
For complete entry see 31.26

28.5   **10 - Methyl - 9 - ((2 - chloroethyl)thio)anthracene**
$C_{18}H_{17}ClS$
J.P.Glusker, D.E.Zacharias   *Acta Cryst. (B),* **28,** 3518, 1972

28.6 8α - Methoxy - 1,1,4aβ - trimethyl - 2β - hydroxy - 1,2,3,4,4a,5,6,7,8,9 - decahydrophenanthrene
$C_{18}H_{28}O_2$
A.Furusaki, N.Hamanaka, T.Matsumoto   *Chem. Letters,* 1041, 1972

28.7   Tricyclo(6.4.0.0^{2,7})dodeca - 2(7),3,5 - triene - 1 - ol p - bromobenzoate
$C_{19}H_{17}BrO_2$
A.Courtois, J.Protas, G.Guillaumet, P.Caubere
C. R. Acad. Sci., Fr., C, **276,** 407, 1973

28.8   DL - 3a,6 - Dimethyl - 6 - (trans - 3 - chlorobut - 2 - ene - 1 - yl) - 2,4,5,6,8,9 - hexahydro - 3H - benz(e)indene - 3,7 - (3aH) - dione
$C_{19}H_{23}ClO_2$
C.S.Yoo, J.Pletcher, M.Sax   *Acta Cryst. (B),* **28,** 2838, 1972
Also classified in 51

28.9 8β - Acetoxy - 1,1,4aβ - trimethyl - 2β - hydroxy - 1,2,3,4,4a,5,6,7,8,9 - decahydrophenanthrene
$C_{19}H_{28}O_3$
A.Furusaki, N.Hamanaka, T.Matsumoto   *Chem. Letters,* 1041, 1972

28.10   9,10,15,16 - Tetradehydro - 11,12,13,14 - tetrahydro - dibenzo(a,c)cyclododecene
$C_{20}H_{16}$
H.Irngartinger   *Chem. Ber.,* **105,** 1184, 1972

28.11   Tricyclo(6.5.0.0^{2,7})trideca - 2,4,6 - triene - 1 - ol p - bromobenzoate
$C_{20}H_{19}BrO_2$
A.Courtois, J.Protas, G.Guillaumet, P.Caubere
C. R. Acad. Sci., Fr., C, **276,** 1171, 1973

28.12   Tricyclo(6.5.0.0^{2,7})tridec - 6 - ene - 1 - ol p - bromobenzoate
$C_{20}H_{23}BrO_2$
A.Courtois, J.Protas, J.-J.Brunet, P.Caubere
C. R. Acad. Sci., Fr., C, **274,** 2162, 1972

28.C   Monobromo - dehydro - bispulegone
$C_{20}H_{29}BrO$
For complete entry see 59.20

28.13   12β - Hydroxy - sandaracopimaric acid
$C_{20}H_{30}O_3$
J.Lapasset, J.Falgueirettes   *Acta Cryst. (B),* **28,** 3321, 1972

28.14   1,2,3,4 - Tetrahydro - 9 - phenanthrol p - bromobenzoate
$C_{21}H_{17}BrO_2$
A.Courtois, J.Protas, J.-J.Brunet, P.Caubere
C. R. Acad. Sci., Fr., C, **275,** 479, 1972

28.15 exo - $1\alpha,7\alpha,2\beta,6\beta$ - **11,11** - **Dimethyltricyclo(5.4.0.0^{2,6})undecan** - **8** - **ol p** - **bromophenylurethane**
$C_{21}H_{26}BrNO_2$
R.M.Bowman, C.Calvo, J.J.McCullough, P.W.Rasmussen, F.F.Snyder
J. Org. Chem., **37**, 2084, 1972

28.16 **5** - **(2′** - **Bromo** - **3′,4′** - **dihydronaphthyl)** - **acenaphthene**
$C_{22}H_{17}Br$
J.Bordner, R.L.Greene, L.A.Jones, R.Watson
Cryst. Struct. Comm., **1**, 393, 1972
Also classified in 24

28.17 **6,7** - **Diphenyl** - **tricyclo(5.3.0.0^{2,6})decan** - **3,10** - **dione**
$C_{22}H_{20}O_2$
M.Magnifico, E.J.O′Connell Junior, A.V.Fratini, C.M.Shaw
J. C. S. Chem. Comm., 1095, 1972

28.18 **cis** - **1** - **(Phenanthr** - **3** - **yl)** - **2** - **(naphth** - **1** - **yl)** - **ethylene**
$C_{26}H_{18}$
G.Tsoucaris, C.de Rango   *Acta Cryst. (A)*, **28**, S28, 1972
Also classified in 24

28.19 **Compound 2**
$C_{26}H_{36}O_4$
G.Stork, J.M.Tabak, J.F.Blount   *J. Amer. Chem. Soc.*, **94**, 4735, 1972

28.C **Fluorenyl** - **lithium bis(quinuclidine)**
$C_{27}H_{35}LiN_2$
For complete entry see 67.20

28.20 **trans** - **1** - **(Phenanthr** - **3** - **yl)** - **2** - **(naphth** - **1** - **yl)** - **ethylene**
$C_{28}H_{18}$
G.Tsoucaris, C.de Rango   *Acta Cryst. (A)*, **28**, S28, 1972
Also classified in 24

HYDROCARBONS (4 FUSED RINGS)

29.C **Tetracyclo(5.5.1.0^{2,6}.0^{10,13})tridecane - 4,8,12 - trione**
$C_{13}H_{14}O_3$
For complete entry see 31.13

29.1 **Spiro(indene - 1,7' - norcaradiene)**
$C_{15}H_{12}$
W.Dreissig, P.Luger, D.Rewicki, C.Tuchscherer
Cryst. Struct. Comm., **2,** 197, 1973

29.2 **Pyrene (neutron study)**
$C_{16}H_{10}$
A.C.Hazell, F.K.Larsen, M.S.Lehmann *Acta Cryst. (B),* **28,** 2977, 1972

29.3 **Pyrene - antimony tribromide complex**
$C_{16}H_{10}$, $2Br_3Sb$
G.Bombieri, G.Peyronel, I.M.Vezzosi *Inorg. Chim. Acta,* **6,** 349, 1972

29.C **Pyrene - 1,3,5 - trinitrobenzene complex**
$C_{16}H_{10}$, $C_6H_3N_3O_6$
For complete entry see 60.29

29.C **Pyrene - 1,2,4,5 - tetracyanobenzene complex (at 290°K)**
$C_{16}H_{10}$, $C_{10}H_2N_4$
For complete entry see 60.30

29.C **Pyrene - 1,2,4,5 - tetracyanobenzene complex (at 178°K)**
$C_{16}H_{10}$, $C_{10}H_2N_4$
For complete entry see 60.31

29.C **Pyrene - 7,7,8,8 - tetracyanoquinodimethane complex (data of Prout and Tickle)**
$C_{16}H_{10}$, $C_{12}H_4N_4$
For complete entry see 60.32

29.C **Pyrene - 7,7,8,8 - tetracyanoquinodimethane complex (data of Wright)**
$C_{16}H_{10}$, $C_{12}H_4N_4$
For complete entry see 60.33

29.4 **2 - Chloro - 4,5,9,10 - tetrahydropyrene**
$C_{16}H_{13}Cl$
C.A.Bear, D.Hall, J.M.Waters, T.N.Waters *J. C. S. Perkin II,* 314, 1973

29.5   **Triphenylene (neutron study)**
9,10 - Benzophenanthrene
$C_{18}H_{12}$
D.W.Jones, J.Yerkess, G.Ferraris   *Acta Cryst. (A)*, **28,** S27, 1972

29.C   **7 - Chloro - 6 - demethyltetracycline hydrochloride trihydrate**
$C_{21}H_{21}ClN_2O_8{}^+$, Cl^- , $3H_2O$
For complete entry see 50.8

29.C   **$4\alpha,8\alpha,10\beta,14\beta$ - Tetramethyl - $\Delta^{9,11}$ - dodecahydro -
cyclopenta(a)phenanthrene - 3,17 - dione**
$C_{21}H_{30}O_2$
For complete entry see 50.9

29.C   **Rearrangement product of a Beyer - 15(16) - en - 12 - one system**
$C_{27}H_{29}BrO_4$
For complete·entry see 59.30

29.C   **N - Bromoacetyl - daunomycin acetone solvate**
$C_{29}H_{30}BrNO_{11}$, C_3H_6O
For complete entry see 50.11

29.6   **2 - Phenyl - 3 - fluoroenylidene - 4 - benzylidene - spiro - cyclobutan - 1,9' -
fluorene**
$C_{42}H_{28}$
W.Dreissig   *Acta Cryst. (A)*, **28,** S4, 1972

HYDROCARBONS (5 OR MORE FUSED RINGS)

30.C **Boron compound**
$C_4H_{12}N^+$, $C_{14}H_{21}B_{10}^-$
For complete entry see 62.12

30.C **Perylene - 7,7,8,8 - tetracyanoquinodimethane complex**
$C_{20}H_{12}$, $C_{12}H_4N_4$
For complete entry see 60.38

30.C **6 - Oxo - 3β,5 - cycloandrostan - 17 - yl acetate**
$C_{21}H_{30}O_3$
For complete entry see 51.20

30.1 **1 - Cyanoacenaphthylene photodimer**
$C_{26}H_{14}N_2$
C.Courseille, B.Busetta, M.Hospital, A.Castellan
Cryst. Struct. Comm., **1,** 337, 1972

30.2 **Hexahelicene**
$C_{26}H_{16}$
C.de Rango, G.Tsoucaris, J.P.Declerq, G.Germain, J.P.Putzeys
Cryst. Struct. Comm., **2,** 189, 1973

30.C **6 - Oxo - 3α,5 - cycloandrostan - 17 - yl p - bromobenzoate**
$C_{26}H_{31}BrO_3$
For complete entry see 51.41

30.3 **2 - Methylhexahelicene**
$C_{27}H_{18}$
G.W.Frank, D.T.Hefelfinger, D.A.Lightner   *Acta Cryst. (B),* **29,** 223, 1973

30.4 **9,10,19,20 - Tetradehydro - tetrabenzo(a,c,g,i)cyclododecene benzene solvate**
$C_{28}H_{16}$, $0.5C_6H_6$
H.Irngartinger   *Chem. Ber.,* **106,** 761, 1973

30.5 **5 - cis - 15 - cis - Tetrabenzo(a,c,g,i)cyclododecene**
$C_{28}H_{20}$
H.Irngartinger   *Eur. Cryst. Meeting,* 1973

30.6   **5 - cis - 15 - trans - Tetrabenzo(a.c.g.i)cyclododecene**
$C_{28}H_{20}$
H.Irngartinger   *Chem. Ber.*, **105,** 2068, 1972

30.7   **trans,trans - Tetrabenzo(a,c,g,i)cyclododecene**
$C_{28}H_{20}$
I.Agranat, M.A.Kraus, E.D.Bergmann, P.J.Roberts, O.Kennard
Tetrahedron Letters, 1265, 1973

30.8   **5 - trans - 15 - trans - Tetrabenzo(a,c,g,i)cyclododecene**
$C_{28}H_{20}$
H.Irngartinger   *Eur. Cryst. Meeting,* 1973

30.C   **(23R) - 23 - Hydroxy - $3\alpha,5\alpha$ - cycloergost - 7 - en - 6 - one**
$C_{28}H_{44}O_2$
For complete entry see 51.43

30.C   **Hopane derivative D**
$C_{29}H_{50}$
For complete entry see 56.3

BRIDGED RING HYDROCARBONS

31.1   anti - 1 - Bromo - tricyclo(5.1.0.0^{3,5})octan - 2,6 - dione
$C_8H_7BrO_2$
J.Heller, A.S.Dreiding, R.Grieb, A.Niggli   *Angew. Chem.*, **84,** 170, 1972
Also classified in 28

31.2   8,8 - Dichloro - tricyclo(3.2.1.0^{1,5})octane (at −40°C)
$C_8H_{10}Cl_2$
K.B.Wiberg, G.J.Burgmaier, K.-W.Shen, S.J.LaPlaca, W.C.Hamilton,
M.D.Newton   *J. Amer. Chem. Soc.*, **94,** 7402, 1972

31.3   Bicyclo(2.2.1)hept - 5 - ene - 2,3 - exo - dicarboxylic anhydride
$C_9H_8O_3$
G.Filippini, C.M.Gramaccioli, C.Rovere, M.Simonetta
Acta Cryst. (B), **28,** 2869, 1972

31.4   7 - syn - Acetoxy - 6 - endo - hydroxy - bicyclo(2.2.1)heptane - 2 - endo -
carboxylic acid lactone
$C_{10}H_{12}O_4$
J.L.Flippen   *Acta Cryst. (B),* **28,** 2046, 1972

31.5   1 - Cyano - tetracyclo(3.3.1.1^{3,7}.0^{3,7})decane
$C_{11}H_{13}N$
C.S.Gibbons, J.Trotter   *Canad. J. Chem.*, **51,** 87, 1973

31.6   1,2,3,4,10,10 - Hexachloro - 1,4,4a,5,8,8a - hexahydro - endo - 1,4 - exo -
5,8 - dimethano - naphthalene
Aldrin
$C_{12}H_8Cl_6$
T.P.DeLacy, C.H.L.Kennard   *J. C. S. Perkin II*, 2153, 1972

31.7   Photoaldrin
$C_{12}H_8Cl_6$
A.A.Khan, W.H.Baur, M.A.Q.Khan   *Acta Cryst. (B),* **28,** 2060, 1972

31.8   1,2,3,4,10,10 - Hexachloro - 6,7 - epoxy - 1,4,4a,5,6,7,8,8a - octahydro -
endo - 1,4 - endo - 5,8 - dimethano - naphthalene
Endrin
$C_{12}H_8Cl_6O$
T.P.DeLacy, C.H.L.Kennard   *J. C. S. Perkin II*, 2153, 1972
Also classified in 38

31.9   **11 - Chloro - 3,8 - methano - (11)annulenone**
$C_{12}H_9ClO$
R.L.Beddoes, O.S.Mills   *Israel J. Chem.*, **10**, 485, 1972

31.10   **S - (−) - p - Bromo - α - phenylethylammonium (+) - 1,6 - methano(10)annulene - 2 - carboxylate**
$C_{12}H_9O_2^-$, $C_8H_{11}BrN^+$
M.V.Stewart, M.G.Newton, R.K.Hill
Amer. Cryst. Assoc., Abstr. Papers (Winter Meeting), 49, 1973
Residue 2 classified in 3

31.11   **5,6 - Dihydroxy - 1,2,3,4,5,6 - hexamethyl - bicyclo(2.1.1)hex - 2 - ene**
$C_{12}H_{20}O_2$
L.A.Paquette, S.A.Lang Junior, S.K.Porter, J.Clardy
Tetrahedron Letters, 3137, 1972

31.12   **11,11 - Dimethyl - tricyclo(4.4.1.0^{1,6})undeca - 2,4,7,9 - tetraene**
$C_{13}H_{14}$
R.Bianchi, G.Morosi, A.Mugnoli, M.Simonetta
Acta Cryst. (B), **29**, 1196, 1973

31.13   **Tetracyclo(5.5.1.0^{2,6}.0^{10,13})tridecane - 4,8,12 - trione**
$C_{13}H_{14}O_3$
T.Akuyama, J.V.Silverton
Amer. Cryst. Assoc., Abstr. Papers (Winter Meeting), 47, 1973
Also classified in 29

31.14   **3 - endo - Phenyl - 2 - endo - norbornanol (neutron study of deuteration)**
$C_{13}H_{16}O$
C.K.Johnson, T.C.Tesch
Amer. Cryst. Assoc., Abstr. Papers (Winter Meeting), 48, 1973

31.15   **Tetracyclo(7.2.1.1^{5,8}.0^{4,9})dodecane - 2,10 - dione**
$C_{13}H_{16}O_2$
M.Przybylska   *Acta Cryst. (B),* **28**, 2814, 1972

31.16   **exo - anti - Tricyclo(3.1.1.0^{2,4})heptan - 6 - yl p - bromobenzoate**
$C_{14}H_{13}BrO_2$
S.Masamune, R.Vukov, M.J.Bennett, J.T.Purdham
J. Amer. Chem. Soc., **94,** 8239, 1972

31.17   **2 - endo - Phenyl - 2,3 - exo - norbornylene carbonate (neutron study of deuteration)**
$C_{14}H_{14}O_4$
C.K.Johnson, T.C.Tesch
Amer. Cryst. Assoc., Abstr. Papers (Winter Meeting), 48, 1973

31.18   **4 - Carboxy - (8) - paracyclophane**
$C_{14}H_{20}O_2$
M.G.Newton, T.J.Walter, N.L.Allinger
Amer. Cryst. Assoc., Abstr. Papers (Winter Meeting), 50, 1973

31.19 Tetracyclo(7.4.1^{3,14}.0.0^{8,15})pentadeca - 4,6,10,12 - tetraene - 2 - one
$C_{15}H_{14}O$
C.Kabuto, Y.Kayama, M.Oda, Y.Kitahara *Chem. Letters,* 885, 1972

31.20 3 - N - Methylamino - 3 - phenyl - 2 - bicyclo(3,2.1)octanone hydrochloride
$C_{15}H_{20}NO^+$, Cl^-
C.L.Stevens, T.A.Treat, P.M.Pillai, W.Schmonsees, M.D.Glick
J. Amer. Chem. Soc., **95,** 1978, 1973

31.21 (2.2)Metacyclophane - 1,9 - diene
$C_{16}H_{12}$
A.W.Hanson, M.Rohrl *Acta Cryst. (B),* **28,** 2032, 1972

31.22 4,6 - (10) - Paracyclophadiyne
$C_{16}H_{16}$
K.Harata, T.Aono, K.Sakabe, N.Sakabe, J.Tanaka
Acta Cryst. (A), **28,** S14, 1972

31.23 (—) - 1,5 - Diamino - 9,10 - ethenoanthracene dihydrobromide monohydrate
(absolute configuration)
$C_{16}H_{16}N_2^{2+}$, $2Br^-$, H_2O
J.Tanaka, C.Katayama, F.Ogura, H.Tatemitsu, M.Nakagawa
J. C. S. Chem. Comm., 21, 1973

31.24 4,8 - Di(carbomethoxy) - tricyclo(6.3.1.0^{1,5})dodecan - 3,12 - dione
$C_{16}H_{20}O_6$
S.Danishefsky, W.E.Hatch, M.Sax, E.Abola, J.Pletcher
J. Amer. Chem. Soc., **95,** 2410, 1973

31.25 1,4,5,8 - Tetramethyl - pentacyclo(6.4.0.0^{2,7}.0^{4,11}.0^{5,10}) dodecane
$C_{16}H_{24}$
J.P.Chesick, J.D.Dunitz, U.V.Gizycki, H.Musso
Chem. Ber., **106,** 150, 1973

31.26 2 - Hydroxy - (4,2,2)propellane p - nitrobenzoate
$C_{17}H_{19}NO_4$
G.W.A.Milne, J.V.Silverton, P.E.Eaton
Amer. Cryst. Assoc., Abstr. Papers (Winter Meeting), 47, 1973
Also classified in 28

31.27 (2.2.2)(1,3,5)Cyclophane - 1,9,17 - triene
$C_{18}H_{12}$
A.W.Hanson, M.Rohrl *Acta Cryst. (B),* **28,** 2287, 1972

31.28 1,6.8,13 - Butane - 1,4 - diylidene(14)annulene
$C_{18}H_{16}$
C.M.Gramaccioli, A.Mugnoli, T.Pilati, M.Raimondi, M.Simonetta
Acta Cryst. (B), **28,** 2365, 1972

31.29   **7 - Methoxycarbonyl - anti - 1,6.8,13 - dimethano(14)annulene**
$C_{18}H_{16}O_2$
C.M.Gramaccioli, A.S.Mimum, A.Mugnoli, M.Simonetta
J. Amer. Chem. Soc., **95**, 3149, 1973

31.30   **Methyl tetracyclo(9.2.2.1^{4,11}.0^{8,16})hexadeca - 4(16),5,7,12,14 - pentaene 1 - carboxylate**
$C_{18}H_{18}O_2$
E.Maverick, S.Smith, K.N.Trueblood   *Acta Cryst. (A)*, **28**, S16, 1972

31.C   **5,7 - (12) - Paracyclophadiyne - tetracyanoethylene complex**
$2C_{18}H_{20}$, C_6N_4
For complete entry see 60.35

31.31   **1,2,3,4 - Tetrachloro - pentacyclo(8.4.2.2^{4,9}.0^{2,10}.0^{3,9}) octadeca - 6,12 - diene**
$C_{18}H_{20}Cl_4$
P.Warner, R.LaRose, C.-M.Lee, J.C.Clardy
J. Amer. Chem. Soc., **94**, 7607, 1972

31.32   **Basketene photodimer**
$C_{20}H_{20}$
N.J.Jones, W.D.Deadman, E.LeGoff   *Tetrahedron Letters*, 2087, 1973

31.C   **Stemodinone**
$C_{20}H_{32}O_2$
For complete entry see 59.21

31.33   **1,5 - Dimethyl - (6,7)benzo - bicyclo(3.2.1)oct - 6 - ene - 8 - ol tosylate**
$C_{21}H_{24}O_3S$
A.Courtois, J.Protas, M.S.Mourad, P.Caubere
C. R. Acad. Sci., Fr., C, **275**, 1017, 1972

31.34   **8 - Acetoxy - 6 - (2,4 - dimethoxy - 5 - bromophenyl) - 3 - methyltricyclo(5.2.1.0^{3,8})decan - 2 - one benzene solvate**
$C_{21}H_{25}BrO_5$, $0.5C_6H_6$
T.R.Kasturi, R.Ramachandra, K.M.Damodaran, K.Vijayan
Tetrahedron Letters, 5059, 1972

31.C   **17 - Ethylenedioxy - 3 - methoxy - 6,7,8 - methylidyne - 1,3,5(10) - estratriene**
$C_{22}H_{26}O_3$
For complete entry see 51.26

31.35   **Dimethyl 2,5 - diphenyl - benzocyclopropen - 1,1 - dicarboxylate**
$C_{23}H_{18}O_4$
E.Carstensen-Oeser, B.Muller, H.Durr   *Angew. Chem.*, **84**, 434, 1972

31.36   **1 - Hydroxy - (2.2)metacyclophane m - bromobenzoyl ester (absolute configuration)**
$C_{23}H_{19}BrO_2$
K.Mislow, M.Brzechffa, H.W.Gschwend, R.T.Puckett
J. Amer. Chem. Soc., **95,** 621, 1973

31.37   **Methyl naphthalene - 2 - carboxylate dimer**
$C_{24}H_{20}O_4$
C.Kowala, G.Sugowdz, W.H.F.Sasse, J.A.Wunderlich
Tetrahedron Letters, 4721, 1972

31.38   **(+) - 2,5 - Dimethoxy - 7 - dimethylamino - tryptycene**
$C_{24}H_{23}NO_2$
N.Sakabe, K.Sakabe, K.Ozeki-Minakata, C.Katayama, J.Tanaka
Acta Cryst. (A), **28,** S28, 1972

31.39   **3 - Methyl - 9,10 - di(hydroxymethyl) - snoutane di - brosylate**
$C_{24}H_{24}Br_2O_6S_2$
L.A.Paquette, R.S.Beckley, D.Truesdell, J.Clardy
Tetrahedron Letters, 4913, 1972

31.40   **(+) - 2,5 - Dimethoxy - 7 - dimethylamino - triptycene hydrobromide (absolute configuration)**
$C_{24}H_{24}NO_2{}^+$, Br^-
N.Sakabe, K.Sakabe, K.Ozeki-Minakata, J.Tanaka
Acta Cryst. (B), **28,** 3441, 1972

31.C   **$14\alpha,17\alpha$ - Etheno - 15,16 - di(trifluoromethyl) - 4,15 - pregnadiene - 3,20 - dione**
$C_{25}H_{26}F_6O_2$
For complete entry see 51.34

31.C   **Aphidicolin bis(acetonide)**
$C_{26}H_{42}O_4$
For complete entry see 50.10

31.C   **Rearrangement product of a Beyer - 15(16) - en - 12 - one system**
$C_{27}H_{29}BrO_4$
For complete entry see 59.30

31.41   **ent - 1α - p - Bromobenzoyloxy - 16S - atis - 13 - en - 2 - one**
$C_{27}H_{33}BrO_3$
M.Laing, K.H.Pegel, L.P.L.Piacenza   *Tetrahedron Letters,* 2393, 1973

31.42   **4,5 - Dihydroxy - 7 - methyl - pentacyclo($7.2.1.1^{2,6}.0^{2,8}.0^{3,12}$) tridecane bis(p - bromobenzoate)**
$C_{28}H_{26}Br_2O_4$
N.Acton, R.J.Roth, T.J.Katz, J.A.K.Frank, C.A.Maier, I.C.Paul
J. Amer. Chem. Soc., **94,** 5446, 1972

31.43 **5a,11a - Dibromo - janusene**
$C_{30}H_{20}Br_2$
W.M.Macintyre, A.H.Tench *J. Org. Chem.*, **38,** 130, 1973

31.44 **Hexa - o - phenylene (centrosymmetric form)**
$C_{36}H_{24}$
H.Irngartinger *Israel J. Chem.*, **10,** 635, 1972

31.45 **Hexa - o - phenylene (orthorhombic form)**
$C_{36}H_{24}$
H.Irngartinger *Acta Cryst. (B),* **29,** 894, 1973

31.C **(2.2) - Paracyclophane bis - O - 18 - crown - 6 ether**
$C_{36}H_{52}O_{12}$
For complete entry see 38.32

31.46 **Quadruple - layered cyclophane (less soluble isomer)**
$C_{40}H_{44}$
H.Mizuno, K.Nishiguchi, T.Otsubo, S.Misumi, N.Morimoto
Tetrahedron Letters, 4981, 1972

HETERO-NITROGEN
(3, 4, 5-MEMBERED MONOCYCLIC)

32.1 **Sodium 4 - amino - 5 - mercapto - 1,2,4 - triazole trihydrate**
$C_2H_3N_4S^-$, Na^+ , $3H_2O$
R.C.Seccombe, J.V.Tillack, C.H.L.Kennard *J. C. S. Perkin II*, 6, 1973

32.2 **3,4,5 - Triamino - 1,2,4 - triazole hydrobromide**
Guanazine hydrobromide
$C_2H_7N_5^+$, Br^-
R.C.Seccombe, C.H.L.Kennard *J. C. S. Perkin II*, 1, 1973

32.3 **trans - 1 - Carbamyl - imidazolidone - 4,5 - diol**
$C_4H_7N_3O_4$
B.S.Hahn, S.Y.Wang, J.L.Flippen, I.L.Karle
J. Amer. Chem. Soc., **95**, 2711, 1973

32.4 **1 - Acetyl - 4 - bromo - pyrazole**
$C_5H_5BrN_2O$
J.Lapasset, A.Escande, J.Falgueirettes *Acta Cryst. (B)*, **28**, 3316, 1972

32.5 **2 - Methyl - 5 - tetrafluorophosphoranyl - pyrrole**
$C_5H_6F_4NP$
M.J.C.Hewson, R.Schmutzler, W.S.Sheldrick
J. C. S. Chem. Comm., 190, 1973
Also classified in 64

32.C **4 - Hydroxy - L - proline (neutron study)**
$C_5H_9NO_3$
For complete entry see 48.27

32.6 **1,4 - Dinitro - 2 - isopropylimidazole**
$C_6H_8N_4O_4$
R.S.Glass, J.F.Blount, D.Butler, A.Perrotta, E.P.Oliveto
Canad. J. Chem., **50**, 3472, 1972

32.7 **1 - (4 - Methyl - 1 - pyrazolin - 3 - yl) - 5,5 - bis(trifluoromethyl) - Δ^2 - 1,2,3 - triazoline**
$C_8H_9F_6N_5$
A.Gieren *Chem. Ber.*, **106**, 288, 1973

32.8   **N - (β - D - Ribofuranosyl)imidazole**
$C_8H_{12}N_2O_4$
M.N.G.James, M.Matsushima   *Acta Cryst. (B)*, **29**, 838, 1973
Also classified in 45

32.9   **1 - (2 - Chloroethyl) - 3 - (4 - carbamoylpyrazol - 3 - yl) - Δ^2 - 1,2,3 - triazolium chloride**
$C_8H_{12}N_6O^+$, Cl^-
J.E.Whinnery, W.H.Watson   *Acta Cryst. (B)*, **28**, 3635, 1972

32.10   **4 - Benzylideneamino - 5 - mercapto - 1,2,4 - triazole**
$C_9H_8N_4S$
R.C.Seccombe, C.H.L.Kennard   *J. C. S. Perkin II*, 9, 1973

32.11   **4 - Amino - 3 - (β - benzoylhydrazino) - 5 - mercapto - 1,2,4 - triazole**
$C_9H_{10}N_6OS$
R.C.Seccombe, C.H.L.Kennard   *J. C. S. Perkin II*, 4, 1973

32.12   **3 - Phenylpyrrolidine - 2,5 - dione (orthorhombic form)**
$C_{10}H_9NO_2$
G.Argay, A.Kalman   *Acta Cryst. (B)*, **29**, 636, 1973

32.13   **2,5 - bis(Ethyleneimino) - 1,4 - benzoquinone (at 300$\pm$5°K)**
$C_{10}H_{10}N_2O_2$
T.Ito, T.Sakurai   *Acta Cryst. (A)*, **28**, S11, 1972
Also classified in 18

32.14   **2,5 - bis(Ethyleneimino) - 1,4 - benzoquinone (at 240$\pm$10°K)**
$C_{10}H_{10}N_2O_2$
T.Ito, T.Sakurai   *Acta Cryst. (A)*, **28**, S11, 1972
Also classified in 18

32.15   **2,5 - bis(Ethyleneimino) - 1,4 - benzoquinone (at 110$\pm$20°K)**
$C_{10}H_{10}N_2O_2$
T.Ito, T.Sakurai   *Acta Cryst. (A)*, **28**, S11, 1972
Also classified in 18

32.16   **3 - (N - Phenyl)amino - pyrrolidine - 2,5 - dione**
$C_{10}H_{10}N_2O_2$
G.Argay, E.Carstensen-Oeser   *Acta Cryst. (B)*, **29**, 1186, 1973

32.17   **Antipyrine**
1 - Phenyl - 2,3 - dimethyl - 5 - pyrazolone
$C_{11}H_{12}N_2O$
T.P.Singh, M.Vijayan   *Acta Cryst. (B)*, **29**, 714, 1973

32.18   **2,5 - bis(Pyrrolidin - 2 - on - 5 - yl)pyrrole monohydrate**
$C_{12}H_{15}N_3O_2$, H_2O
G.D.Andreetti, G.Bocelli, L.Cavalca, P.Sgarabotto
Abstr. Ital.-Yug. Congr, 121, 1973

32.19   **1 - (2' - Carboxyphenyl) - 2 - acetyl - 5 - methyl - 3 - pyrazolone**
$C_{13}H_{12}N_2O_4$
G.D.Andreetti, G.Bocelli, L.Cavalca, P.Sgarabotto
Gazz. Chim. Ital., **102**, 106, 1972

32.C   **1 - Morpholinomethyl - 3 - methyl - 3 - phenylpyrrolidin - 2,5 - dione**
$C_{16}H_{20}N_2O_3$
For complete entry see 40.30

32.20   **2 - (p - Fluorophenyl) - 3 - (cyclohexylimino) - 4,4 - bis(trifluoromethyl) - 1 - azetine**
$C_{17}H_{15}F_7N_2$
A.Gieren   *Eur. Cryst. Meeting*, 1973

32.21   **1,3 - bis(4 - (2' - Bromoethoxy)phenyl) - 5 - tetrazolone**
$C_{17}H_{16}Br_2N_4O_3$
Y.Iitaka, S.Uchiyama, Z.Tamura   *Chem. Pharm. Bull.*, **20**, 1181, 1972
Also classified in 17

32.C   **cis - (2 - Methoxycarbonyl - 3,3 - dimethyl - 8 - oxo - 7 - phthalimido - 4,5 - dithia - 1 - aza - bicyclo(4.2.0)octane)**
$C_{17}H_{16}N_2O_5S_2$
For complete entry see 41.39

32.22   **4 - Hydroxy - 4,5 - dimethyl - 3,5 - diphenyl - pyrrolidin - 2 - one (stereoisomer i)**
$C_{18}H_{19}NO_2$
L.Fanfani, A.Nunzi, P.F.Zanazzi, A.R.Zanzari
Abstr. Ital.-Yug. Congr, 128, 1973

32.23   **2 - (Diphenylamino - aminylio) - 3,5 - dioxo - 4 - phenyl - 1,2,4 - triazolidine - 1 - ide**
$C_{20}H_{15}N_5O_2$
J.E.Weidenborner, E.Fahr, M.J.Richter, K.-H.Koch
Angew. Chem., **85**, 229, 1973
Also classified in 9

32.24   **5,5' - Di(ethoxycarbonyl) - 3,3',4,4' - tetraethyl - dipyrrol - 2 - yl - methane**
$C_{23}H_{34}N_2O_4$
R.Bonnett, M.B.Hursthouse, S.Neidle   *J. C. S. Perkin II*, 1335, 1972

32.25   **1 - p - Iodophenyl - 3 - carbomethoxy - 4 - phenacylidene - 5 - hydroxy - 5 - phenylpyrazoline**
$C_{25}H_{19}IN_2O_4$
G.D.Andreetti, G.Bocelli, L.Cavalca, P.Sgarabotto
Gazz. Chim. Ital., **102**, 355, 1972

HETERO-NITROGEN
(6-MEMBERED MONOCYCLIC)

33.1 **6 - Azauracil**
$C_3H_3N_3O_2$
P.Singh, D.J.Hodgson   *J. C. S. Chem. Comm.*, 439, 1973

33.2 **Cyclotrimethylene - trinitramine**
$C_3H_6N_6O_6$
C.S.Choi, E.Prince   *Acta Cryst. (B),* **28,** 2857, 1972

33.3 **4,5 - Dichloro - 3,6 - pyridazinedione (low angle data)**
Dichloromaleic hydrazide
$C_4H_2Cl_2N_2O_2$
T.Ottersen   *Acta Chem. Scand.,* **27,** 797, 1973

33.4 **4,5 - Dichloro - 3,6 - pyridazinedione (high angle data)**
Dichloromaleic hydrazide
$C_4H_2Cl_2N_2O_2$
T.Ottersen   *Acta Chem. Scand.,* **27,** 797, 1973

33.5 **6 - Azathymine**
$C_4H_5N_3O_2$
P.Singh, D.J.Hodgson   *J. C. S. Chem. Comm.*, 439, 1973

33.6 **5 - Chloro - 2 - pyridone**
C_5H_4ClNO
A.Kvick, S.S.Booles   *Acta Cryst. (B),* **28,** 3405, 1972

33.7 **2,5 - Dichloro - 3 - methoxypyrazine**
$C_5H_4Cl_2N_2O$
D.R.Carter, F.P.Boer   *J. Heterocycl. Chem.,* **9,** 335, 1972

33.C **4 - Nitropyridine - N - oxide - hydroquinone complex**
$C_5H_4N_2O_3$, $C_6H_6O_2$
For complete entry see 60.2

33.8 **Pyridine iodomonochloride**
C_5H_5ClIN
C.Romming   *Acta Chem. Scand.,* **26,** 1555, 1972

33.9   **4 - Chloropyridinium hexachlorostannate(iv)**
$2C_5H_5ClN^+$, Cl_6Sn^{2-}
R.C.Gearhart, T.B.Brill, W.A.Welsh, R.H.Wood
J. C. S. Dalton, 359, 1973

33.10   **Pyridinium hexaiodo - penta - argentate (at −30°C)**
$C_5H_6N^+$, $5Ag^+$, $6I^-$
S.Geller, B.B.Owens   *J. Phys. Chem. Solids*, **33**, 1241, 1972

33.11   **Pyridinium hexaiodo - penta - argentate (at 23°C)**
$C_5H_6N^+$, $5Ag^+$, $6I^-$
S.Geller, B.B.Owens   *J. Phys. Chem. Solids*, **33**, 1241, 1972

33.12   **Pyridinium hexaiodo - penta - argentate (at 40°C)**
$C_5H_6N^+$, $5Ag^+$, $6I^-$
S.Geller, B.B.Owens   *J. Phys. Chem. Solids*, **33**, 1241, 1972

33.13   **Pyridinium hexaiodo - penta - argentate (at 55°C)**
$C_5H_6N^+$, $5Ag^+$, $6I^-$
S.Geller, B.B.Owens   *J. Phys. Chem. Solids*, **33**, 1241, 1972

33.14   **Pyridinium hexaiodo - penta - argentate (at 80°C)**
$C_5H_6N^+$, $5Ag^+$, $6I^-$
S.Geller, B.B.Owens   *J. Phys. Chem. Solids*, **33**, 1241, 1972

33.15   **Pyridinium hexaiodo - penta - argentate (at 125°C)**
$C_5H_6N^+$, $5Ag^+$, $6I^-$
S.Geller, B.B.Owens   *J. Phys. Chem. Solids*, **33**, 1241, 1972

33.16   **Pyridinium hexachloroantimonate(v)**
$C_5H_6N^+$, Cl_6Sb^-
S.K.Porter, R.A.Jacobson   *Cryst. Struct. Comm.*, **1**, 431, 1972

33.17   **Pyridinium bromo - niobium chloride**
$2C_5H_6N^+$, $Br_{12}Cl_6Nb_6^{2-}$
B.Spreckelmeyer, H.G.von Schnering   *Z. Anorg. Allg. Chem.*, **386**, 27, 1971

33.18   **Pyridinium chloro - bromo - niobium bromide**
$2C_5H_6N^+$, $Br_{12}Cl_6Nb_6^{2-}$
B.Spreckelmeyer, H.G.von Schnering   *Z. Anorg. Allg. Chem.*, **386**, 27, 1971

33.19   **Pyridinium chloro - niobium chloride**
$2C_5H_6N^+$, $Cl_{18}Nb_6^{2-}$
B.Spreckelmeyer, H.G.von Schnering   *Z. Anorg. Allg. Chem.*, **386**, 27, 1971

33.20   **Piperidine hydrochloride**
$C_5H_{12}N^+$, Cl^-
J.K.Dattagupta, N.N.Saha   *Stockholm Symposium*, 89, 1973

33.C   **Piperidinium tetrakis(benzoylacetonato) europate**
$C_5H_{12}N^+$, $C_{40}H_{36}EuO_8^-$
For complete entry see 77.21

33.21 **bis(Piperidinium) pentabromoantimonate(iii)**
$2C_5H_{12}N^+$, Br_5Sb^{2-}
H.A.Abdel-Rehim, E.A.Meyers *Cryst. Struct. Comm.*, **2,** 45, 1973

33.22 **Di(pyridinium) octafluoro - di - μ - oxo - arsenate**
$2C_6H_6N^+$, $As_2F_8O_2^{2-}$
W.Haase *Chem. Ber.*, **106,** 734, 1973

33.23 **3 - Thioamido - pyridine**
$C_6H_6N_2S$
G.R.Form, E.S.Raper, T.C.Downie *Acta Cryst. (B)*, **29,** 776, 1973

33.24 **2 - Picolinium tetraiodoantimonate(iii)**
$C_6H_8N^+$, I_4Sb^-
H.A.Abdel-Rehim, E.A.Meyers *Cryst. Struct. Comm.*, **2,** 121, 1973

33.25 **4 - Methylpyridinium hexabromoantimonate(v) tribromide**
$2C_6H_8N^+$, Br_6Sb^- , Br_3^-
S.L.Lawton, D.M.Hoh, R.C.Johnson, A.S.Knisely
Inorg. Chem., **12,** 277, 1973

33.26 **1,2 - Dimethyl - 3,6 - pyridazinedione**
N,N - Dimethylmaleic hydrazide
$C_6H_8N_2O_2$
T.Ottersen *Acta Chem. Scand.*, **27,** 835, 1973

33.27 **Chinchomeronic acid**
$C_7H_5NO_4$
F.Takusagawa, K.Hirotsu, A.Shimada *Acta Cryst. (A)*, **28,** S15, 1972

33.28 **Dinicotinic acid**
$C_7H_5NO_4$
F.Takusagawa, K.Hirotsu, A.Shimada *Acta Cryst. (A)*, **28,** S15, 1972

33.29 **Quinolinic acid**
$C_7H_5NO_4$
F.Takusagawa, K.Hirotsu, A.Shimada *Acta Cryst. (A)*, **28,** S15, 1972

33.30 **Dipicolinic acid monohydrate**
$C_7H_5NO_4$, H_2O
F.Takusagawa, K.Hirotsu, A.Shimada *Acta Cryst. (A)*, **28,** S15, 1972

33.31 **3,5 - Dichloro - 2,6 - dimethyl - 4 - pyridinol**
Clopidol
$C_7H_7Cl_2NO$
F.P.Boer *Acta Cryst. (B)*, **28,** 3200, 1972

33.32 **2 - Formylpyridine selenosemicarbazide**
$C_7H_8N_4Se$
A.Conde, A.Lopez-Castro, R.Marquez *Acta Cryst. (B)*, **28,** 3464, 1972
Also classified in 8, 9

33.33 **1 - Methylnicotinamide aden - 9 - yl acetate dihydrate**
$C_7H_9N_2O^+$, $C_7H_6N_5O_2^-$, $2H_2O$
D.Voet *J. Amer. Chem. Soc.*, **95,** 3763, 1973
Residue 2 classified in 44

33.C **Thymine N,N - diethylmelamine monohydrate**
$C_7H_{14}N_6$, $C_5H_6N_2O_2$, H_2O
For complete entry see 60.4

33.34 **Pyridoxal hydrochloride**
$C_8H_{10}NO_3^+$, Cl^-
T.Fujiwara, Y.Izumi, K.Tomita *Acta Cryst. (A)*, **28,** S49, 1972

33.35 **Pyridoxal phosphate monohydrate**
$C_8H_{10}NO_6P$, H_2O
T.Fujiwara, Y.Izumi, K.Tomita *Acta Cryst. (A)*, **28,** S49, 1972
Residue 1 also classified in 46

33.C **L - Mimosine**
β - N - (3 - Hydroxy - 4 - pyridone) - α - amino - propionic acid
$C_8H_{10}N_2O_4$
For complete entry see 48.53

33.36 **Pyridoxamine dihydrochloride monohydrate**
$C_8H_{14}N_2O_2^{2+}$, $2Cl^-$, H_2O
T.Fujiwara, Y.Izumi, K.Tomita *Acta Cryst. (A)*, **28,** S49, 1972

33.37 **Pyridoxol 5' - methylphosphonate**
$C_9H_{14}NO_4P$
F.E.Cole, B.Lachmann, W.Korytnyk *J. Heterocycl. Chem.*, **9,** 1129, 1972
Also classified in 64

33.38 **2,2,6,6 - Tetramethyl - piperidine - 1 - oxy - piperidine (tetragonal form)**
$C_9H_{18}NO$
A.Capiomont, D.Bordeaux, J.Lajzerowicz
C. R. Acad. Sci., Fr., C, **275,** 317, 1972

33.39 **2,2' - Bipyridine**
$C_{10}H_8N_2$
K.Nakatsu, H.Yoshioka, M.Matsui, S.Koda, S.Ooi
Acta Cryst. (A), **28,** S24, 1972

33.40 **2,2' - Bipyridinium bromide**
$C_{10}H_{10}N_2^{2+}$, $2Br^-$
K.Nakatsu, H.Yoshioka, M.Matsui, S.Koda, S.Ooi
Acta Cryst. (A), **28,** S24, 1972

33.41 **2,2' - Bipyridinium chloride**
$C_{10}H_{10}N_2^{2+}$, $2Cl^-$
K.Nakatsu, H.Yoshioka, M.Matsui, S.Koda, S.Ooi
Acta Cryst. (A), **28,** S24, 1972

33.42   **3 - Deaza - uridine**
$C_{10}H_{13}NO_6$
C.H.Schwalbe, W.Saenger   *Acta Cryst. (B)*, **29**, 61, 1973
Also classified in 45

33.43   **2 - Butyl - 4 - thiocarbamoyl - pyridine**
$C_{10}H_{14}N_2S$
J.-C.Colleter, M.Gadret, M.Goursolle, F.Leroy
C. R. Acad. Sci., Fr., C, **274**, 1803, 1972

33.44   **N,N' - bis - (3 - Bromopropionyl)piperazine**
$C_{10}H_{16}Br_2N_2O_2$
L.Olansky, J.W.Moncrief   *Acta Cryst. (B)*, **29**, 357, 1973

33.45   **2,5 - Dihydroxyphenyl - pyridinium bromide monohydrate**
$C_{11}H_{10}NO_2^+$, Br^- , H_2O
N.Shamala, K.Venkatesan   *Cryst. Struct. Comm.*, **2**, 239, 1973
Residue 1 also classified in 17

33.C   **Diaquo - magnesium picolinate**
$C_{12}H_{12}MgN_2O_6$
For complete entry see 67.8

33.46   **bis(N,N - Cyclopentamethylene - thiocarbonyl)disulfide**
Piperidylthiuram disulfide
$C_{12}H_{20}N_2S_4$
M.F.Dix, A.D.Rae   *Cryst. Struct. Comm.*, **2**, 159, 1973
Also classified in 11

33.47   **2,3 - Di(2 - pyridyl) - 2,3 - butanediol**
$C_{14}H_{16}N_2O_2$
J.N.Brown, R.M.Jenevein, J.H.Stocker, L.M.Trefonas
J. Org. Chem., **37**, 3712, 1972

33.C   **1,4 - Dihydro - 1 - methyl - 4 - (2' - methyl - 5' - phenylthiopyran - 4' - ylidene) pyridine - S,S - dioxide chloroform solvate**
$C_{14}H_{17}NO_3S$, $CHCl_3$
For complete entry see 39.26

33.48   **(±) - α - Promedol alcohol**
(±) - α - 1,2a,5e - Trimethyl - 4e - phenylpiperidin - 4a - ol
$C_{14}H_{21}NO$
F.R.Ahmed, W.H.DeCamp   *Acta Cryst. (B)*, **28**, 3489, 1972

33.49   **(±) - β - Promedol alcohol (rhombohedral form)**
(±) - β - 1,2,5 - Trimethyl - 4 - phenylpiperidin - 4 - ol
$C_{14}H_{21}NO$
W.H.DeCamp, F.R.Ahmed   *Acta Cryst. (B)*, **28**, 3484, 1972

33.50 **Chlorpheniramine maleate**
$C_{16}H_{20}ClN_2{}^+$, $C_4H_3O_4{}^-$
M.N.G.James, G.J.B.Williams
Amer. Cryst. Assoc., Abstr. Papers (Winter Meeting), 85, 1973
Residue 2 classified in 2

33.C **1,2 - Dihydro - 1 - methyl - 2 - (5' - phenylthiopyran - 2' - ylidene) pyridine S,S - dioxide**
$C_{17}H_{15}NO_2S$
For complete entry see 39.33

33.C **Tyrosine pyridoxyl phosphate heptahydrate**
$C_{17}H_{21}N_2O_8P$, $7H_2O$
For complete entry see 48.73

33.C **Neuroleptic drug i**
$C_{18}H_{18}ClN_3O$
For complete entry see 40.32

33.C **Neuroleptic drug ii**
$C_{18}H_{19}ClN_4$
For complete entry see 36.23

33.C **Neuroleptic drug iii**
$C_{18}H_{19}ClN_4$
For complete entry see 36.24

33.51 **1,4 - bis(β - Pyridyl - 2 - vinyl) - benzene**
$C_{20}H_{16}N_2$
H.Nakanishi, K.Ueno, M.Hasegawa, Y.Sasada   *Chem. Letters*, 301, 1972
Also classified in 19

33.C **2,4,6 - Triphenylverdazyl**
$C_{20}H_{17}N_4$
For complete entry see 12.12

33.C **Methixene hydrochloride monohydrate**
$C_{20}H_{24}NS^+$, Cl^- , H_2O
For complete entry see 39.44

33.52 **Trihexyphenidyl**
syn - α - Cyclohexyl - α - phenyl - 1 - piperidine - propanol
$C_{20}H_{31}NO$
N.Camerman, A.Camerman   *J. Amer. Chem. Soc.*, **94,** 8553, 1972
Also classified in 21

33.53 **1 - Benzyl - 4 - (2,6 - dioxo - 3 - phenyl - 3 - piperidyl)piperidine**
Benzitimide
$C_{23}H_{26}N_2O_2$
M.H.J.Koch, O.Dideberg   *Acta Cryst. (B)*, **29**, 369, 1973

33.54   **4,5,6 - Tri - (p - methoxyphenyl) - 1,2,3 - triazine**
$C_{24}H_{21}N_3O_3$
E.Oeser, L.Schiele   *Chem. Ber.*, **105**, 3704, 1972
A so classified in 17

33.55   **N,N' - Dibenzyl - 4,4' - bipyridylium tetra(7,7,8,8 - tetracyanoquinodimethane)**
$C_{24}H_{22}N_2{}^{2+}$, $2C_{12}H_4N_4{}^-$, $2C_{12}H_4N_4$
T.Sundaresan, S.C.Wallwork   *Acta Cryst. (B)*, **28**, 2474, 1972
Residue 2 classified in 7

33.C   **2 - Phenyl - 4 - benzyl - 5,6 - dihydro - 5,6 - tetramethylene - 6 - piperidino - 1,3,4 - oxadiazine**
$C_{25}H_{29}N_3O_2$
For complete entry see 40.36

33.C   **Di - (2,2,6,6 - tetramethyl - 4 - piperidinyl - 1 - oxyl)suberate**
$C_{26}H_{46}N_2O_6$
For complete entry see 1.37

HETERO-NITROGEN
(7- AND HIGHER-MEMBERED MONOCYCLIC)

34.C Cyclo - di - β - alanyl
$C_6H_{10}N_2O_2$
For complete entry see 48.39

34.1 syn - N - Methyl - thiocapryllactam
$C_9H_{17}NS$
J.L.Flippen *Acta Cryst. (B)*, **28**, 3618, 1972

34.2 1 - Tosyl - 1,2 - diazepine (orthorhombic form)
$C_{12}H_{12}N_2O_2S$
R.Allmann, A.Frankowski, J.Streith *Tetrahedron*, **28**, 581, 1972

34.3 1,3,5,7 - Tetra - acetyl - 1,3,5,7 - tetra - azacyclo - octane
$C_{12}H_{20}N_4O_4$
C.S.Choi, J.E.Abel, B.Dickens, J.M.Stewart *Acta Cryst. (B)*, **29**, 651, 1973

34.4 N - Phenoxycarbonyl - azepine
$C_{13}H_{11}NO_2$
H.J.Lindner, B.von Gross *Chem. Ber.*, **105**, 434, 1972

34.5 syn - N - Methyl - thiolauryllactam
$C_{13}H_{25}NS$
J.L.Flippen *Acta Cryst. (B)*, **28**, 3618, 1972

34.C Cyclo - (tri - L - prolyl)
$C_{15}H_{21}N_3O_3$
For complete entry see 48.71

34.C Cyclo - (L - prolyl - L - prolyl - L - hydroxyprolyl)
$C_{15}H_{21}N_3O_4$
For complete entry see 48.72

34.C 10 - Chloro - 2,3,5,6,7,11b - hexahydro - 3 - hydroximino - 7 - methyl - 11b - phenyl - oxazolo(3,2 - d)benzo - 1,4 - diazepine
$C_{18}H_{18}ClN_3O_2$
For complete entry see 40.33

34.6 3 - (Diphenyl - hydroxy - methyl) - 3H - azepine
$C_{19}H_{17}NO$
H.J.Lindner, B.von Gross *Chem. Ber.*, **106**, 1033, 1973

34.7   **3,5,7 - Triphenyl - 4H - 1,2 - diazepine picrate**
$C_{23}H_{19}N_2^+$, $C_6H_2N_3O_7^-$
R.Gerdil   *Helv. Chim. Acta,* **55,** 2159, 1972
Residue 2 classified in 15, 17

34.8   **3,5 - Diphenyl - 2,4,6,7 - tetrakis(methoxycarbonyl) - 3H - azepine**
$C_{26}H_{23}NO_8$
E.Carstensen-Oeser   *Chem. Ber.,* **105,** 982, 1972

34.9   **4,6 - Diphenyl - 2,3,5,7 - tetrakis(methoxycarbonyl) - 3H - azepine**
$C_{26}H_{23}NO_8$
E.Carstensen-Oeser   *Chem. Ber.,* **105,** 982, 1972

HETERO-NITROGEN (2 FUSED RINGS)

35.1 **Allopurinol**
$C_5H_4N_4O$
P.Prusiner, M.Sundaralingam *Acta Cryst. (B)*, **28**, 2148, 1972

35.2 **5H - Pyrazolo - (3,4 - d)pyrimidine - 4 - thione**
$C_5H_4N_4S$
M.Gadret, M.Goursolle, J.-M.Leger
C. R. Acad. Sci., Fr., C, **276**, 1007, 1973

35.3 **3 - Methoxy - 1H - pyrazolo(4,3 - e) - 1,2,4 - triazine**
$C_5H_5N_5O$
H.J.Lindner, G.Schaden *Chem. Ber.*, **105**, 1949, 1972

35.4 **Benzimidazole**
$C_7H_6N_2$
C.J.Dik-Edixhoven, H.Schenk, H.van der Meer
Cryst. Struct. Comm., **2**, 23, 1973

35.5 **2,3 - Diazanaphthalene**
Phthalazine
$C_8H_6N_2$
C.Huiszoon, B.W.van de Waal, A.B.van Egmond, S.Harkema
Acta Cryst. (B), **28**, 3415, 1972

35.6 **5 - Chloro - 7 - iodo - 8 - quinolinol**
C_9H_5ClINO
J.-M.Leger, P.Marsau, J.Housty *C. R. Acad. Sci., Fr., C,* **274**, 1991, 1972

35.7 **6 - Mercapto - 3 - phenyl - s - triazolo(4,3 - b) - s - tetrazine pyridine solvate**
$C_9H_6N_6S$, C_5H_5N
R.C.Seccombe, C.H.L.Kennard *J. C. S. Perkin II,* 11, 1973

35.8 **N - Ethanol - β - isatoxime**
$C_{10}H_{10}N_2O_3$
F.Plana, M.Font-Altaba, C.Miravitlles *Acta Cryst. (A)*, **28**, S23, 1972

35.9 **1 - (β - D - Ribofuranosyl)imidazo(4,5 - b)pyrazine 4 - oxide**
$C_{10}H_{12}N_4O_5$
F.E.Cole *Amer. Cryst. Assoc., Abstr. Papers (Winter Meeting),* 61, 1973
Also classified in 45

35.C **Tetracyanoquinodimethane - 1 - methyl - N - ethylbenzimidazolium complex acetonitrile solvate**
$C_{10}H_{13}N_2^+$, $C_{12}H_4N_4$, $C_{12}H_4N_4^-$, C_2H_3N
For complete entry see 7.9

35.10 **1,2,3 - Trimethyl - benzimidazolium tetracyanoquinodimethanide**
$C_{10}H_{13}N_2^+$, $C_{12}H_4N_4^-$
D.Chasseau, J.Gaultier, C.Hauw, M.Schvoerer
C. R. Acad. Sci., Fr., C, **275**, 1491, 1972
Residue 2 classified in 7, 12

35.11 **2 - Methoxy - 4 - hydroxy - 5 - oxo - benz(f)azepine**
$C_{11}H_9NO_3$
W.A.Denne, M.F.Mackay *Tetrahedron,* **28**, 1795, 1972

35.12 **2 - (1 - (E) - Hydroxyiminoethyl) - 3 - acetylimidazo(1,2 - a)pyridine**
$C_{11}H_{11}N_3O_2$
N.W.Alcock, B.T.Golding, D.R.Hall, U.Horn
J. Amer. Chem. Soc., **94**, 8610, 1972

35.13 **Tubercidin**
$C_{11}H_{14}N_4O_4$
R.M.Stroud *Acta Cryst. (B),* **29**, 690, 1973
Also classified in 45

35.14 **Tubercidin**
$C_{11}H_{14}N_4O_4$
J.Abola, M.Sundaralingam *Acta Cryst. (B),* **29**, 697, 1973
Also classified in 45

35.C **Tetracyanoquinodimethane - 1,2 - dimethyl - N - ethylbenzimidazolium complex acetonitrile solvate**
$C_{11}H_{15}N_2^+$, $C_{12}H_4N_4$, $C_{12}H_4N_4^-$, C_2H_3N
For complete entry see 7.10

35.15 **3 - Carbethoxy - 4 - oxo - 6 - methyl - homopyrimidazole**
$C_{12}H_{12}N_2O_3$
K.Sasvari, J.C.Horvai, K.Simon *Acta Cryst. (B),* **28**, 2405, 1972

35.16 **N,N - Dimethyltryptamine (form II)**
$C_{12}H_{16}N_2$
G.Falkenberg *Acta Cryst. (B),* **28**, 3075, 1972
Also classified in 3

35.17 **Bufotenine**
5 - Hydroxy - N,N - dimethyltryptamine
$C_{12}H_{16}N_2O$
G.Falkenberg *Acta Cryst. (B),* **28**, 3219, 1972
Also classified in 3

35.18   3 - Carboethoxy - 4 - oxo - 6 - methyl - 6,7,8,9 -
tetrahydrohomopyrimidazole
$C_{12}H_{16}N_2O_3$
K.Sasvari, K.Simon   *Acta Cryst. (B)*, **29**, 1245, 1973

35.19   2 - Phenyl - pyrimido(1,2 - b)pyridazin - 4 - one
$C_{13}H_9N_3O$
L.Golic, S.Kavcic   *Abstr. Ital.-Yug. Congr,* 104, 1973

35.20   2 - (4 - O - Acetyl - 2,3 - dideoxy - β - L - glycero - pent - 2 - enopyranosyl) -
5,6 - dichloro - benzotriazole
$C_{13}H_{11}Cl_2N_3O_3$
J.Lopez de Lerma, S.Martinez-Carrera, S.Garcia-Blanco
Acta Cryst. (B), **29**, 537, 1973
Also classified in 38

35.21   N - Tosyl - 2 - trichloromethyl - 4,5 - epoxy - piperidine
$C_{13}H_{14}Cl_3NO_3S$
E.R.Hessling, J.G.H.de Jong, A.R.Overbeek, H.Schenk
Cryst. Struct. Comm., **2**, 247, 1973
Also classified in 38

35.22   **Melatonin**
N - Acetyl - 5 - methoxy - tryptamine
$C_{13}H_{16}N_2O_2$
A.Wakahara, T.Fujiwara, K.-I.Tomita   *Chem. Letters,* 1139, 1972
Also classified in 3

35.23   1,6 - Dimethyl - 3 - carbethoxy - 4 - oxo - 1,6,7,8,10 - hexahydro -
homopyrimidazole
$C_{13}H_{20}N_2O_3$
K.Simon, K.Sasvari   *Cryst. Struct. Comm.*, **1**, 419, 1972

35.24   cis - 1 - (6 - Acetoxymethyltetrahydro - 2 - pyranyl) - 5,6 -
dichlorobenzotriazole
$C_{14}H_{15}Cl_2N_3O_3$
J.Fayos, S.Garcia-Blanco   *Acta Cryst. (B)*, **28**, 2863, 1972
Also classified in 38

35.25   7 - Acetyl - 3,5,5,9,9 - pentamethyl - 1,6 - diazabicyclo(4.3.0)nona - 3,7 -
dien - 2 - one
$C_{14}H_{20}N_2O_2$
A.C.Day, A.N.McDonald, B.F.Anderson, T.J.Bartczak, O.J.R.Hodder
J. C. S. Chem. Comm., 247, 1973

35.26   5 - Chloro - 3 - cyano - 2 - dimethylamino - 3 - phenyl - 3H - indole
$C_{17}H_{14}ClN_3$
N.W.Gilman, J.F.Blount, L.H.Sternbach   *J. Org. Chem.*, **37**, 3201, 1972

35.27   5 - Chloro - 2 - dimethylamino - 3 - methoxyiminomethyl - 3 - phenyl - 3H - indole
$C_{18}H_{18}ClN_3O$
N.W.Gilman, J.F.Blount, L.H.Sternbach   *J. Org. Chem.*, **37**, 3201, 1972

35.28   7 - Chloro - 4 - (4' - diethylamino - 1 - methylbutyl - amino)quinoline
$C_{18}H_{26}ClN_3$
C.Courseille, B.Busetta, M.Hospital   *Cryst. Struct. Comm.*, **2**, 283, 1973
Also classified in 3

35.29   7 - Chloro - 4 - (4 - diethylamino - 1 - methylbutyl - amino) - quinoline sulfate monohydrate
$C_{18}H_{28}ClN_3{}^{2+}$, O_4S^{2-} , H_2O
J.-M.Leger, J.-P.Bideau   *C. R. Acad. Sci., Fr., C,* **275**, 313, 1972

35.30   1,1,1',1' - tetrakis(Trifluoromethyl) - 3,3' - bi(1H - isoindole)
$C_{20}H_8F_{12}N_2$
A.Gieren, K.Burger, K.Einhellig   *Angew. Chem.*, **85**, 170, 1973

35.31   2,3 - Diphenylindole
$C_{20}H_{15}N$
B.Schmelter, H.Bradaczek, P.Luger   *Acta Cryst. (B)*, **29**, 971, 1973

35.32   Dimethyl 1,2 - bis - (4 - bromophenyl) - 3,5 - dimethyl - 4 - oxo - 6,7 - diazabicyclo(3.2.0)hept - 2 - ene - 6,7 - dicarboxylate
$C_{23}H_{20}Br_2N_2O_5$
P.C.Chieh, D.Mackay, L.Wong   *J. C. S. Perkin II*, 2094, 1972

35.33   8 - (3 - (p - Fluorobenzoyl)propyl) - 1 - phenyl - 1,3,8 - triaza - spiro(4,5)deca - 4 - one
$C_{23}H_{26}FN_3O_2$
M.H.J.Koch   *Acta Cryst. (B)*, **29**, 379, 1973

35.34   5,6 - bis(Methoxycarbonyl) - 7 - p - bromophenyl - 3 - methyl - 4 - phenylpyrrolo(1,2 - b)pyridazin - 2 - one
$C_{24}H_{19}BrN_2O_5$
J.A.Moore, R.C.Gearhart, O.S.Rothenberger, P.C.Thorstenson,
R.H.Wood   *J. Org. Chem.*, **37**, 3774, 1972

35.35   5,5',6,6' - Tetrachloro - 1,1',3,3' - tetraethyl - benzimidazolocarbocyanine iodide methanol solvate
$C_{25}H_{27}Cl_4N_4{}^+$, I^- , $2CH_4O$
D.L.Smith, H.R.Luss   *Acta Cryst. (B)*, **28**, 2793, 1972

35.36   5,5',6,6' - Tetrachloro - 1,1',3,3' - tetraethyl - benzimidazolocarbocyanine iodide acetonitrile solvate
$C_{25}H_{27}Cl_4N_4{}^+$, I^- , C_2H_3N
D.L.Smith, H.R.Luss   *Acta Cryst. (B)*, **28**, 2793, 1972

35.37   **1,1′ - Azo - 2 - phenylimidazo(1,2 - a)pyridinium dibromide**
$C_{26}H_{20}N_6^{2+}$, $2Br^-$
D.J.Pointer, J.B.Wilford   *J. C. S. Perkin II,* 2259, 1972
Residue 1 also classified in 9

35.38   **1,1 - bis(5′ - Methyl - 2′ - phenyl - 1′ - benzoyl - indolizin - 2′ - yl) - ethylene**
$C_{46}H_{34}N_2O_2$
E.Oeser   *Chem. Ber.,* **105,** 2351, 1972

HETERO-NITROGEN
(MORE THAN 2 FUSED RINGS)

36.1   **Allyl azide dimer**
$C_6H_{10}N_6$
J.C.Pezzullo, E.R.Boyko *J. Org. Chem.*, **38,** 168, 1973

36.2   **4,6 - Dimethyl - 4,6 - diaza - 9,11 - dioxa - tricyclo(6.3.0.0^{2,7})undecan - 3,5,10 - trione**
$C_9H_{10}N_2O_5$
C.Pascard-Billy *Acta Cryst. (B)*, **29,** 521, 1973
Also classified in 38

36.3   **cis - anti - 6 - Methyluracil dimer**
$C_{10}H_{12}N_4O_4$
L.M.Amzel, H.P.Avey, L.N.Becka, R.J.Poljak, M.N.Khattak, S.Y.Wang
Nature New Biology, **238,** 204, 1972
Also classified in 44

36.4   **2 - Bromo - 1 - methyl - benzo(c)pyrazolo(1.2 - a)pyrazole - 3,9 - dione**
$C_{11}H_7BrN_2O_2$
I.Sotofte *Acta Chem. Scand.*, **27,** 661, 1973

36.5   **10 - Methyl - isoalloxazine silver nitrate**
$C_{11}H_8N_4O_2$, Ag^+ , NO_3^-
C.J.Fritchie Junior, G.D.Sproul, T.D.Wade *Acta Cryst. (A)*, **28,** S48, 1972

36.6   **bis(10 - Methylisoalloxazine) - sesqui(silver perchlorate)**
$4C_{11}H_8N_4O_2$, $3Ag^+$, $3ClO_4^-$
C.J.Fritchie Junior, G.D.Sproul, T.D.Wade *Acta Cryst. (A)*, **28,** S48, 1972

36.7   **Salicylato - (1,10 - phenanthroline) thallium(i)**
$C_{12}H_8N_2$, $C_7H_5O_3^-$, Tl^+
D.L.Hughes, M.R.Truter *J. C. S. Dalton,* 2214, 1972

36.C   **7,7,8,8 - Tetracyanoquinodimethane - 1,10 - phenanthroline complex**
$C_{12}H_8N_2$, $C_{12}H_4N_4$
For complete entry see 60.14

36.C   **7,7,8,8 - Tetracyanoquinodimethane - phenazine**
$C_{12}H_8N_2$, $C_{12}H_4N_4$
For complete entry see 60.15

36.8 **1,10 - Phenanthrolinium hydrogen sulfate**
$C_{12}H_9N_2^+$, HO_4S^-
K.Nakatsu, H.Yoshioka, M.Matsui, S.Koda, S.Ooi
Acta Cryst. (A), **28,** S24, 1972

36.9 **Phenanthrolium pentachloromanganate(iii)**
$C_{12}H_{10}N_2^{2+}$, Cl_5Mn^{2-}
M.Matsui, S.Koda, S.Ooi, H.Kuroya, I.Bernal *Chem. Letters,* 51, 1972

36.10 **1,1' - Ethylene - 2,2' - bipyridylium 7,7,8,8 - tetracyanoquinodimethane**
$C_{12}H_{12}N_2^{2+}$, $2C_{12}H_4N_4^-$
T.Sundaresan, S.C.Wallwork *Acta Cryst. (B),* **28,** 3065, 1972
Residue 2 classified in 7

36.11 **4,4a,9,9a,14,14a - Hexahydro - 3H,8H,13H - tripyridazino(1,6 - a.1',6' - c,1'',6'' - e) - s - triazine**
4,5 - Dihydropyridazine trimer
$C_{12}H_{18}N_6$
B.K.Bandlish, J.N.Brown, J.W.Timberlake, L.M.Trefonas
J. Org. Chem., **38,** 1102, 1973

36.12 **Phenanthridine**
$C_{13}H_9N$
P.Roychowdhury *Acta Cryst. (B),* **29,** 1362, 1973

36.13 **Chromophore I (at 5°C)**
$C_{13}H_{12}N_3O_2^+$, Cl^- , $2H_2O$
K.Sasaki, Y.Hirata *J. C. S. Perkin II,* 485, 1973

36.14 **Lumiflavinium chloride - hydroquinone complex**
$C_{13}H_{13}N_4O_2^+$, $C_6H_6O_2$, Cl^-
R.Karlsson *Acta Cryst. (B),* **28,** 2358, 1972
Residue 2 classified in 17

36.15 **Chloro - zoanthoxanthin trihydrate**
$C_{13}H_{14}ClNS$, $3H_2O$
L.Cariello, S.Crescenzi, G.Prota, F.Giordano, L.Mazzarella
J. C. S. Chem. Comm., 99, 1973

36.16 **5 - Dimethylamino - 12 - chloro - 12 - methyl - 4,6,11,13 - tetra - aza - tricyclo(8.3.0.0^{3,7})trideca - 1,3,5,7,9 - pentane trihydrate**
$C_{13}H_{14}ClN_5$, $3H_2O$
S.Capasso, F.Giordano, L.Mazzarella *Abstr. Ital.-Yug. Congr,* 221, 1973

36.17 **1,10 - Dichloro - 3,8 - dimethyl - 4,7 - phenanthroline**
$C_{14}H_{10}Cl_2N_2$
F.H.Herbstein, M.Kapon, D.Rabinovich *Israel J. Chem.,* **10,** 537, 1972

36.18 **9 - Bromo - 3,7,8,10 - tetramethyl - isoalloxazine monohydrate**
$C_{14}H_{13}BrN_4O_2$, H_2O
M.von Glehn, R.Norrestam *Acta Chem. Scand.,* **26,** 1490, 1972

36.C **7,7,8,8 - Tetracyanoquinodimethane - N,N' - dimethyldihydrophenazine (photographic data)**
$C_{14}H_{14}N_2$, $C_{12}H_4N_4$
For complete entry see 60.17

36.C **7,7,8,8 - Tetracyanoquinodimethane - N,N' - dimethyldihydrophenazine (diffractometer data)**
$C_{14}H_{14}N_2$, $C_{12}H_4N_4$
For complete entry see 60.18

36.19 **6,7 - Dihydro - 6 - methyl - 5H - dibenz(c,e)azepin - 5 - one**
$C_{15}H_{13}NO$
G.H.Wahl Junior, K.J.Wildonger, J.Bordner
Cryst. Struct. Comm., **2**, 267, 1973

36.20 **1,3,7,8,10 - Pentamethyl - isoalloxazinium iodide monohydrate**
$C_{15}H_{17}N_4O_2^+$, I^- , H_2O
R.Norrestam, L.Torbjornsson, F.Muller
Acta Chem. Scand., **26**, 2441, 1972

36.21 **Naphtho(2,1 - c)cinnoline**
$C_{16}H_{10}N_2$
A.K.Bhuiya, E.Stanley, J.Chan   *Z. Kristallogr.*, **135**, 408, 1972

36.C **10 - Aza - 19 - nor - 5β,9β - androst - 8(14) - en - 3,17 - dione**
$C_{17}H_{23}NO_2$
For complete entry see 51.1

36.22 **10 - Bromo - 2,3,5,6,7 - 11b - hexahydro - 2 - methyl - 11b - phenylbenzo(6,7) - 1,4 - diazepine(4,5 - b) - oxazol - 6 - one ethanol solvate**
$C_{18}H_{17}BrN_2O_2$, $2C_2H_6O$
S.Sato, N.Sakurai, T.Miyadera, C.Tamura, R.Tachikawa
Chem. Pharm. Bull., **19**, 2501, 1971
Residue 1 also classified in 40

36.23 **Neuroleptic drug ii**
$C_{18}H_{19}ClN_4$
T.J.Petcher, H.P.Weber   *Stockholm Symposium*, 38, 1973
Also classified in 33

36.24 **Neuroleptic drug iii**
$C_{18}H_{19}ClN_4$
T.J.Petcher, H.P.Weber   *Stockholm Symposium*, 38, 1973
Also classified in 33

36.25 **6,6,12,12 - tetrakis(Trifluoromethyl) - 6,12 - dihydro - dibenzo(c,h)(1,5)naphthyridine**
$C_{20}H_8F_{12}N_2$
A.Gieren, K.Burger, K.Einhellig   *Angew. Chem.*, **85**, 171, 1973

36.26   **(4aRS,4bRS,13bRS) - 12 - Bromo - 1 - (p - bromophenyl) - 1,4a,4b,5,6,13b - hexahydro - 4H - dipyridazino(1,6 - a.4,3 - c)quinoline**
$C_{20}H_{18}Br_2N_4$
H.Hjeds, I.K.Larsen *Acta Chem. Scand.*, **25,** 2394, 1971

36.27   **2,7 - Diamino - 9 - phenyl - 10 - methyl - phenanthridinium bromide monohydrate**
$C_{20}H_{18}N_3^+$, Br^- , H_2O
C.Courseille, B.Busetta, M.Hospital *C. R. Acad. Sci., Fr., C,* **275,** 95, 1972

36.28   **8 - Methoxy - N - tosyl - 3 - trichloromethyl - 2,3,4,4a,5,6 - hexahydrobenzo(f)quinoline**
$C_{22}H_{18}Cl_3NO_3S$
H.van der Meer *Rec. Trav. Chim. Pays-Bas,* **92,** 210, 1973

36.29   **4 - p - Bromobenzylidene - 9 - p - bromophenyl - 10a - hydroxy - 2,7 - dimethyl - 2,7 - diaza - 10 - oxa - 1,2,3,4,5,6,7,8,8a,10a - decahydroanthracene monohydrate**
$C_{26}H_{28}Br_2N_2O_2$, H_2O
C.P.Huber *Acta Cryst. (B),* **29,** 1046, 1973
Residue 1 also classified in 38

36.30   **6,7 - Diphenyldibenzo(e,g)(1,4)diazocine**
$C_{26}H_{28}N_2$
C.J.Finder, M.G.Newton, N.L.Allinger
Amer. Cryst. Assoc., Abstr. Papers (Winter Meeting), 50, 1973

36.31   **10,10' - Dimethyl - 9,10,9',10' - tetrahydro - 9,9' - biacridyl**
$C_{28}H_{24}N_2$
J.Preuss, A.Gieren, W.Hoppe, V.Zanker *Ann. Chem.*, 221, 1973

36.C   **Kinamycin C p - bromobenzoate benzene solvate (absolute configuration)**
$C_{31}H_{23}BrN_2O_{11}$, C_6H_6
For complete entry see 50.14

36.32   **4' - Methylene - 1,2 - di - m - bromophenyl - 1',2',6',7' - tetraphenyl - spiro(pyrazolidin - 4,8' - (8H,4H) - benzo(1,2 - c.4,5 - c)dipyrazoline) - 3,5,3',5' - tetraone**
$C_{47}H_{30}Br_2N_6O_4$
L.Fanfani, A.Nunzi, P.F.Zanazzi, A.R.Zanzari
Abstr. Ital.-Yug. Congr, 135, 1973

HETERO-NITROGEN
(BRIDGED RING SYSTEMS)

37.C   **bis(1,1,1,5,5,5 - Hexafluoropentaen - 2,4 - dionato) copper(ii) - 1,4 - diazabicyclo(2.2.2)octane complex**
$C_6H_{12}N_2$, $C_{10}H_2CuF_{12}O_4$
For complete entry see 60.12

37.1   **Hexamethylenetetramine**
$C_6H_{12}N_4$
E.D.Stevens, H.Hope
Amer. Cryst. Assoc., Abstr. Papers (Winter Meeting), 38, 1973

37.2   **(R) - (−) - 3 - Acetoxy - quinuclidine methiodide (absolute configuration)**
$C_{10}H_{18}NO_2^+$, I^-
R.W.Baker, P.J.Pauling   *J. C. S. Perkin II,* 2340, 1972

37.3   **1 - Azabicyclo(3.3.3)undecane hydrochloride**
$C_{10}H_{20}N^+$, Cl^-
A.H.-J.Wang, R.J.Missavage, S.R.Byrn, I.C.Paul
J. Amer. Chem. Soc., **94,** 7100, 1972

37.4   **2,9 - Diacetyl - 9 - azabicyclo(4.2.1)non - 2 - ene (absolute configuration)**
$C_{12}H_{17}NO_2$
C.S.Huber   *Acta Cryst. (B),* **28,** 2577, 1972

37.5   **exo - 2 - Methoxy - 3 - aza - 4 - keto - 7,8 - benzobicyclo(4.2.1)nonene**
$C_{13}H_{15}NO$
H.L.Ammon, P.H.Mazzocchi, W.J.Kopecky Junior, H.J.Tamburin,
P.H.Watts Junior   *J. Amer. Chem. Soc.,* **95,** 1968, 1973

37.6   **N - p - Bromobenzyl - exo - 2,3 - aziridinobicyclo(2.2.1)heptane**
$C_{14}H_{14}BrNO$
E.M.Gopalakrishna   *Acta Cryst. (B),* **28,** 2754, 1972

37.7   **7,8,9,10 - Tetrahydro - 6,10 - propano - 6H - cyclohepta(b)quinoxaline**
$C_{16}H_{18}N_2$
R.Hafter, J.Murray-Rust, P.Murray-Rust, W.Parker
J. C. S. Chem. Comm., 1127, 1972

37.8   N - n - Butyl - 2,3 - dimethyl - 1,4,endo - (3',4' - pyrrolidino) - 1,4 -
dihydronaphthalene hydrochloride
$C_{20}H_{28}N^+$, Cl^-
P.Argos, R.Clayton   *Acta Cryst. (B)*, **29,** 910, 1973

37.9   Decahydro - 7,14a,7a,14 - ethanediylidenenaphtho(1,8 - de.4,5 - d'e')
bisazocine - 4,6,11,13 - (1H,7H,8H,14H) - tetrone dihydrate
$C_{22}H_{26}N_2O_4$, $2H_2O$
T.Iwakuma, H.Nakai, O.Yonemitsu, D.S.Jones, I.L.Karle, B.Witkop
J. Amer. Chem. Soc., **94,** 5136, 1972

37.10   4,11 - Diacetyl - dodecahydro - 7H - 1,7,8a - ethanylylidene - 8,14 -
methanocyclopropa(1,6) - benzo(1,2 - d.4,3 - d') - bisazocine -
3,12,15,17(4H,9H) - tetrone
$C_{24}H_{26}N_2O_6$
T.Iwakuma, H.Nakai, O.Yonemitsu, D.S.Jones, I.L.Karle, B.Witkop
J. Amer. Chem. Soc., **94,** 5136, 1972

37.C   Fluorenyl - lithium bis(quinuclidine)
$C_{27}H_{35}LiN_2$
For complete entry see 67.20

HETERO-OXYGEN

38.1 **Furan (tetragonal form, at 123°K)**
C_4H_4O
R.Fourme *Acta Cryst. (B)*, **28,** 2984, 1972

38.2 **Furan (orthorhombic form, at 152°K)**
C_4H_4O
R.Fourme *Acta Cryst. (B)*, **28,** 2984, 1972

38.C **3,4 - Epoxysulfolane**
$C_4H_6O_3S$
For complete entry see 39.3

38.3 **2 - Formyl - 4 - bromofuran**
$C_5H_3BrO_2$
C.Riche *Acta Cryst. (B)*, **29,** 756, 1973

38.C **β - Chloroglutaric acid anhydride**
$C_5H_5ClO_3$
For complete entry see 1.12

38.C **Furamide - oxalic acid complex**
$2C_5H_5NO_2 , C_2H_2O_4$
For complete entry see 60.3

38.C **Calcium(+) - allo - hydroxycitrate lactone tetrahydrate**
Hibiscus acid calcium salt tetrahydrate
$C_6H_4O_7^{2-} , Ca^{2+} , 4H_2O$
For complete entry see 2.26

38.C **Potassium hydrogen (+) - isocitrate lactone**
$C_6H_5O_6^- , K^+$
For complete entry see 2.29

38.C **γ - Hydroxy - L - isoleucine - lactone hydrobromide (absolute configuration)**
$C_6H_{12}NO_2^+ , Br^-$
For complete entry see 48.40

38.C **DL - Methylkasugaminide selenate monohydrate methanol solvate**
$C_7H_{18}N_2O_2^{2+} , O_4Se^{2-} , H_2O , CH_4O$
For complete entry see 50.3

38.4 **Coumarin**
$C_9H_6O_2$
E.Gavuzzo, F.Mazza, E.Giglio *Eur. Cryst. Meeting,* 1973

38.C **4,6 - Dimethyl - 4,6 - diaza - 9,11 - dioxa - tricyclo(6.3.0.0^{2,7})undecan - 3,5,10 - trione**
$C_9H_{10}N_2O_5$
For complete entry see 36.2

38.5 **9,9 - Dichloro - trans,trans - bicyclo(6.1.0)non - 4 - ene oxide**
$C_9H_{12}Cl_2O$
W.A.Szabo, M.F.Betkouski, J.A.Deyrup, M.Mathew, G.J.Palenik
J. C. S. Perkin II, 339, 1973

38.6 **2,5 - Dimethyl - 3,4 - diacetylfuran**
$C_{10}H_{12}O_3$
S.Baggio, L.M.Amzel *J. Cryst. Mol. Struct.,* **3,** 115, 1973

38.C **Sceleratinic acid**
$C_{10}H_{13}ClO_4$
For complete entry see 59.5

38.7 **1 - Hydroxymethyl - 4 - methyl - 7 - oxabicyclo(4.3.0)non - 4 - en - 8 - one**
$C_{10}H_{14}O_3$
M.Currie *J. C. S. Perkin II,* 240, 1973

38.8 **1,3,8,10 - Tetraoxacyclotetradecane**
$C_{10}H_{20}O_4$
I.W.Bassi, R.Scordamaglia, L.Fiore *J. C. S. Perkin II,* 1726, 1972

38.C **anti - 4 - Phenyl - 6,7 - epoxy - 2 - oxa - 3 - aza - bicyclo(3.2.0)hepta - 3 - ene**
$C_{11}H_9NO_2$
For complete entry see 40.22

38.C **syn - 4 - Phenyl - 6,7 - epoxy - 2 - oxa - 3 - aza - bicyclo(3.2.0)hepta - 3 - ene**
$C_{11}H_9NO_2$
For complete entry see 40.23

38.C **Ranuncoside monohydrate**
$C_{11}H_{16}O_8 , H_2O$
For complete entry see 45.23

38.9 **Octachlorodibenzo - p - dioxin**
$C_{12}Cl_8O_2$
M.A.Neuman, P.P.North, F.P.Boer *Acta Cryst. (B),* **28,** 2313, 1972

38.10 **2,8 - Dichlorodibenzo - p - dioxin**
$C_{12}H_6Cl_2O_2$
F.P.Boer, M.A.Neuman, O.Aniline *Acta Cryst. (B),* **28,** 2878, 1972

38.C   **1,2,3,4,10,10 - Hexachloro - 6,7 - epoxy - 1,4,4a,5,6,7,8,8a - octahydro - endo - 1,4 - endo - 5,8 - dimethano - naphthalene**
Endrin
$C_{12}H_8Cl_6O$
For complete entry see 31.8

38.11   **Dibenzofuran**
$C_{12}H_8O$
A.Banerjee   *Indian J. Phys.*, **46,** 481, 1972

38.12   **Dibenzo - p - dioxin**
$C_{12}H_8O_2$
M.Senma, Z.Taira, T.Taga, K.Osaki   *Cryst. Struct. Comm.*, **2,** 311, 1973

38.C   **7,7,8,8 - Tetracyanoquinodimethane - dibenzo - p - dioxin**
$C_{12}H_8O_2$, $C_{12}H_4N_4$
For complete entry see 60.16

38.C   **Xanthotoxin**
8 - Methoxy - 3,'2' - 6,7 - furocoumarin
$C_{12}H_8O_4$
For complete entry see 59.6

38.13   **syn - 8,8 - Dichloro - 4 - phenyl - 3,5 - dioxabicyclo(5.1.0)octane**
$C_{12}H_{12}Cl_2O_2$
G.R.Clark, G.J.Palenik   *J. C. S. Perkin II*, 194, 1973

38.14   **1 - (p - Bromophenyl) - 1,2 - epoxycyclohexane**
$C_{12}H_{13}BrO$
S.Merlino, G.Lami, B.Macchia, F.Macchia, L.Monti
J. Org. Chem., **37,** 703, 1972

38.C   **1,2,4,5 - Di - O - isopropylidene - β - fructopyranose**
$C_{12}H_{20}O_6$
For complete entry see 45.25

38.C   **2 - (4 - O - Acetyl - 2,3 - dideoxy - β - L - glycero - pent - 2 - enopyranosyl) - 5,6 - dichloro - benzotriazole**
$C_{13}H_{11}Cl_2N_3O_3$
For complete entry see 35.20

38.C   **N - Tosyl - 2 - trichloromethyl - 4,5 - epoxy - piperidine**
$C_{13}H_{14}Cl_3NO_3S$
For complete entry see 35.21

38.C   **4,9 - Dimethoxy - 7 - methyl - 5H - furo(3,2 - g)benzopyran - 5 - one**
Khellin
$C_{14}H_{12}O_5$
For complete entry see 59.9

38.15 DL - 4 - Methoxy - 6 - styryl - 5,6 - dihydro - α - pyrone
Kavain
$C_{14}H_{14}O_3$
A.Yoshino, W.Nowacki *Z. Kristallogr.*, **136**, 66, 1972

38.C cis - 1 - (6 - Acetoxymethyltetrahydro - 2 - pyranyl) - 5,6 -
dichlorobenzotriazole
$C_{14}H_{15}Cl_2N_3O_3$
For complete entry see 35.24

38.16 4 - Methoxy - 6 - (1',2' - dihydrostyryl) - 5,6 - dihydro - α - pyrone
Dihydrokavain
$C_{14}H_{16}O_3$
P.Engel, W.Nowacki *Z. Kristallogr.*, **136**, 453, 1972

38.17 3 - Bromo - 4'a,10'a - dihydrospiro(2,5 - cyclohexadiene - 1,3' - cyclo - octa -
as - trioxin) - 4 - one
$C_{14}H_{17}BrO_4$
E.J.Gardner, R.H.Squire, R.C.Elder, R.M.Wilson
J. Amer. Chem. Soc., **95**, 1693, 1973
Also classified in 22

38.18 2 - (4 - Bromophenyl) - r - 2,4,4,c - 6 - tetramethyl - 1,3 - dioxan
$C_{14}H_{19}BrO_2$
G.M.Kellie, P.Murray-Rust, F.G.Riddell *J. C. S. Perkin II*, 2384, 1972

38.19 4 - Bromo - 8 - oxa - tetracyclo(9.2.2.1^{2,10}.0^{7,16})hexadeca -
2,4,6,10(16),12,14 - hexaene - 9 - one
$C_{15}H_9BrO_2$
T.Asao, N.Morita, C.Kabuto, Y.Kitahara *Tetrahedron Letters*, 4379, 1972

38.20 11 - Carbomethoxy - 11 - hydroxy - 6 - methoxy - 8 - methyl - 3,12 - dioxa -
tricyclo(8.3.0.0^{4,9})trideca - 1(10),4(9),5,7 - tetraene - 2,13 - dione
$C_{15}H_{12}O_8$
M.F.C.Ladd, D.V.Povey, R.Thomas *J. C. S. Chem. Comm.*, 333, 1973

38.21 4 - Methoxy - 6 - (5',6' - dioxymethylene - styryl) - 5,6 - dihydro - α - pyrone
Methysticin
$C_{15}H_{14}O_5$
P.Engel, W.Nowacki *Z. Kristallogr.*, **136**, 437, 1972

38.22 2 - Bromobenzo(b)indeno(1,2 - e)pyran
$C_{16}H_9BrO$
M.F.C.Ladd, D.C.Povey *J. Cryst. Mol. Struct.*, **2**, 243, 1972

38.23 6,7 - Dihydro - 6,6 - dimethyl - 3 - phenylbenzofuran - 2,4(3H,5H) - dione 2 -
oxime
$C_{16}H_{17}NO_3$
N.D.Jones, M.O.Chaney *Acta Cryst. (B)*, **28**, 3190, 1972

38.C   **Anthrotaxin monohydrate**
$C_{17}H_{16}O_6$, H_2O
For complete entry see 59.12

38.C   **Axivalin monohydrate**
$C_{17}H_{22}O_5$, H_2O
For complete entry see 59.13

38.24   **3 - Hydroxy - 5 - methyl - 7 - methyl - 10 - isopropyl - 6,7,8,9,10,11 - hexahydrodibenzofuran**
$C_{17}H_{24}O_2$
E.Foresti Serantoni, R.Mongiorgi, L.Riva di Sanseverino
Abstr. Ital.-Yug. Congr, 116, 1973

38.C   **Phenylsulfoxonium iodide**
$C_{18}H_{13}OS^+$, I^-
For complete entry see 42.11

38.C   **Usnic acid (at −110°C)**
$C_{18}H_{16}O_7$
For complete entry see 59.14

38.C   **Mesylmethynolide**
$C_{18}H_{30}O_7S$
For complete entry see 59.15

38.C   **Strigol**
$C_{19}H_{22}O_6$
For complete entry see 59.16

38.C   **Averufin**
$C_{20}H_{16}O_7$
For complete entry see 59.17

38.C   **Triptolide (absolute configuration)**
$C_{20}H_{24}O_6$
For complete entry see 59.18

38.C   **Tripdiolide**
$C_{20}H_{24}O_7$
For complete entry see 59.19

38.25   **2,3.11,12 - Dicyclohexyl - 1,4,7,10,13,16 - hexaoxacyclo - octane sodium bromide dihydrate**
$C_{20}H_{36}NaO_6^+$, Br^- , $2H_2O$
D.E.Fenton, M.Mercer, M.R.Truter
Biochem. Biophys. Res. Comm., **48,** 10, 1972
Residue 1 also classified in 67

38.C μ - 1,4 - Dioxane - bis(di(tetrahydrofuran) lithium) di(trichloro -
cyclopentadienyl - chromium)
$C_{20}H_{40}Li_2O_6{}^{2+}$, $2C_5H_5Cl_3Cr^-$
For complete entry see 67.16

38.26   14 - (aj) - Dibenzoxanthenium chloride acetic acid solvate
$C_{21}H_{13}O^+$, Cl^- , $C_2H_4O_2$
G.D.Andreetti, G.Bocelli, P.Sgarabotto   *Cryst. Struct. Comm.*, **2**, 91, 1973

38.27   14 - (aj) - Dibenzoxanthene
$C_{21}H_{14}O$
G.D.Andreetti, A.Manfredotti, P.Sgarabotto, A.Villa
Eur. Cryst. Meeting, 1973

38.C   Steroid Nic - 10
$C_{21}H_{22}O_4$
For complete entry see 51.15

38.28   2,2 - Di - (p - ethoxyphenyl) - 3,3 - dimethyl - oxetan
$C_{21}H_{26}O_3$
G.Holan, C.Kowala, J.A.Wunderlich   *J. C. S. Chem. Comm.*, 34, 1973
Also classified in 17

38.C   2' - Bromopodophyllotoxin ethyl acetate solvate (absolute configuration)
$C_{22}H_{21}BrO_8$, $0.5C_4H_8O_2$
For complete entry see 59.24

38.C   5' - Demethoxy - β - peltatin A methyl ether
$C_{22}H_{22}O_7$
For complete entry see 59.25

38.C   Episteganol (absolute configuration)
$C_{22}H_{22}O_8$
For complete entry see 59.26

38.C   17 - Ethylenedioxy - 3 - methoxy - 6,7,8 - methylidyne - 1,3,5(10) -
estratriene
$C_{22}H_{26}O_3$
For complete entry see 51.26

38.C   Wortmannin
$C_{23}H_{24}O_8$
For complete entry see 59.27

38.C   19,19 - Dimethoxy - 17β - acetoxy - 4 - androsten - 3 - one irradiation
product
$C_{23}H_{34}O_5$
For complete entry see 51.30

38.29   **2,2' - o - Xylenoxydiphenylbutadiyne**
$C_{24}H_{16}O_2$
T.Taga   *Acta Cryst. (A)*, **28**, S29, 1972

38.C   **Leuco - thelephoric acid hexamethyl ether**
(2,3,6,8,9,12 - Hexamethoxybenzo - bis(1,2 - b.4,5 - b')benzofuran)
$C_{24}H_{22}O_8$
For complete entry see 59.28

38.C   **3,20 - bis(Ethylenedioxy) - pregna - 5,7 - diene**
$C_{25}H_{36}O_4$
For complete entry see 51.37

38.C   **4 - p - Bromobenzylidene - 9 - p - bromophenyl - 10a - hydroxy - 2,7 - dimethyl - 2,7 - diaza - 10 - oxa - 1,2,3,4,5,6,7,8,8a,10a - decahydroanthracene monohydrate**
$C_{26}H_{28}Br_2N_2O_2$, H_2O
For complete entry see 36.29

38.C   **1,2.5,6 - Di - O - isopropylidene - 3,4 - di - O - tosyl - L - chiro - inositol**
$C_{26}H_{32}O_{10}S_2$
For complete entry see 21.17

38.C   **Aphidicolin bis(acetonide)**
$C_{26}H_{42}O_4$
For complete entry see 50.10

38.30   **2,6 - Diphenyl - 4 - (p - bromophenyl) - 5 - nitro - 2,6 - diethoxy - 5,6 - dihydro - 2H - pyran**
$C_{27}H_{26}BrNO_5$
C.L.Pedersen, O.Buchard, S.Larsen, K.J.Watson
Tetrahedron Letters, 2195, 1973

38.C   **Dothistromin bromoethyl ether tetra - acetate (absolute configuration)**
$C_{28}H_{23}BrO_{13}$
For complete entry see 59.31

38.31   **Potassium iodide bis(2,3,5,6,8,9,11,12 - octahydro - 1,4,7,10,13 - benzo - pentaoxacyclopentadecine)**
Potassium iodide bis(benzo - 15 - crown - 5)
$C_{28}H_{40}KO_{10}^+$, I^-
P.R.Mallinson, M.R.Truter   *J. C. S. Perkin II*, 1818, 1972
Residue 1 also classified in 67

38.C   **N - Bromoacetyl - daunomycin acetone solvate**
$C_{29}H_{30}BrNO_{11}$, C_3H_6O
For complete entry see 50.11

38.C   **Wortmannin p - bromobenzoate (absolute configuration)**
$C_{30}H_{26}BrO_9$
For complete entry see 59.32

38.C   **Cesium chlorothircolide methyl ester trihydrate (absolute configuration)**
$C_{30}H_{37}O_8^-$, Cs^+ , $3H_2O$
For complete entry see 50.13

38.C   **5 - Bromo - antibiotic X - 537A hemihydrate**
$C_{34}H_{53}BrO_8$, $0.5H_2O$
For complete entry see 50.15

38.32   **(2.2) - Paracyclophane bis - O - 18 - crown - 6 ether**
$C_{36}H_{52}O_{12}$
K.Parker, R.C.Helgeson, E.Maverick, K.N.Trueblood
Amer. Cryst. Assoc., Abstr. Papers (Winter Meeting), 49, 1973
Also classified in 31

38.C   **Sodium - bis(tetrahydrofuran) - bis(N,N' - ethylenebis(salicylideneiminato) cobalt(ii)) tetraphenylborate**
$C_{40}H_{44}Co_2N_4NaO_6^+$, $C_{24}H_{20}B^-$
For complete entry see 67.22

38.C   **Nonactin**
$C_{40}H_{64}O_{12}$
For complete entry see 50.16

38.C   **Tetranactin - potassium thiocyanate (form ii)**
$C_{44}H_{72}KO_{12}^+$, CNS^-
For complete entry see 50.17

38.C   **Tetranactin - potassium thiocyanate (form i)**
$C_{44}H_{72}KO_{12}^+$, CNS^-
For complete entry see 50.18

38.C   **Tetranactin - sodium thiocyanate (form ii)**
$C_{44}H_{72}NaO_{12}^+$, CNS^-
For complete entry see 50.19

38.C   **Tetranactin**
$C_{44}H_{72}O_{12}$
For complete entry see 50.20

38.C   **Tetranactin - rubidium thiocyanate (form i)**
$C_{44}H_{72}O_{12}Rb^+$, CNS^-
For complete entry see 50.22

38.C   **Oligomycin B**
$C_{45}H_{72}O_{12}$
For complete entry see 50.23

38.C   **Kidamycin derivative di - iodide methanol solvate monohydrate**
$C_{48}H_{64}N_2O_{13}^{2+}$, $2I^-$, $4CH_4O$, H_2O
For complete entry see 50.25

38.C **Antibiotic A204A silver salt acetone solvate (absolute configuration)**
$C_{49}H_{83}O_{17}{}^-$, Ag^+ , C_3H_6O
For complete entry see 50.26

38.C **Antibiotic A204A**
$C_{49}H_{84}O_{17}$
For complete entry see 50.28

38.C **bis(Thiocyanato - cesium - (6,7,9,10,17,18,20,21 - octahydro - 7R,9R,18S,20S - tetramethyldibenzo(b,k)(1,4,7,10,13,16)hexa - oxacyclo - octadecin))**
$C_{50}H_{64}Cs_2N_2O_{12}S_2$
For complete entry see 67.24

HETERO-SULPHUR AND HETERO-SELENIUM

39.1   **1,3 - Dithiolane - 2 - thione**
Ethylene trithiocarbonate
$C_3H_4S_3$
B.Klewe, H.M.Seip   *Acta Chem. Scand.*, **26**, 1860, 1972

39.2   **3 - Chlorthietane 1,1 - dioxide**
$C_3H_5ClO_2S$
G.D.Andreetti, G.Bocelli, P.Sgarabotto   *Cryst. Struct. Comm.*, **2**, 323, 1973

39.3   **3,4 - Epoxysulfolane**
$C_4H_6O_3S$
D.E.Sands   *Acta Cryst. (B)*, **28**, 2463, 1972
Also classified in 38

39.4   **1,3,5,7 - Tetrathiocane**
$C_4H_8S_4$
G.W.Frank, O.J.Degen, F.A.L.Anet   *J. Amer. Chem. Soc.*, **94**, 4792, 1972

39.5   **6a - Thiathiophthene (photographic data)**
$C_5H_4S_3$
L.K.Hansen, A.Hordvik   *Acta Chem. Scand.*, **27**, 411, 1973

39.6   **3 - Acetoxythietane 1,1 - dioxide**
$C_5H_8O_4S$
G.D.Andreetti, G.Bocelli, P.Sgarabotto   *Cryst. Struct. Comm.*, **1**, 423, 1972

39.C   **2,2' - Bi - 1,3 - dithiole 7,7,8,8 - tetracyanoquinodimethanide**
$C_6H_4S_4^+$, $C_{12}H_4N_4^-$
For complete entry see 60.6

39.7   **2 - Methyl - 6a - thiathiophthene**
$C_6H_6S_3$
A.Hordvik, L.J.Saethre   *Acta Chem. Scand.*, **26**, 1729, 1972

39.8   **1,4 - Dithionia - bicyclo(2.2.2)octane tetrachlorozincate**
$C_6H_{12}S_2^{2+}$, Cl_4Zn^{2-}
E.Deutsch   *J. Org. Chem.*, **37**, 3481, 1972

121

39.9   **2,4,6 - Trimethyl - 1,3,5 - triselenane**
Triselenoparaldehyde
$C_6H_{12}Se_3$
G.Valle, A.Del Pra, M.Mammi   *Cryst. Struct. Comm.*, **2**, 169, 1973

39.10   **3 - Carboxy - 2 - acetyl - thiophene**
$C_7H_6O_3S$
M.Griffe, F.Durant, A.F.Pieret   *Bull. Soc. Chim. Belges*, **81**, 319, 1972

39.11   **3 - Carboxy - 4 - acetyl - thiophene**
$C_7H_6O_3S$
M.Griffe, F.Durant, A.F.Pieret   *Bull. Soc. Chim. Belges*, **81**, 319, 1972

39.12   **2 - Carboxy - 3 - acetyl - thiophene**
$C_7H_6O_3S$
M.Griffe, F.Durant, A.F.Pieret   *Bull. Soc. Chim. Belges*, **81**, 319, 1972

39.13   **3,4 - Dimethyl - 6a - selenathiophthene**
$C_7H_8S_2Se$
A.Hordvik, T.S.Rimala, L.J.Saethre   *Acta Chem. Scand.*, **27**, 360, 1973

39.C   **Trinitrobenzene - benzothiophene complex**
C_8H_6S , $C_6H_3N_3O_6$
For complete entry see 60.5

39.14   **3 - Methoxyimino - 1,2 - benzodithiol - 1,1 - dioxide**
$C_8H_7NO_3S_2$
M.B.Ferrari, L.C.Capacchi, G.G.Fava   *Cryst. Struct. Comm.*, **2**, 185, 1973

39.15   **4,8,9,10 - Tetrachloro - 2,6 - dithia - adamantane**
$C_8H_8Cl_4S_2$
T.Higuchi, W.Nowacki   *Z. Kristallogr.*, **135**, 56, 1972

39.16   **3,4 - Trimethylene - 6a - selenaselenophthene**
$C_8H_8Se_3$
A.Hordvik, J.A.Porten   *Acta Chem. Scand.*, **27**, 485, 1973

39.C   **3 - Ethyl - 5 - (2' - (1',3' - dithiolanylidene)) rhodanine**
$C_8H_9NOS_4$
For complete entry see 41.12

39.17   **L - α - (p - Chlorobenzene - sulfonamido) - β - propionothiolactone**
$C_9H_8ClNO_3S_2$
I.Milinovic, A.Bezjak   *Abstr. Ital.-Yug. Congr*, 97, 1973

39.18   **2,5 - bis(Dicyanomethylene) - 2,5 - dihydrothiophene**
$C_{10}H_2N_4S$
B.Aurivillius   *Acta Chem. Scand.*, **26**, 3612, 1972

39.19   **Pyromellitic dithioanhydride**
$C_{10}H_2O_4S_2$
I.V.Bulgarovskaya, Z.V.Zvonkova   *Kristallografija*, **17**, 292, 1972
Also classified in 13

39.C   **8,3′ - S - Cycloadenosine monohydrate**
$C_{10}H_{11}N_5O_3S$, H_2O
For complete entry see 47.14

39.C   **8,2′ - S - Cycloadenosine - 3′,5′ - cyclic monophosphate trihydrate**
$C_{10}H_{11}N_5O_5PS$, $3H_2O$
For complete entry see 47.15

39.20   **3,5 - bis(N,N - Diethylimmonium) - 1,2,4 - trithiolane tetraiododi - μ - iododimercurate**
$C_{10}H_{20}N_2S_3^{2+}$, $Hg_2I_6^{2-}$
P.T.Beurskens, W.P.J.H.Bosman, J.A.Cras
J. Cryst. Mol. Struct., **2**, 183, 1972

39.21   **Diethyldithiocarbamato - selenium**
$C_{10}H_{20}N_2S_4Se$
A.Conde, A.Lopez-Castro, R.Marquez
An. R. Soc. Esp. Fis. Quim., A, **68**, 59, 1972

39.22   **1H,3H - Naphtho(1,8)thiopyran**
$C_{12}H_{10}S$
B.-M.Lunden   *Acta Cryst. (B)*, **29**, 1219, 1973

39.23   **2 - (2 - (3 - Thiacyclohexylidene)ethyl) - 2 - methyl - cyclopentane - 1,3 - dione**
$C_{13}H_{18}O_2S$
A.R.Overbeek, H.Schenk   *Cryst. Struct. Comm.*, **2**, 29, 1973
Also classified in 20

39.24   **4 - t - Butyl - 2 - (2,2 - dimethyl - 1 - methylthiopropylidene) - 1,3 - dithiole**
$C_{13}H_{22}S_3$
G.Roelofsen, J.A.Kanters   *Cryst. Struct. Comm.*, **2**, 95, 1973

39.25   **5H,8H - Dibenzo(d,f)(1,2)dithiocin**
$C_{14}H_{12}S_2$
G.H.Wahl Junior, J.Bordner, D.N.Harpp, J.G.Gleason
J. C. S. Chem. Comm., 985, 1972

39.26   **1,4 - Dihydro - 1 - methyl - 4 - (2′ - methyl - 5′ - phenylthiopyran - 4′ - ylidene) pyridine - S,S - dioxide chloroform solvate**
$C_{14}H_{17}NO_3S$, $CHCl_3$
G.D.Andreetti, G.Bocelli, P.Sgarabotto   *J. C. S. Perkin II*, 1189, 1973
Residue 1 also classified in 33

39.27 **2 - (5 - Phenyl - 1,2 - dithiole - 3 - ylio)phenolate**
$C_{15}H_{10}OS_2$
E.C.Llaguno, I.C.Paul, R.Pinel, Y.Mollier *Tetrahedron Letters*, 4687, 1972
Also classified in 17

39.28 **cis - 1,1 - Dioxy - 2,2 - diphenyl - 3,4 - dichloro - thietane**
$C_{15}H_{12}Cl_2O_2S$
S.Kumakura, T.Shimozawa *Acta Cryst. (A)*, **28**, S29, 1972

39.29 **cis - 2,2 - Diphenyl - 3,4 - dichloro - thietane**
$C_{15}H_{12}Cl_2S$
S.Kumakura, T.Shimozawa *Acta Cryst. (A)*, **28**, S29, 1972

39.30 **Selenoindigo**
$C_{16}H_8O_2Se_2$
P.Susse, G.Kunz, W.Luttke *Naturwissenschaften*, **60**, 4, 1973

39.31 **1 - Cyano - 2 - hydroxy - 3,4 - epithiobutane - α - naphthylurethane**
$C_{16}H_{14}N_2O_2S$
R.B.Bates, R.A.Grady, T.C.Sneath *J. Org. Chem.*, **37**, 2145, 1972
Also classified in 24

39.32 **2,5 - Diphenyl - 6a - selenathiophthene**
$C_{17}H_{12}S_2Se$
A.Hordvik, T.S.Rimala, L.J.Saethre *Acta Chem. Scand.*, **26**, 2139, 1972

39.33 **1,2 - Dihydro - 1 - methyl - 2 - (5' - phenylthiopyran - 2' - ylidene) pyridine S,S - dioxide**
$C_{17}H_{15}NO_2S$
G.D.Andreetti, G.Bocelli, P.Sgarabotto *J. C. S. Perkin II*, 1189, 1973
Also classified in 33

39.34 **9 - Isobutyl - thioxanthene 10 - dioxide**
$C_{17}H_{18}O_2S$
S.S.C.Chu *Amer. Cryst. Assoc., Abstr. Papers (Winter Meeting)*, 86, 1973

39.35 **9 - Isobutyl - thioxanthene**
$C_{17}H_{18}S$
S.S.C.Chu *Amer. Cryst. Assoc., Abstr. Papers (Winter Meeting)*, 86, 1973

39.36 **5,6,11,12 - Tetradehydro - 7,10 - dihydro - 8,9 - dithiadibenzo(a,c)cyclododecene**
$C_{18}H_{12}S_2$
H.Irngartinger *Chem. Ber.*, **106**, 751, 1973

39.C **Phenylsulfoxonium iodide**
$C_{18}H_{13}OS^+$, I^-
For complete entry see 42.11

39.37   **2 - N,N - Diethylamino - 3 - carbomethoxynaphtho(2,3 - b)thiophene - 4,9 - dione**
$C_{18}H_{17}NO_4S$
C.S.Gibbons, J.A.Lerbscher, J.Trotter   *J. Cryst. Mol. Struct.*, **2,** 235, 1972

39.38   **2,5 - Diphenyl - 3,4 - dimethylene - 6a - thiathiophthene**
$C_{19}H_{14}S_3$
B.Birknes, A.Hordvik, L.J.Saethre   *Acta Chem. Scand.*, **27,** 382, 1973

39.39   **2,5 - Diphenyl - 3,4 - dimethyl - 6a - thiathiophthene**
$C_{19}H_{16}S_3$
A.Hordvik, O.Sjolset, L.J.Saethre   *Acta Chem. Scand.*, **27,** 379, 1973

39.40   **2 - (p - Dimethylanilino) - 4 - phenyl - 6a - thiathiophthene**
$C_{19}H_{17}NS_3$
A.Hordvik, L.J.Saethre   *Acta Chem. Scand.*, **26,** 3114, 1972
Also classified in 16

39.41   **2 - (2 - p - Methoxyphenyl - 2 - methylthiovinyl) - 3,4 - trimethylene - 5 - methylthio - 1,6,6a - thiathiophthene**
$C_{19}H_{20}OS_5$
J.Sletten   *Acta Chem. Scand.*, **27,** 229, 1973

39.42   **2,5 - Diphenyl - 3,4 - trimethylene - 6a - thiathiophthene**
$C_{20}H_{16}S_3$
B.Birknes, A.Hordvik, L.J.Saethre   *Acta Chem. Scand.*, **26,** 2140, 1972

39.43   **2 - (Morpholino - carboxamide) - 4 - methyl - 5 - methylthio - 2H - thiopyran - 3 - carboxylic acid p - bromobenzyl ester**
$C_{20}H_{22}BrNO_4S_2$
A.E.Smith, R.Kalish, E.J.Smutny   *Acta Cryst. (B)*, **28,** 3494, 1972
Also classified in 40

39.44   **Methixene hydrochloride monohydrate**
$C_{20}H_{24}NS^+$, Cl^- , H_2O
S.S.C.Chu   *Acta Cryst. (B)*, **28,** 3625, 1972
Residue 1 also classified in 33

39.45   **N - (3 - Phenyl - 2 - benzo(b)thienyl)thiobenzamide**
$C_{21}H_{15}NS_2$
G.Argay, A.Kalman   *Cryst. Struct. Comm.*, **2,** 19, 1973

HETERO-(NITROGEN AND OXYGEN)

40.1   **Methyl 2 - bromo - 5 - oxo - Δ - 1,3,4 - oxadiazolin - 4 - carboxylate**
$C_4H_3BrN_2O_4$
D.Seyferth, H.Shih, M.D.LaPrade   *J. C. S. Chem. Comm.*, 1036, 1972

40.2   **Muscimol hemihydrate**
3 - Hydroxy - 5 - aminomethyl - isoxazole hydrate
$C_4H_6N_2O_2$, $0.5H_2O$
L.Brehm   *Stockholm Symposium,* 37, 1973

40.3   **3 - Methyl - 4 - furoxan - carboxylic acid hydrazide**
$C_4H_6N_4O_3$
G.Germain, D.Viterbo   *Cryst. Struct. Comm.*, **1**, 411, 1972

40.4   **4 - Methyl - 3 - furoxan - carboxylic acid hydrazide**
$C_4H_6N_4O_3$
M.Calleri, G.Chiari, D.Viterbo   *Cryst. Struct. Comm.*, **1**, 407, 1972

40.5   **Morpholinium 7,7,8,8 - tetracyanoquinodimethane**
$C_4H_{10}NO^+$, $C_{12}H_4N_4^-$
T.Sundaresan, S.C.Wallwork   *Acta Cryst. (B)*, **28**, 3507, 1972
Residue 2 classified in 7

40.C   **αS,5S - α - Amino - 3 - chloro - 2 - isoxazoline - 5 - acetic acid (absolute configuration)**
$C_5H_7ClN_2O_3$
For complete entry see 50.1

40.6   **1,3,5,7 - Tetra - oxa - 9 - azacyclodecan - 10 - one**
$C_5H_9NO_5$
D.Kobelt, E.F.Paulus   *Acta Cryst. (B)*, **29**, 633, 1973

40.7   **Benzoxazoline - 2 - thione**
C_7H_5NOS
P.Groth   *Acta Chem. Scand.*, **27**, 945, 1973

40.8   **Benzoxazolin - 2 - one**
$C_7H_5NO_2$
P.Groth   *Acta Chem. Scand.*, **27**, 945, 1973

40.9 3 - Methyl - benzoxazoline - 2 - thione
C_8H_7NOS
P.Groth, K.Davidkov, D.Simov *Acta Chem. Scand.*, **26**, 1931, 1972

40.10 3 - Methyl - benzoxazolin - 2 - one
$C_8H_7NO_2$
P.Groth *Acta Chem. Scand.*, **27**, 945, 1973

40.11 syn - 5,10 - Dimethyl - 3,8 - dioxa - 4,9 - diaza - tricyclo(5.3.0.0^{2,6})deca - 4,9 - diene
$C_8H_{10}N_2O_2$
G.D.Andreetti, G.Bocelli, P.Sgarabotto *Cryst. Struct. Comm.*, **2**, 115, 1973

40.12 2,4 - (β,β - Dimethyl - trimethylene) - (1,2,5)oxathiazolo(2,3 - b)(1,2,5) oxathiazole - 7 - S^{iv}
$C_8H_{10}N_2O_2S$
E.C.Llaguno, I.C.Paul *Tetrahedron Letters*, 1565, 1973
Also classified in 41

40.13 Morpholine biguanide hydrobromide
$C_8H_{14}N_5O^+$, Br^-
R.Handa, N.N.Saha *Acta Cryst. (B)*, **29**, 554, 1973
Residue 1 also classified in 8

40.14 3 - Methyl - 4 - phenylsulfonyl - furoxan
$C_9H_8N_2O_4S$
G.Chiari, D.Viterbo, A.G.Manfredotti, C.Guastini
Abstr. Ital.-Yug. Congr, 110, 1973

40.15 5,5 - Dimethyl - 2 - p - bromophenylimino - Δ^3 - 1,3,4 - oxadiazoline
$C_{10}H_{10}BrN_3O$
A.Prakash, C.Calvo, A.M.Cameron, J.Warkentin
J. Cryst. Mol. Struct., **3**, 71, 1973

40.C Anhydro - bromo - nitro - camphane
$C_{10}H_{13}BrNO$
For complete entry see 52.2

40.16 bis(4 - Morpholinethiocarbonyl) - trisulfide
$C_{10}H_{16}N_2O_2S_5$
S.Husebye *Acta Chem. Scand.*, **27**, 756, 1973
Also classified in 11

40.17 3 - Phenyl - 5 - chloro - 7 - bromoisoxazolo(4,5 - d)pyrimidine
$C_{11}H_5BrClN_3O$
B.Bovio, S.Locchi *J. Cryst. Mol. Struct.*, **2**, 251, 1972
Also classified in 44

40.18 3 - Bromo - 2 - methyl - pyrazolo(a)(3,1)benzoxazin - 5 - one
$C_{11}H_7BrN_2O_2$
I.Sotofte *Acta Chem. Scand.*, **27**, 661, 1973

40.19   **4 - Phenyl - $6\alpha,7\beta$ - dibromo - 2,3 - oxabicyclo(3.2.0)hepta - $1\beta,5\beta,6\beta,7\alpha$ - 3 - ene**
$C_{11}H_9Br_2NO$
M.B.Ferrari, G.F.Gasparri, M.A.Pellinghelli
Abstr. Ital.-Yug. Congr, 118, 1973

40.20   **4 - Phenyl - $6\alpha,7\beta$ - dibromo - 2,3 - oxabicyclo(3.2.0)hepta - $1\alpha,5\alpha,6\beta,7\alpha$ - 3 - ene**
$C_{11}H_9Br_2NO$
M.B.Ferrari, G.F.Gasparri, M.A.Pellinghelli
Abstr. Ital.-Yug. Congr, 118, 1973

40.21   **4 - Phenyl - 6,7 - dichloro - 2,3 - oxabicyclo(3.2.0)hept - 3 - ene**
$C_{11}H_9Cl_2NO$
A.C.Bonamartini, C.G.Palmieri   *Abstr. Ital.-Yug. Congr,* 124, 1973

40.22   **anti - 4 - Phenyl - 6,7 - epoxy - 2 - oxa - 3 - aza - bicyclo(3.2.0)hepta - 3 - ene**
$C_{11}H_9NO_2$
M.B.Ferrari, G.G.Fava, M.Nardelli, M.A.Pellinghelli
Eur. Cryst. Meeting, 1973
Also classified in 38

40.23   **syn - 4 - Phenyl - 6,7 - epoxy - 2 - oxa - 3 - aza - bicyclo(3.2.0)hepta - 3 - ene**
$C_{11}H_9NO_2$
M.B.Ferrari, G.F.Gasparri, M.Nardelli, M.A.Pellinghelli
Eur. Cryst. Meeting, 1973
Also classified in 38

40.24   **Acenaphthofuroxan**
$C_{12}H_6N_2O_2$
M.Calleri, G.Chiari, D.Viterbo   *Cryst. Struct. Comm.,* **2,** 335, 1973

40.25   **2 - Methyl - 3 - phenyl - 4 - (N - methyl - N - hydroxyamidin) - isoxazolin - 5 - one hydrobromide**
$C_{12}H_{14}N_3O_3^+$, Br^-
L.Fanfani, A.Nunzi, P.F.Zanazzi, A.R.Zanzari
Acta Cryst. (B), **28,** 2598, 1972

40.26   **7,10 - Dichloroanthraquinoneoxadiazole**
7,10 - Dichloroanthra(1,2 - c) - (1,2,5)oxadiazole - 6,11 - dione
$C_{14}H_4Cl_2N_2O_3$
O.A.Mikhno, L.A.Chetkina, Z.I.Ezhkova   *Kristallografija,* **17,** 297, 1972

40.27   **Anthraquinoneoxadiazole**
Anthra(1,2 - c)(1,2,5)oxadiazole - 6,11 - dione
$C_{14}H_6N_2O_3$
L.A.Chetkina, S.L.Ginzburg, M.G.Neigauz, G.A.Gol'der
Zh. Strukt. Khim., **13,** 91, 1972
Also classified in 26

40.28   2 - Benzylimino - benzoxazoline
$C_{14}H_{12}N_2O$
P.Groth *Acta Chem. Scand.*, **27**, 945, 1973

40.C   Lithium(4,7,13,19 - tetra - oxa - 1,10 - diaza - bicyclo(8.5.5)eicosane) iodide
$C_{14}H_{28}LiN_2O_4^+$, I^-
For complete entry see 67.10

40.29   2 - Benzylimino - 3 - methyl - benzoxazoline
$C_{15}H_{14}N_2O$
P.Groth *Acta Chem. Scand.*, **27**, 945, 1973

40.30   1 - Morpholinomethyl - 3 - methyl - 3 - phenylpyrrolidin - 2,5 - dione
$C_{16}H_{20}N_2O_3$
G.Argay, J.Seres *Acta Cryst. (B)*, **29**, 1146, 1973
Also classified in 32

40.C   10 - Bromo - 2,3,5,6,7 - 11b - hexahydro - 2 - methyl - 11b -
phenylbenzo(6,7) - 1,4 - diazepine(4,5 - b) - oxazol - 6 - one ethanol solvate
$C_{18}H_{17}BrN_2O_2$, $2C_2H_6O$
For complete entry see 36.22

40.31   2 - Phenyl - 4 - (p - bromophenyl) - 5,6 - dimethyl - 5,6 - dihydro - 4H - 1,3 -
oxazine
$C_{18}H_{18}BrNO$
F.Garbassi, L.Giarda *Acta Cryst. (B)*, **29**, 1190, 1973

40.32   Neuroleptic drug i
$C_{18}H_{18}ClN_3O$
T.J.Petcher, H.P.Weber *Stockholm Symposium*, 38, 1973
Also classified in 33

40.33   10 - Chloro - 2,3,5,6,7,11b - hexahydro - 3 - hydroximino - 7 - methyl - 11b -
phenyl - oxazolo(3,2 - d)benzo - 1,4 - diazepine
$C_{18}H_{18}ClN_3O_2$
R.Jaunin, W.E.Oberhansli, J.Hellerbach *Helv. Chim. Acta*, **55**, 2975, 1972
Also classified in 34

40.34   3,8 - Dihydro - 3,3,8,8 - tetramethyl - 6 - phenoxyacetamido - 1 - oxo -
oxazolo(4,3 - c) - (1,4)thiazinium chloride
$C_{18}H_{21}N_2O_4S^+$, Cl^-
R.Thomas, D.J.Williams *J. C. S. Chem. Comm.*, 226, 1973
Residue 1 also classified in 41

40.C   Cesium(4,7,13,16,21,24 - hexa - oxa - 1,10 - diaza -
bicyclo(8.8.8)hexaeicosane) thiocyanate monohydrate
$C_{18}H_{36}CsN_2O_6^+$, CNS^- , H_2O
For complete entry see 67.12

40.C   **Potassium(4,7,13,16,21,24 - hexa - oxa - 1,10 - diaza - bicyclo(8.8.8)hexaeicosane)iodide**
$C_{18}H_{36}KN_2O_6{}^+$, I^-
For complete entry see 67.13

40.C   **Sodium(4,7,13,16,21,24 - hexa - oxa - 1,10 - diaza - bicyclo(8.8.8)hexaeicosane)iodide**
$C_{18}H_{36}N_2NaO_6{}^+$, I^-
For complete entry see 67.14

40.C   **Rubidium(4,7,13,16,21,24 - tetra - oxa - 1,10 - diaza - bicyclo(8.8.8)hexaeicosane) thiocyanate monohydrate**
$C_{18}H_{36}N_2O_6Rb^+$, CNS^- , H_2O
For complete entry see 67.15

40.C   **2 - (Morpholino - carboxamide) - 4 - methyl - 5 - methylthio - 2H - thiopyran - 3 - carboxylic acid p - bromobenzyl ester**
$C_{20}H_{22}BrNO_4S_2$
For complete entry see 39.43

40.35   **10,21 - Diethyl - 3,14 - dioxa - 1,12 - diazatricyclo(16,4,0,0^{7,12})docosane**
$C_{22}H_{42}N_2O_2$
D.A.Whiting, R.Cahill, T.A.Crabb   *J. C. S. Chem. Comm.*, 1307, 1972

40.36   **2 - Phenyl - 4 - benzyl - 5,6 - dihydro - 5,6 - tetramethylene - 6 - piperidino - 1,3,4 - oxadiazine**
$C_{25}H_{29}N_3O_2$
E.F.Sorantoni, R.Mongiorgi, L.Riva di Sanseverino
Acta Cryst. (A), **28**, S24, 1972
Also classified in 33

40.C   **4,4,6 - Trimethyl - 1 - oxa - 3 - azacyclohex - 2 - enylcarbenyl phosphonium chloride**
$C_{26}H_{29}NOP^+$, Cl^-
For complete entry see 64.38

HETERO-(NITROGEN AND SULPHUR)

41.1   **2 - Imino - 4 - thiazolidinone**
$C_3H_4N_2OS$
V.Amirthalingam, K.V.Muralidharan *Acta Cryst. (B)*, **28**, 2421, 1972

41.2   **3 - Acetamido - 1,2,4 - dithiazole - 5 - thione**
$C_4H_4N_2OS_3$
G.Eide, A.Hordvik, L.J.Saethre *Acta Chem. Scand.*, **26**, 2140, 1972

41.3   **S - Thiazolidine - 4 - carboxylic acid**
$C_4H_7NO_2S$
M.Goodman, V.Chen, E.Benedetti, C.Pedone, P.Corradini
Biopolymers, U. S. A., **11**, 1779, 1972

41.4   **2 - Amino - 4 - thiazolidinone - 5 - acetic acid**
$C_5H_6N_2O_3S$
V.Amirthalingam, K.V.Muralidharan *Acta Cryst. (B)*, **28**, 2417, 1972

41.5   **3 - Methyl - 4 - oxo - 1,3 - thiazine - 2 - thione**
$C_5H_7NOS_2$
V.Amirthalingam, V.S.Jakkal *Acta Cryst. (B)*, **28**, 2612, 1972

41.C   **Chondrine**
$C_5H_9NO_3S$
For complete entry see 48.28

41.6   **N,N' - Dimethyl - 3,6 - epitetrathio - piperazin - 2,5 - dione**
$C_6H_8N_2O_2S_4$
B.R.Davis, I.Bernal, U.Schmidt *Angew. Chem.*, **84**, 640, 1972

41.7   **3,5 - Diacetylamino - 1,2,4 - thiadiazole**
$C_6H_8N_4O_2S$
S.Sato, T.Kinoshita, T.Hata, C.Tamura *Acta Cryst. (A)*, **28**, S25, 1972

41.8   **2,3,4 - Trimethyl - thiazolium bromide**
$C_6H_{10}NS^+$, Br^-
G.Pepe, M.Pierrot *Acta Cryst. (B)*, **28**, 2118, 1972

41.9   **5 - Acetylimino - 4 - methyl - 3 - methylamino - Δ^2 - 1,2,4 - thiadiazoline**
$C_6H_{10}N_4OS$
S.Sato, T.Kinoshita, T.Hata, C.Tamura *Acta Cryst. (A)*, **28**, S25, 1972

41.C   **N - Ethyl - 2 - methyl - thiazoline - bis(7,7,8,8 - tetracyanoquinodimethane)**
$C_6H_{12}NS^+$, $C_{12}H_4N_4^-$, $C_{12}H_4N_4$
For complete entry see 60.8

41.10   **Dihydro - 6 - chloro - 7 - sulfamoyl - benzo - 1,2,4 - thiadizine 1,1 - dioxide**
Hydrochlorothiazide
$C_7H_8ClN_3O_4S_2$
L.Dupont, O.Dideberg   *Acta Cryst. (B)*, **28**, 2340, 1972

41.11   **2 - Imino - 3 - methyl - 5 - (methoxycarbonyl - methylene) - 1,3 - thiazolidin - 4 - one**
$C_7H_8N_2O_3S$
J.F.B.Mercer, G.M.Priestley, R.N.Warrener, E.Adman, L.H.Jensen
Synthetic Comm., **2**, 35, 1972

41.12   **3 - Ethyl - 5 - (2' - (1',3' - dithiolanylidene)) rhodanine**
$C_8H_9NOS_4$
H.R.Luss, D.L.Smith   *Acta Cryst. (B)*, **29**, 998, 1973
Also classified in 39

41.C   **2,4 - (β,β - Dimethyl - trimethylene) - (1,2,5)oxathiazolo(2,3 - b)(1,2,5) oxathiazole - 7 - S^{iv}**
$C_8H_{10}N_2O_2S$
For complete entry see 40.12

41.13   **N - Chloroacetyl - chlorothiazide**
6 - Chloro - 7 - N - chloroacetylsulfamoyl - 1,2,4 - benzothiadiazine 1,1 - dioxide
$C_9H_6Cl_2N_3O_5S_2$
B.Magnusson, O.Lindqvist   *Acta Chem. Scand.*, **26**, 1411, 1972

41.14   **1,2 - Benzidothiazol - 3 - yl methyl ketoxime**
$C_9H_8N_2OS$
A.Braibanti, M.A.Pellinghelli, A.Tiripicchio, M.T.Camellini
Acta Cryst. (B), **29**, 43, 1973

41.15   **2 - Amino - 5 - phenyl - thiazolin - 4 - one (form ii)**
$C_9H_8N_2OS$
J.-P.Mornon, R.Bally   *Acta Cryst. (B)*, **28**, 2074, 1972

41.C   **Homosulfanilamide - sulfathiazole complex**
$C_9H_8N_3O_2S_2^-$, $C_7H_{11}N_2O_2S^+$
For complete entry see 11.5

41.16   **3,5 - bis((Dimethylthiocarbamoyl) - imino) - 4 - methyl - 1,2,4 - dithiazolidine**
$C_9H_{15}N_5S_4$
J.E.Oliver, J.L.Flippen, J.Karle   *J. C. S. Chem. Comm.*, 1153, 1972

41.17   2,4 - Di - isopropyl - 3 - methyl - thiazolium bromide trihydrate
$C_{10}H_{18}NS^+$, Br^- , $3H_2O$
G.Pepe, M.Pierrot   *Acta Cryst. (B)*, **28**, 2118, 1972

41.18   D(−) - Luciferin (absolute configuration)
$C_{11}H_8N_2O_3S_2$
D.Dennis, R.H.Stanford Junior   *Acta Cryst. (B)*, **29**, 1053, 1973

41.19   6 - Phenyl - imidazo(2,1 - b)thiazole
$C_{11}H_8N_2S$
L.Cavalca, P.Domiano, A.Musatti   *Cryst. Struct. Comm.*, **1**, 345, 1972

41.20   2,3,5,6 - Tetrahydro - 6 - phenyl - imidazo(2,1 - b)thiazole
$C_{11}H_{12}N_2S$
A.L.Spek   *Cryst. Struct. Comm.*, **1**, 309, 1972

41.21   2,3,6,7 - Tetrahydro - 6 - phenyl - 5H - imidazo(2,1 - b)thiazolium chloride
$C_{11}H_{13}N_2S^+$, Cl^-
R.W.Baker, P.J.Pauling   *J. C. S. Perkin II*, 203, 1973

41.22   Thiamine chloride monohydrate
$C_{12}H_{17}N_4OS^+$, Cl^- , H_2O
J.Pletcher, M.Sax, S.Sengupta, J.Chu, C.S.Yoo
Acta Cryst. (B), **28**, 2928, 1972
Residue 1 also classified in 44

41.23   1 - (2 - Thiazolylazo) - 6 - bromo - 2 - naphthol
$C_{13}H_8BrN_3OS$
M.Kurahashi, A.Kawase, K.Hirotsu, M.Fukuyo, A.Shimada
Bull. Chem. Soc. Jap., **45**, 1940, 1972
Also classified in 24

41.24   3 - Formyl - 4 - (N - 4' - bromophenylcarbamoyl) - 2,2 -
dimethylthiazolidine - 1 - oxide
$C_{13}H_{15}BrN_2O_3S$
B.W.Bycroft, C.Fouweather, T.J.King   *Acta Cryst. (A)*, **28**, S26, 1972

41.25   1,1' - Diethyl - 2,2' - thiazolinecarbocyanine iodide
$C_{13}H_{21}N_2S_2^+$, I^-
T.E.Borowiak, N.G.Bokii, Yu.T.Struchkov
Zh. Strukt. Khim., **13**, 480, 1972

41.26   3,5 - bis((Dimethylthiocarbamoyl) - imino) - 4 - phenyl - 1,2,4 -
dithiazolidine
$C_{14}H_{17}N_5S_4$
J.E.Oliver, J.L.Flippen, J.Karle   *J. C. S. Chem. Comm.*, 1153, 1972

41.27   cis - 2,3 - (3' - Cyclohexanon - 1',2' - ylene) - 5 - methyl - 8 - ethoxydihydro -
thiazolo(3,2 - a)pyridinium bromide
$C_{14}H_{18}NOS^+$, Br^-
P.Groth   *Acta Chem. Scand.*, **26**, 3131, 1972

41.C   **3,1' - Anhydro - 2 - (4,6 - di - O - acetyl - 2,3 - dideoxy - α - D - ribopyranose) - 3 - hydroxy - 5 - methyl - 2H - 1,2,6 - thiadiazine - 1,1 - dioxide**
$C_{14}H_{18}N_2O_9S$
For complete entry see 47.30

41.28   **bis(1,1,3,3 - Tetramethylbutyl)thiadiaziridine 1,1 - dioxide**
$C_{14}H_{28}N_2O_2S$
L.M.Trefonas, L.D.Cheung   *J. Amer. Chem. Soc.*, **95**, 636, 1973

41.29   **Tri(pyridine) - 2,6',2',6''.2',6 - trisulfide**
$C_{15}H_9N_3S_3$
P.Groth   *Acta Chem. Scand.*, **27**, 5, 1973

41.30   **2,5 - Diphenyl - 3,4 - diaza - 6a - thiathiophthene**
$C_{15}H_{10}N_2S_3$
A.Hordvik, L.M.Milje   *Acta Chem. Scand.*, **27**, 510, 1973

41.31   **2 - Phenylamino - 5 - phenyl - thiazolin - 4 - one ethanol solvate**
$C_{15}H_{11}N_2OS$, $0.5C_2H_6O$
G.Lepicard, J.Delettre, J.-P.Mornon
C. R. Acad. Sci., Fr., C, **276**, 657, 1973

41.32   **2 - Phenylamino - 5 - phenyl - thiazolin - 4 - one**
$C_{15}H_{12}N_2OS$
R.Bally, J.-P.Mornon   *Acta Cryst. (B)*, **29**, 1157, 1973

41.33   **2 - Chloro - 6,11 - dihydro - 6H - dibenzo(b,5)(1,4)thiazocine - 12 - carboxamide**
$C_{15}H_{13}ClN_2OS$
H.L.Yale, F.Sowinski, E.R.Spitzmiller   *J. Heterocycl. Chem.*, **9**, 899, 1972

41.34   **Phenothiazine - 10 - propionic acid**
$C_{15}H_{13}NO_2S$
M.C.Malmstrom, A.W.Cordes   *J. Heterocycl. Chem.*, **9**, 325, 1972

41.35   **2 - Phenylimino - 3 - methyl - 5 - phenyl - thiazolidin - 4 - one**
$C_{16}H_{14}N_2OS$
R.Bally, J.-P.Mornon   *Acta Cryst. (B)*, **29**, 1160, 1973

41.C   **Phenoxymethyl - anhydropenicillin ethanol solvate (absolute configuration)**
$C_{16}H_{16}N_2O_4S$, C_2H_6O
For complete entry see 50.6

41.36   **2,4 - Dimethyl - 3,5 - diphenylimino - 1,2,4 - thiadiazolidine**
$C_{16}H_{16}N_4S$
S.Sato, T.Kinoshita, T.Hata, C.Tamura   *Acta Cryst. (A)*, **28**, S25, 1972

41.37   **Methylene blue pentahydrate**
3,7 - bis(Dimethylamino)phenazothionium chloride pentahydrate
$C_{16}H_{18}N_3S^+$, Cl^- , $5H_2O$
H.E.Marr III, J.M.Stewart, M.F.Chiu   *Acta Cryst. (B)*, **29**, 847, 1973

41.38   **Methylene blue thiocyanate**
$C_{16}H_{18}N_3S^+$, CNS^-
A.Kahn-Harari, R.E.Ballard, E.K.Norris   *Acta Cryst. (B)*, **29**, 1124, 1973

41.C   **Dimethyl - thiacyanine bis(7,7,8,8 - tetracyanoquinodimethane)**
$C_{17}H_{15}N_2S_2^+$, $C_{12}H_4N_4^-$, $C_{12}H_4N_4$
For complete entry see 60.34

41.39   **cis - (2 - Methoxycarbonyl - 3,3 - dimethyl - 8 - oxo - 7 - phthalimido - 4,5 - dithia - 1 - aza - bicyclo(4.2.0)octane)**
$C_{17}H_{16}N_2O_5S_2$
S.Kukolja, P.V.Demarco, N.D.Jones, M.O.Chaney, J.W.Paschal
J. Amer. Chem. Soc., **94**, 7592, 1972
Also classified in 32

41.C   **6 - Methoxy phenoxymethyl anhydropenicillin**
$C_{17}H_{18}N_2O_5S$
For complete entry see 50.7

41.40   **10 ׃ (2' - Dimethylamino - propyl) - phenothiazine hydrobromide**
$C_{17}H_{21}N_2S^+$, Br^-
P.Marsau, B.Busetta   *Acta Cryst. (B)*, **29**, 986, 1973
Residue 1 also classified in 3

41.41   **2,4 - bis - (1,2 - Benzisothiazol - 3 - yl) - 3 - aminocrotononitrile dimethyl - formamide solvate**
$C_{18}H_{12}N_4S_2$, C_3H_7NO
E.Gaetani, T.Vitali, A.Mangia, M.Nardelli, G.Pelizzi
J. C. S. Perkin II, 2125, 1972

41.42   **2 - (2,6 - Dimethyl - phenyl) - imino - 3 - (2 - chlorobenzoyl) - thiazolidine**
$C_{18}H_{17}ClN_2OS$
G.Argay, A.Kalman, G.Toth, L.Toldy   *Tetrahedron Letters*, 3179, 1972

41.43   **2,2' - bis(2,3 - Dimethyl - benzothiazoline)**
$C_{18}H_{20}N_2S_2$
E.Miler-Srenger   *Acta Cryst. (B)*, **29**, 1119, 1973

41.C   **3,8 - Dihydro - 3,3,8,8 - tetramethyl - 6 - phenoxyacetamido - 1 - oxo - oxazolo(4,3 - c) - (1,4)thiazinium chloride**
$C_{18}H_{21}N_2O_4S^+$, Cl^-
For complete entry see 40.34

41.44   **3 - Methoxy - 10 - (3' - dimethylammonium - propyl) - phenothiazine hydrogen maleate**
$C_{18}H_{23}N_2OS^+$, $C_4H_3O_4^-$
P.Marsau, J.Gauthier   *Acta Cryst. (B)*, **29,** 992, 1973
Residue 1 also classified in 3; residue 2 classified in 2

41.45   **10 - (2' - Trimethylammonium - n - propyl) - phenothiazine methyl sulfate**
$C_{18}H_{23}N_2S^+$, $CH_3O_4S^-$
P.Marsau, Y.Cam   *Acta Cryst. (B)*, **29,** 980, 1973
Residue 1 also classified in 3; residue 2 classified in 11

41.C   **Dimethyl - thiacarbocyanine bis(7,7,8,8 - tetracyanoquinodimethane)**
$C_{19}H_{17}N_2S_2^+$, $C_{12}H_4N_4^-$, $C_{12}H_4N_4$
For complete entry see 60.36

41.46   **3,3' - Diethyl - thiacyanine bromide**
$C_{19}H_{19}N_2S_2^+$, Br^-
K.Nakatsu, H.Yoshioka, T.Aoki   *Chem. Letters*, 339, 1972

41.C   **Diethyl - thiacyanine bis(7,7,8,8 - tetracyanoquinodimethane)**
$C_{19}H_{19}N_2S_2^+$, $C_{12}H_4N_4^-$, $C_{12}H_4N_4$
For complete entry see 60.37

41.C   **Diethyl - thiacarbocyanine bis(7,7,8,8 - tetracyanoquinodimethane) (triclinic form)**
$C_{21}H_{21}N_2S_2^+$, $C_{12}H_4N_4^-$, $C_{12}H_4N_4$
For complete entry see 60.39

41.C   **7 - Methylthio - 7 - phenylacetamido - deacetoxycephalosporanic acid t - butyl ester**
$C_{21}H_{26}N_2O_4S_2$
For complete entry see 59.23

41.47   **3,5 - Epidithio - 2,5 - diphenyl - 2,4 - pentadienylidene - aniline**
$C_{23}H_{17}NS_3$
F.Leung, S.C.Nyburg   *Canad. J. Chem.*, **50,** 324, 1972

41.48   **2,4 - Diphenyl - 5 - (3' - phenyl - 2' - thiabutene - 4' - yl) - isothiazolium iodide**
$C_{24}H_{20}NS_2^+$, I^-
F.Leung, S.C.Nyburg   *Canad. J. Chem.*, **50,** 324, 1972

41.49   **2 - Phenyl - 3 - methyl - 5 - (1' - phenyl - 2' - benzyl - 2' - thiolato - ethylenyl) - 1,3,4 - thiadiazole**
$C_{24}H_{20}N_2S_2$
J.L.Flippen   *Acta Cryst. (B)*, **28,** 2749, 1972

41.50   **5 - Isopropyl - 2,3,5 - triphenyl - thiazolidin - 4 - one 1,1 - dioxide**
$C_{24}H_{23}NO_3S$
J.M.Decazes, J.L.Luche, H.B.Kagan, R.Parthasarathy, J.Ohrt
Tetrahedron Letters, 3633, 1972

41.C   Chaetocin (absolute configuration)
$C_{30}H_{28}N_6O_6S_4$
For complete entry see 50.12

HETERO-MIXED MISCELLANEOUS

42.1 **1,5 - Dihydroxy - 9 - oxa - 3,7 - dithiabicyclo(3.3.1)nonane - 3 - oxide**
$C_6H_{10}O_4S_2$
S.Abrahamsson, G.Rehnberg *Acta Chem. Scand.*, **26,** 3309, 1972

42.2 **2,4,6 - Trimethyl - 1,3,5 - dioxaselenane**
Monoselenoparaldehyde
$C_6H_{12}O_2Se$
G.Valle, V.Busetti, M.Mammi *Cryst. Struct. Comm.*, **2,** 167, 1973

42.C **3,7 - Dicyano - 3,5,7 - triaza - 1 - phosphabicyclo(3.3.1)nonane**
$C_7H_{10}N_5P$
For complete entry see 64.8

42.C **N,N - bis(2' - Chloroethyl) - N',O - trimethylene - phosphodiamide**
$C_7H_{14}Cl_2N_2O_2P$
For complete entry see 64.9

42.3 **4 - Phenyl - 1,3,2 - oxathiazole - 5 - one**
$C_8H_5NO_2S$
G.D.Andreetti, G.Bocelli, L.Cavalca, P.Sgarabotto
Gazz. Chim. Ital., **102,** 23, 1972

42.4 **2,4 - (β,β - Dimethyl - trimethylene) - (1,2,5 - oxaselenazolo(2,3 - b) (1,2,5)oxaselenazole) - 7 - Se(iv)**
$C_8H_{10}N_2O_2Se$
E.C.Llaguno, I.C.Paul *J. C. S. Perkin II*, 2001, 1972

42.5 **1 - Acetoxy - 1,2 - benziodoxolin - 3 - one**
$C_9H_7IO_4$
J.Z.Gougoutas, J.C.Clardy *J. Solid State Chem.*, **4,** 226, 1972

42.6 **2 - p - Nitrophenyl - 1,3 - oxathiane**
$C_{10}H_{11}NO_3S$
N.de Wolf, G.C.Verschoor, C.Romers *Acta Cryst. (B)*, **28,** 2424, 1972

42.7 **2,7 - Dichloro - selenophenazine**
$C_{12}H_7Cl_2NSe$
F.Bernier, A.Conde, A.Lopez-Castro, R.Marquez
Acta Cryst. (A), **28,** S31, 1972

42.8   1 - (2' - Iodobenzoyloxy) - 1,2 - benziodoxolin - 3 - one (α form)
$C_{14}H_8I_2O_4$
J.Z.Gougoutas, J.C.Clardy   *J. Solid State Chem.*, **4**, 230, 1972

42.9   1 - (2' - Iodobenzoyloxy) - 1,2 - benziodoxolin - 3 - one (β form)
$C_{14}H_8I_2O_4$
J.Z.Gougoutas, J.C.Clardy   *J. Solid State Chem.*, **4**, 230, 1972

42.10   bis - (2 - Carboxyphenyl)sulfur dihydroxide dilactone
$C_{14}H_8O_4S$
A.Kalman, K.Sasvari, I.Kapovits   *Acta Cryst. (B)*, **29**, 355, 1973

42.11   Phenylsulfoxonium iodide
$C_{18}H_{13}OS^+$, I^-
A.I.Gusev, Yu.T.Struchkov   *Zh. Strukt. Khim.*, **12**, 1120, 1971
Residue 1 also classified in 38, 39

42.C   3,5 - Di - (t - butyl) - 1 - phenylphosphoryl - 4 - aza - 2,5 - cyclohexadiene
$C_{18}H_{26}NOP$
For complete entry see 64.28

42.C   2 - Cyclohexyl - 3 - phenyl - 2 - aza - 3 - phospha - bicyclo(2.2.2)octane - 3 - thione
$C_{18}H_{26}NPS$
For complete entry see 64.29

42.C   Caged adamantanoid oxyphosphorane compound
$C_{19}H_{19}N_2O_4P$
For complete entry see 64.30

42.12   3,3,5,6 - Tetraphenyl - 2,3 - dihydro - 1,4 - oxathiin - 2 - one - 4,4 - dioxide
$C_{28}H_{20}O_4S$
N.Yasuoka, N.Kasai, M.Tanaka, T.Nagai, N.Tokura
Acta Cryst. (B), **28**, 3393, 1972

42.13   2,3,6,7 - Tetrahydro - 3,4,4,7 - tetraphenyl - 3 - oxo - 1,5,2,7 - thiaoxadiazepin - 6 - ylidene diphenylmethane benzene solvate
$C_{40}H_{30}N_2O_2S$, $0.5C_6H_6$
N.Yasuoka, Y.Kai, N.Kasai, T.Minami, K.Yamataka, Y.Ohshiro,
T.Agawa   *Acta Cryst. (A)*, **28**, S16, 1972

BARBITURATES

43.1   **Lithium purpurate dihydrate**
Lithium 5,5' - nitrilo - dibarbiturate dihydrate
$C_8H_4N_5O_6^-$, Li^+ , $2H_2O$
H.B.Burgi, S.Djuric, M.Dobler, J.D.Dunitz
Helv. Chim. Acta, **55,** 1771, 1972

43.C   **Strontium violurate tetrahydrate**
$C_8H_4N_6O_8Sr$, $4H_2O$
For complete entry see 67.4

43.C   **9 - Ethyladenine - 5,5 - diethylbarbituric acid complex**
$C_8H_{12}N_2O_3$, $C_7H_9N_5$
For complete entry see 60.10

43.2   **5,5 - Diethylbarbituric acid potassium tri - iodide complex**
$2C_8H_{12}N_2O_3$, K^+ , I_3^-
A.H.Robbins, B.M.Craven
Amer. Cryst. Assoc., Abstr. Papers (Winter Meeting), 87, 1973

43.3   **1 - Methyl - 5,5 - diethylbarbituric acid**
Metharbital
$C_9H_{14}N_2O_3$
H.Wunderlich   *Acta Cryst. (B),* **29,** 168, 1973

43.C   **9 - Ethyladenine - 5 - isopropyl - 5 - bromoallylbarbituric acid complex**
$C_{10}H_{15}BrN_2O$, $C_7H_9N_5$
For complete entry see 60.11

43.C   **Amobarbital - salicylamide complex**
$C_{11}H_{18}N_2O_3$, $C_7H_7NO_2$
For complete entry see 60.13

PYRIMIDINES AND PURINES

44.1   **Cytosine**
$C_4H_5N_3O$
R.J.McClure, B.M.Craven   *Acta Cryst. (B)*, **29**, 1234, 1973

44.2   **Cytosine monohydrate**
$C_4H_5N_3O$, H_2O
R.J.McClure, B.M.Craven   *Acta Cryst. (B)*, **29**, 1234, 1973

44.3   **bis(Cytosine)resorcinic acid monohydrate**
$C_4H_5N_3O$, $C_7H_5O_4^-$, $C_4H_6N_3O^+$, H_2O
C.Tamura, T.Hata, S.Sato   *Acta Cryst. (A)*, **28**, S46, 1972
Residue 1 also classified in 17

44.4   **Cytosine - N - benzoylglycine complex monohydrate**
$C_4H_5N_3O^+$, $C_9H_8NO_3^-$, H_2O
C.Tamura, T.Hata, S.Sato, N.Sakurai
Bull. Chem. Soc. Jap., **45**, 3254, 1972
Residue 2 classified in 48

44.5   **1 - Methyl - 5 - bromouracil**
$C_5H_5BrN_2O_2$
H.Mizuno, K.Morita, T.Fujiwara, K.-I.Tomita   *Chem. Letters*, 965, 1972

44.C   **Thymine N,N - diethylmelamine monohydrate**
$C_5H_6N_2O_2$, $C_7H_{14}N_6$, H_2O
For complete entry see 60.4

44.C   **bis(Adeninium) tetra - aquo - bis(adenine) cobalt(ii) sulfate hexahydrate**
$2C_5H_6N_5^+$, $C_{10}H_{18}CoN_{10}O_4^{2+}$, $2O_4S^{2-}$, $6H_2O$
For complete entry see 83.26

44.6   **Guanine picrate monohydrate**
$C_5H_6N_5O^+$, $C_6H_2N_3O_7^-$, H_2O
C.E.Bugg, U.Thewalt, E.Subramanian   *Stockholm Symposium*, 81, 1973
Residue 2 classified in 17, 15

44.7   **Thioguanine picrate monohydrate**
$C_5H_6N_5S^+$, $C_6H_2N_3O_7^-$, H_2O
C.E.Bugg, U.Thewalt, E.Subramanian   *Stockholm Symposium*, 81, 1973
Residue 2 classified in 17, 15

44.8 Adenine - N^1 - oxide sulfuric acid
$C_5H_7N_5O^{2+}$, O_4S^{2-}
P.Prusiner, M.Sundaralingam *Acta Cryst. (B)*, **28**, 2142, 1972

44.9 cis - Thymine glycol
$C_5H_8N_2O_4$
J.L.Flippen, J.H.Konnert
Amer. Cryst. Assoc., Abstr. Papers (Winter Meeting), 59, 1973

44.10 1 - Ethyl - 5 - bromouracil (form i)
$C_6H_7BrN_2O_2$
H.Mizuno, T.Fujiwara, K.Tomita *Bull. Chem. Soc. Jap.*, **45,** 905, 1972

44.11 1 - Ethyl - 5 - bromouracil (form ii)
$C_6H_7BrN_2O_2$
T.Tsukihara, T.Ashida, M.Kakudo *Bull. Chem. Soc. Jap.*, **45,** 909, 1972

44.12 N^6 - Methyladenine
$C_6H_7N_5$
H.Sternglanz, C.E.Bugg
Amer. Cryst. Assoc., Abstr. Papers (Winter Meeting), 61, 1973

44.13 1 - Methylthymine - 9 - methyladenine complex (neutron study)
$C_6H_8N_2O_2$, $C_6H_7N_5$
T.F.Koetzle, M.N.Frey, M.S.Lehmann, W.C.Hamilton
Amer. Cryst. Assoc., Abstr. Papers (Winter Meeting), 93, 1973

44.C 1 - Methylnicotinamide aden - 9 - yl acetate dihydrate
$C_7H_6N_5O_2^-$, $C_7H_9N_2O^+$, $2H_2O$
For complete entry see 33.33

44.14 6 - Methyl - uracil - 5 - acetic acid
$C_7H_8N_2O_4$
R.Destro, R.E.Marsh *Acta Cryst. (B)*, **28,** 2971, 1972

44.C 9 - Ethyladenine - 5,5 - diethylbarbituric acid complex
$C_7H_9N_5$, $C_8H_{12}N_2O_3$
For complete entry see 60.10

44.C 9 - Ethyladenine - 5 - isopropyl - 5 - bromoallylbarbituric acid complex
$C_7H_9N_5$, $C_{10}H_{15}BrN_2O$
For complete entry see 60.11

44.C 6 - Azauridine
$C_8H_{11}N_3O_6$
For complete entry see 47.1

44.C 6 - Azauridine - 5' - phosphoric acid
$C_8H_{12}N_3O_9P$
For complete entry see 47.2

44.C $O^2,2'$ **- Cyclouridine**
$C_9H_{10}N_2O_5$
For complete entry see 47.3

44.C **5 - Chloro - 2' - deoxyuridine**
$C_9H_{11}ClN_2O_5$
For complete entry see 47.4

44.C **Sodium β - cytidine 2',3' - cyclic phosphate dihydrate**
$C_9H_{11}N_3O_7P^-$, Na^+ , $2H_2O$
For complete entry see 47.5

44.C **5 - Iodo - 2' - deoxycytidine**
$C_9H_{12}IN_3O_4$
For complete entry see 47.6

44.C **Deoxyuridine**
$C_9H_{12}N_2O_5$
For complete entry see 47.7

44.C **Uracil - β - D - arabinofuranoside**
$C_9H_{12}N_2O_6$
For complete entry see 47.8

44.C **Cyclocytidine hydrochloride**
2,2' - Anhydro - 1 - β - D - arabinofuranosylcytosine hydrochloride
$C_9H_{12}N_3O_4^+$, Cl^-
For complete entry see 47.9

44.C **1 - β - D - Arabinofuranosyl - cytosine**
$C_9H_{13}N_3O_5$
For complete entry see 47.10

44.C **Pyrazomycin monohydrate**
$C_9H_{13}N_3O_6$, H_2O
For complete entry see 50.5

44.C **5,6 - Dihydro - 2 - thiouridine**
$C_9H_{14}N_2O_5S$
For complete entry see 47.11

44.C **1 - (β - D - Arabinofuranosyl)cytosine hydrochloride (absolute configuration)**
$C_9H_{14}N_3O_5^+$, Cl^-
For complete entry see 47.12

44.C **Cytidine 2' - phosphate trihydrate**
Cytidylic acid a trihydrate
$C_9H_{14}N_3O_8P$, $3H_2O$
For complete entry see 47.13

143

44.C   **8,3′ - S - Cycloadenosine monohydrate**
$C_{10}H_{11}N_5O_3S$, H_2O
For complete entry see 47.14

44.C   **8,2′ - S - Cycloadenosine - 3′,5′ - cyclic monophosphate trihydrate**
$C_{10}H_{11}N_5O_5PS$, $3H_2O$
For complete entry see 47.15

44.15   **Potassium N - (purine - 6 - yl carbamoyl) - L - threonine tetrahydrate**
$C_{10}H_{11}N_6O_4^-$, K^+ , $4H_2O$
R.Parthasarathy, J.Ohrt, G.B.Chheda
Amer. Cryst. Assoc., Abstr. Papers (Winter Meeting), 62, 1973
Residue 1 also classified in 48

44.C   **cis - anti - 6 - Methyluracil dimer**
$C_{10}H_{12}N_4O_4$
For complete entry see 36.3

44.C   **Inosine (orthorhombic form)**
$C_{10}H_{12}N_4O_5$
For complete entry see 47.16

44.16   **N⁶ - (Δ² - Isopentenyl)adenine**
$C_{10}H_{13}N_5$
H.Sternglanz, U.Thewalt, C.E.Bugg   *Stockholm Symposium*, 81, 1973

44.C   **Adenosine (absolute configuration)**
$C_{10}H_{13}N_5O_4$
For complete entry see 47.17

44.C   **9 - β - D - Arabinofuranosyl - adenine**
$C_{10}H_{13}N_5O_4$
For complete entry see 47.18

44.C   **6 - Thioguanosine monohydrate**
$C_{10}H_{13}N_5O_4S$, H_2O
For complete entry see 47.19

44.C   **Sodium 2′ - deoxyadenosine - 5′ - phosphate hexahydrate**
$C_{10}H_{13}N_5O_6P^-$, Na^+ , $6H_2O$
For complete entry see 47.20

44.C   **6 - Methyluridine**
$C_{10}H_{14}N_2O_6$
For complete entry see 47.21

44.C   **Adenosine hydrochloride**
$C_{10}H_{14}N_5O_4^+$, Cl^-
For complete entry see 47.22

44.C   **Thioguanosine monohydrate**
$C_{10}H_{14}N_5O_4S$, H_2O
For complete entry see 47.23

44.C   **Adenosine triphosphate disodium salt trihydrate**
$C_{10}H_{14}N_5O_{13}P_3{}^{2-}$, $2Na^+$, $3H_2O$
For complete entry see 47.24

44.C   **3 - Methyl - cytidine methosulfate**
$C_{10}H_{16}N_3O_5{}^+$, $CH_3O_4S^-$
For complete entry see 47.25

44.C   **3 - Phenyl - 5 - chloro - 7 - bromoisoxazolo(4,5 - d)pyrimidine**
$C_{11}H_5BrClN_3O$
For complete entry see 40.17

44.C   **5' - Methylene - adenosine - 3',5' - cyclic monophosphonate monohydrate**
$C_{11}H_{14}N_5O_5P$, H_2O
For complete entry see 47.26

44.17   **4 - (β - D - Ribopyranosyl) - amino - 8 - amino - pyrimido(4,5 - d)pyrimidine**
$C_{11}H_{14}N_6O_4$
H.M.Berman, W.C.Hamilton, R.Rousseau
Amer. Cryst. Assoc., Abstr. Papers (Winter Meeting), 95, 1973
Also classified in 45

44.18   **6 - Cyclohexyladenine - iodine complex**
$C_{11}H_{15}N_5$, I_2
D.van der Helm, S.D.Christian, L.-N.Lin
J. Amer. Chem. Soc., **95**, 2409, 1973
Residue 1 also classified in 21

44.C   **2,5' - Anhydro - 2',3' - isopropylidene - cyclouridine**
$C_{12}H_{14}N_2O_5$
For complete entry see 47.27

44.C   **Thiamine chloride monohydrate**
$C_{12}H_{17}N_4OS^+$, Cl^- , H_2O
For complete entry see 41.22

44.C   **N_2 - Dimethylguanosine**
$C_{12}H_{17}N_5O_5$
For complete entry see 47.28

44.C   **1,7 - Dimethylguanosine iodide**
$C_{12}H_{18}N_5O_5{}^+$, I^-
For complete entry see 47.29

44.19   **1,1' - Trimethylene - bis(thymine)**
$C_{13}H_{16}N_4O_4$
J.A.K.Frank, I.C.Paul   *J. Amer. Chem. Soc.*, **95**, 2324, 1973

44.C **2' - O - Tetrahydropyranyl - adenosine**
$C_{15}H_{21}N_5O_5$
For complete entry see 47.31

44.C **N - 5 - L - Phenylalanyl - 5 - amino - uridine**
$C_{18}H_{22}N_4O_7$
For complete entry see 47.32

44.C **Sodium adenylyl - 3',5' - uridine phosphate hydrate**
$C_{19}H_{23}N_7O_{12}P^-$, Na^+ , $6.5H_2O$
For complete entry see 47.33

44.C **Sodium guanylyl - 3',5' - cytidine nonahydrate**
$C_{19}H_{24}N_8O_{12}P^-$, Na^+ , $9H_2O$
For complete entry see 47.34

44.C **Sodium thymidylyl - 5',3' - thymidylate - 5' hydrate**
$C_{20}H_{27}N_4O_{15}P_2^-$, Na^+ , $12H_2O$
For complete entry see 47.35

CARBOHYDRATES

45.1 β - **D** - **Ribopyranoside**
$C_5H_{10}O_5$
V.James *Stockholm Symposium,* 100, 1973

45.2 β - **D** - **Ribopyranoside (neutron study)**
$C_5H_{10}O_5$
V.James *Stockholm Symposium,* 100, 1973

45.3   **D** - **iso** - **Ascorbic acid**
$C_6H_8O_6$
N.Azarnia, H.M.Berman, R.D.Rosenstein *Acta Cryst. (B),* **28,** 2157, 1972

45.4   **1,6** - **Anhydro** - β - **D** - **mannofuranose**
$C_6H_{10}O_5$
J.Lechat, G.A.Jeffrey *Acta Cryst. (B),* **28,** 3410, 1972

45.5   **Trisodium 6** - **phosphogluconate dihydrate**
$C_6H_{10}O_{10}P^{3-}$, $3Na^+$, $2H_2O$
G.D.Smith, C.N.Caughlan, A.Fitzgerald, J.P.Ashmore, K.A.Kerr
Amer. Cryst. Assoc., Abstr. Papers (Winter Meeting), 52, 1973
Residue 1 also classified in 46

45.6   **Potassium D** - **gluconate monohydrate (absolute configuration)**
$C_6H_{11}O_7^-$, K^+ , H_2O
G.A.Jeffrey, E.J.Fasiska *Carbohyd. Res.,* **21,** 187, 1972

45.7   **myo** - **Inositol calcium bromide pentahydrate**
$C_6H_{12}CaO_6^{2+}$, $2Br^-$, $5H_2O$
W.J.Cook, C.E.Bugg
Amer. Cryst. Assoc., Abstr. Papers (Winter Meeting), 58, 1973
Residue 1 also classified in 67

45.8   **Methyl 5** - **thio** - α - **D** - **ribopyranoside (absolute configuration)**
$C_6H_{12}O_4S$
R.L.Girling, G.A.Jeffrey *Acta Cryst. (B),* **29,** 1102, 1973

45.9   **Methyl 5** - **thio** - β - **D** - **ribopyranoside (absolute configuration)**
$C_6H_{12}O_4S$
R.L.Girling, G.A.Jeffrey *Acta Cryst. (B),* **29,** 1102, 1973

45.10   **Methyl 1 - thio - α- D - vibopgvanside (absolute configuration)**
$C_6H_{12}O_4S$
R.L.Girling, G.A.Jeffrey   *Acta Cryst. (B)*, **29**, 1006, 1973

45.11   **Methyl β - D - ribopyranoside**
$C_6H_{12}O_5$
V.J.James, J.D.Stevens   *Carbohyd. Res.*, **21**, 334, 1972

45.12   **α - D - Glucose monohydrate (absolute configuration)**
$C_6H_{12}O_6$, H_2O
E.Hough, S.Neidle, D.Rogers, P.G.H.Troughton
Acta Cryst. (B), **29**, 365, 1973

45.13   **α - Galactose calcium bromide trihydrate**
$C_6H_{12}CaO_6{}^{2+}$, $2Br^-$, $3H_2O$
W.J.Cook, C.E.Bugg
Amer. Cryst. Assoc., Abstr. Papers (Winter Meeting), 58, 1973
Residue 1 also classified in 67

45.14   **Phosphogluconic acid**
$C_6H_{13}O_{10}P$
C.N.Caughlan, C.K.Wang, A.Fitzgerald   *Acta Cryst. (A)*, **28**, S40, 1972
Also classified in 46

45.15   **β - D - Galactosamine hydrochloride**
$C_6H_{14}NO_5{}^+$, Cl^-
M.Takai, S.Watanabe, T.Ashida, M.Kakudo
Acta Cryst. (B), **28**, 2370, 1972

45.16   **β - D - Mannofuranose - calcium chloride tetrahydrate**
$(C_6H_{18}CaO_9{}^{2+})_n$, $2nCl^-$, nH_2O
D.C.Craig, N.C.Stephenson, J.D.Stevens   *Carbohyd. Res.*, **22**, 494, 1972
Residue 1 also classified in 67

45.17   **Methyl 3,6 - anhydro - α - D - glucopyranoside**
$C_7H_{12}O_5$
B.Lindberg, B.Lindberg, S.Svensson   *Acta Chem. Scand.*, **27**, 373, 1973

45.18   **Methyl 3,6 - anhydro - α - D - galactoside**
$C_7H_{12}O_5$
J.W.Campbell, M.M.Harding   *J. C. S. Perkin II*, 1721, 1972

45.C   **6 - Azauridine**
$C_8H_{11}N_3O_6$
For complete entry see 47.1

45.C   **N - (β - D - Ribofuranosyl)imidazole**
$C_8H_{12}N_2O_4$
For complete entry see 32.8

45.C   **6 - Azauridine - 5′ - phosphoric acid**
$C_8H_{12}N_3O_9P$
For complete entry see 47.2

45.C   **o - (β - D - Xylopyranosyl) - L - serine**
$C_8H_{15}NO_7$
For complete entry see 48.55

45.C   **O^2,2′ - Cyclouridine**
$C_9H_{10}N_2O_5$
For complete entry see 47.3

45.C   **5 - Chloro - 2′ - deoxyuridine**
$C_9H_{11}ClN_2O_5$
For complete entry see 47.4

45.C   **Sodium β - cytidine 2′,3′ - cyclic phosphate dihydrate**
$C_9H_{11}N_3O_7P^-$, Na^+ , $2H_2O$
For complete entry see 47.5

45.C   **5 - Iodo - 2′ - deoxycytidine**
$C_9H_{12}IN_3O_4$
For complete entry see 47.6

45.C   **Deoxyuridine**
$C_9H_{12}N_2O_5$
For complete entry see 47.7

45.C   **Uracil - β - D - arabinofuranoside**
$C_9H_{12}N_2O_6$
For complete entry see 47.8

45.C   **Cyclocytidine hydrochloride**
2,2′ - Anhydro - 1 - β - D - arabinofuranosylcytosine hydrochloride
$C_9H_{12}N_3O_4^+$, Cl^-
For complete entry see 47.9

45.C   **1 - β - D - Arabinofuranosyl - cytosine**
$C_9H_{13}N_3O_5$
For complete entry see 47.10

45.C   **Pyrazomycin monohydrate**
$C_9H_{13}N_3O_6$, H_2O
For complete entry see 50.5

45.C   **5,6 - Dihydro - 2 - thiouridine**
$C_9H_{14}N_2O_5S$
For complete entry see 47.11

45.C   1 - (β - **D** - **Arabinofuranosyl)cytosine hydrochloride (absolute configuration)**
$C_9H_{14}N_3O_5{}^+$, Cl^-
For complete entry see 47.12

45.C   **Cytidine 2' - phosphate trihydrate**
Cytidylic acid a trihydrate
$C_9H_{14}N_3O_8P$, $3H_2O$
For complete entry see 47.13

45.19   **Methyl 6 - O - acetyl -** β - **D - galactopyranoside**
$C_9H_{16}O_7$
B.Lindberg, P.J.Garegg, C.-G.Swahn *Acta Chem. Scand.*, **27,** 380, 1973

45.C   **8,3' - S - Cycloadenosine monohydrate**
$C_{10}H_{11}N_5O_3S$, H_2O
For complete entry see 47.14

45.C   **8,2' - S - Cycloadenosine - 3',5' - cyclic monophosphate trihydrate**
$C_{10}H_{11}N_5O_5PS$, $3H_2O$
For complete entry see 47.15

45.C   1 - (β - **D - Ribofuranosyl)imidazo(4,5 - b)pyrazine 4 - oxide**
$C_{10}H_{12}N_4O_5$
For complete entry see 35.9

45.C   **Inosine (orthorhombic form)**
$C_{10}H_{12}N_4O_5$
For complete entry see 47.16

45.C   **3 - Deaza - uridine**
$C_{10}H_{13}NO_6$
For complete entry see 33.42

45.C   **Adenosine (absolute configuration)**
$C_{10}H_{13}N_5O_4$
For complete entry see 47.17

45.C   9 - β - **D - Arabinofuranosyl - adenine**
$C_{10}H_{13}N_5O_4$
For complete entry see 47.18

45.C   **6 - Thioguanosine monohydrate**
$C_{10}H_{13}N_5O_4S$, H_2O
For complete entry see 47.19

45.C   **Sodium 2' - deoxyadenosine - 5' - phosphate hexahydrate**
$C_{10}H_{13}N_5O_6P^-$, Na^+ , $6H_2O$
For complete entry see 47.20

45.C   **6 - Methyluridine**
$C_{10}H_{14}N_2O_6$
For complete entry see 47.21

45.C   **Adenosine hydrochloride**
$C_{10}H_{14}N_5O_4^+$, Cl^-
For complete entry see 47.22

45.C   **Thioguanosine monohydrate**
$C_{10}H_{14}N_5O_4S$, H_2O
For complete entry see 47.23

45.C   **Adenosine triphosphate disodium salt trihydrate**
$C_{10}H_{14}N_5O_{13}P_3^{2-}$, $2Na^+$, $3H_2O$
For complete entry see 47.24

45.C   **3 - Methyl - cytidine methosulfate**
$C_{10}H_{16}N_3O_5^+$, $CH_3O_4S^-$
For complete entry see 47.25

45.C   **Tubercidin**
$C_{11}H_{14}N_4O_4$
For complete entry see 35.13

45.C   **Tubercidin**
$C_{11}H_{14}N_4O_4$
For complete entry see 35.14

45.C   **5' - Methylene - adenosine - 3',5' - cyclic monophosphonate monohydrate**
$C_{11}H_{14}N_5O_5P$, H_2O
For complete entry see 47.26

45.C   **4 - (β - D - Ribopyranosyl) - amino - 8 - amino - pyrimido(4,5 - d)pyrimidine**
$C_{11}H_{14}N_6O_4$
For complete entry see 44.17

45.20   **Tri - O - acetyl - β - D - arabinopyranosyl bromide**
$C_{11}H_{15}BrO_7$
P.W.R.Corfield, J.D.Mokren, P.L.Durette, D.Horton
Carbohyd. Res., **23,** 158, 1972

45.21   **Tri - O - acetyl - β - D - xylopyranosyl - chloride**
$C_{11}H_{15}ClO_7$
J.Hjortas   *Eur. Cryst. Meeting,* 1973

45.22   **2,3,4 - Tri - O - acetyl - α - D - xylopyranoside**
$C_{11}H_{16}O_8$
V.James   *Stockholm Symposium,* 100, 1973

45.23   **Ranuncoside monohydrate**
$C_{11}H_{16}O_8$, H_2O
R.A.Mariezcurrena, S.E.Rasmussen   *Acta Cryst. (B)*, **29**, 1030, 1973
Residue 1 also classified in 38

45.C   **2,5' - Anhydro - 2',3' - isopropylidene - cyclouridine**
$C_{12}H_{14}N_2O_5$
For complete entry see 47.27

45.C   **N_2 - Dimethylguanosine**
$C_{12}H_{17}N_5O_5$
For complete entry see 47.28

45.24   **Calcium ascorbate dihydrate**
$C_{12}H_{18}CaO_{14}$, $2H_2O$
R.A.Hearn, C.E.Bugg
Amer. Cryst. Assoc., Abstr. Papers (Winter Meeting), 58, 1973
Residue 1 also classified in 67

45.C   **1,7 - Dimethylguanosine iodide**
$C_{12}H_{18}N_5O_5{}^+$, I^-
For complete entry see 47.29

45.25   **1,2,4,5 - Di - O - isopropylidene - β - fructopyranose**
$C_{12}H_{20}O_6$
S.Takagi, R.Shiono, R.D.Rosenstein   *Acta Cryst. (B)*, **29,** 1177, 1973
Also classified in 38

45.26   **Calcium lactobionate bromide tetrahydrate (absolute configuration)**
$C_{12}H_{21}O_{12}{}^-$, Ca^{2+}, Br^-, $4H_2O$
W.J.Cook, C.E.Bugg   *Acta Cryst. (B)*, **29,** 215, 1973

45.27   **Lactose calcium bromide heptahydrate**
$C_{12}H_{22}CaO_{12}{}^{2+}$, $2Br^-$, $7H_2O$
W.J.Cook, C.E.Bugg
Amer. Cryst. Assoc., Abstr. Papers (Winter Meeting), 58, 1973
Residue 1 also classified in 67

45.28   **Sucrose (low angle data only)**
$C_{12}H_{22}O_{11}$
J.C.Hanson, L.C.Sieker, L.H.Jensen   *Acta Cryst. (B)*, **29,** 797, 1973

45.29   **Sucrose (high angle data only)**
$C_{12}H_{22}O_{11}$
J.C.Hanson, L.C.Sieker, L.H.Jensen   *Acta Cryst. (B)*, **29,** 797, 1973

45.30   **Sucrose (neutron study,refinement)**
$C_{12}H_{22}O_{11}$
G.M.Brown, H.A.Levy   *Acta Cryst. (B)*, **29,** 790, 1973

45.31 **Sucrose (low and high angle data)**
$C_{12}H_{22}O_{11}$
J.C.Hanson, L.C.Sieker, L.H.Jensen *Acta Cryst. (B)*, **29**, 797, 1973

45.32 **Lactose calcium bromide complex heptahydrate (absolute configuration)**
$C_{12}H_{22}O_{11}$, Ca^{2+}, $2Br^-$, $7H_2O$
C.E.Bugg *J. Amer. Chem. Soc.*, **95**, 908, 1973

45.33 **Lactose - calcium chloride heptahydrate**
$C_{12}H_{22}O_{11}$, Ca^{2+}, $2Cl^-$, $7H_2O$
W.J.Cook, C.E.Bugg *Acta Cryst. (B)*, **29**, 907, 1973

45.34 **Isomaltulose monohydrate**
$C_{12}H_{22}O_{11}$, H_2O
W.Dreissig, P.Luger *Acta Cryst. (B)*, **29**, 514, 1973

45.35 α,α **- Trehalose dihydrate (data of Rohrer and Berking)**
$C_{12}H_{22}O_{11}$, $2H_2O$
G.M.Brown, D.C.Rohrer, B.Berking, C.A.Beevers, R.O.Gould, R.Simpson
Acta Cryst. (B), **28**, 3145, 1972

45.36 α,α **- Trehalose dihydrate (data of Brown)**
$C_{12}H_{22}O_{11}$, $2H_2O$
G.M.Brown, D.C.Rohrer, B.Berking, C.A.Beevers, R.O.Gould, R.Simpson
Acta Cryst. (B), **28**, 3145, 1972

45.37 α,α **- Trehalose dihydrate**
$C_{12}H_{22}O_{11}$, $2H_2O$
T.Taga, M.Senma, K.Osaki *Acta Cryst. (B)*, **28**, 3258, 1972

45.38 α,α **- Trehalose dihydrate (data of Beevers et al.)**
$C_{12}H_{22}O_{11}$, $2H_2O$
G.M.Brown, D.C.Rohrer, B.Berking, C.A.Beevers, R.O.Gould, R.Simpson
Acta Cryst. (B), **28**, 3145, 1972

45.39 α **- Trehalose calcium bromide dihydrate**
$C_{12}H_{22}O_{12}$, Ca^{2+}, $2Br^-$, $2H_2O$
W.J.Cook, C.E.Bugg
Amer. Cryst. Assoc., Abstr. Papers (Winter Meeting), 58, 1973.
Residue 1 also classified in 67

45.40 **1,2,3,4 - Tetra - O - acetyl -** α **- D - arabinopyranoside**
$C_{13}H_{15}O_9$
V.James *Stockholm Symposium*, 100, 1973

45.41 **1,2,3,4 - Tetra - O - acetyl -** β **- D - arabinopyranoside**
$C_{13}H_{18}O_9$
V.James *Stockholm Symposium*, 100, 1973

45.42   1,2,3,4 - Tetra - O - acetyl - β - D - arabinopyranoside (neutron study)
$C_{13}H_{18}O_9$
V.James   *Stockholm Symposium*, 100, 1973

45.43 α - Methyl - 3,2' - anhydro - 4 - (3' - desoxy - α - L - rhamnosyl) - 2 - desoxy - L - fucoside
$C_{13}H_{22}O_6$
W.Richle, E.K.Winkler, D.M.Hawley, M.Dobler, W.Keller-Schierlein
Helv. Chim. Acta, **55,** 467, 1972

45.C   3,1' - Anhydro - 2 - (4,6 - di - O - acetyl - 2,3 - dideoxy - α - D - ribopyranose) - 3 - hydroxy - 5 - methyl - 2H - 1,2,6 - thiadiazine - 1,1 - dioxide
$C_{14}H_{18}N_2O_8S$
For complete entry see 47.30

45.44   Ethyl 2,3.4,5 - di - O - isopropylidene - 1 - thio - β - D - glucoseptanoside
$C_{14}H_{24}O_5S$
J.P.Beale, N.C.Stephenson, J.D.Stevens   *Acta Cryst. (B)*, **28,** 3115, 1972

45.C   2' - O - Tetrahydropyranyl - adenosine
$C_{15}H_{21}N_5O_5$
For complete entry see 47.31

45.45   Methyl 2,3,4,5 - tetra - O - acetyl - α - D - glucoseptanoside
$C_{15}H_{22}O_{10}$
J.F.McConnell, J.D.Stevens   *Acta Cryst. (A)*, **28,** S21, 1972

45.46   o - (4 - O - Methyl - α - D - glucopyranosyluronic acid) - (1 - 2) - O - β - D - xylopyranosyl - (1 - 4) - D - xylopyranose trihydrate (at $-193°C$)
$C_{17}H_{28}O_{15}$, $3H_2O$
R.A.Moran, G.F.Richards   *Carbohyd. Res.*, **25,** 250, 1972

45.C   N - 5 - L - Phenylalanyl - 5 - amino - uridine
$C_{18}H_{22}N_4O_7$
For complete entry see 47.32

45.47   3 - O - (p - Bromobenzoyl) - α - D - xylopyranose 1,2,4 - orthobenzoate
$C_{19}H_{15}BrO_6$
L.G.Vorontsova, A.F.Bochkov, I.V.Obruchnikov, V.I.Andrianov,
B.L.Tarnopolsky   *Carbohyd. Res.*, **23,** 326, 1972

45.48   Methyl - 2,3 - anhydro - 4,6 - benzylidene - α - D - mannopyranoside
$C_{19}H_{16}O_5$
A.-M.Pilotti, B.Stensland   *Acta Cryst. (B)*, **28,** 2821, 1972

45.C   Sodium adenylyl - 3',5' - uridine phosphate hydrate
$C_{19}H_{23}N_7O_{12}P^-$, Na^+ , $6.5H_2O$
For complete entry see 47.33

45.49   **5 - Brosyl - 3 - deoxy - 3 - C - (R) - (ethoxycarbonylformamido)methyl - 1,2 O - iso - propylidene - α - D - ribofuranose**
$C_{19}H_{24}BrNO_9S$
J.Coetzer, A.Jordaan, G.J.Lourens, M.J.Nolte
Acta Cryst. (B), **28**, 3537, 1972

45.C   **Sodium guanylyl - 3′,5′ - cytidine nonahydrate**
$C_{19}H_{24}N_8O_{12}P^-$, Na^+ , $9H_2O$
For complete entry see 47.34

45.C   **Sodium thymidylyl - 5′,3′ - thymidylate - 5′ hydrate**
$C_{20}H_{27}N_4O_{15}P_2^-$, Na^+ , $12H_2O$
For complete entry see 47.35

45.C   **Datiscoside bis(p - iodobenzoate) dihydrate (absolute configuration)**
$C_{52}H_{60}I_2O_{14}$, $2H_2O$
For complete entry see 59.34

PHOSPHATES

46.1   **Ammonium dimethylphosphate**
$C_2H_6O_4P^-$, H_3N^+
F.Garbassi, L.Giarda   *Abstr. Ital.-Yug. Congr,* 66, 1973

46.2   **Disodium DL - glycerol 3 - phosphate hexahydrate**
$C_3H_7O_6P^{2-}$, $2Na^+$, $6H_2O$
R.H.Fenn, G.E.Marshall   *Biochem. J.,* **130,** 1, 1972

46.C   **Silver diethylphosphate**
$(C_4H_{10}AgO_4P)_n$
For complete entry see 84.2

46.C   **Propylguanidinium diethylphosphate**
$C_4H_{10}O_4P^-$, $C_4H_{12}N_3^+$
For complete entry see 8.13

46.C   **Arginium diethylphosphate**
$C_4H_{10}O_4P^-$, $C_6H_{15}N_4O_2^+$
For complete entry see 48.51

46.C   **Putrescinium di(diethylphosphate)**
Tetramethylenediammonium di(diethylphosphate)
$2C_4H_{10}O_4P^-$, $C_4H_{14}N_2^{2+}$
For complete entry see 3.14

46.C   **Calcium phosphoryl choline chloride**
$C_5H_{13}NO_4P^-$, Ca^{2+} , Cl^-
For complete entry see 59.1

46.3   **L - α - Glycerylphosphoryl - ethanolamine monohydrate (absolute configuration)**
$C_5H_{14}NO_6P$, H_2O
G.T.DeTitta, B.M.Craven   *Acta Cryst. (B),* **29,** 1354, 1973
Residue 1 also classified in 3

46.4   **Mono - (p - nitrophenyl) phosphate**
$C_6H_6NO_6P$
C.N.Caughlan, C.K.Wang, A.Fitzgerald   *Acta Cryst. (A),* **28,** S40, 1972
Also classified in 15

46.5   **Monophenyl phosphate**
$C_6H_7O_4P$
C.N.Caughlan, C.K.Wang, A.Fitzgerald *Acta Cryst. (A)*, **28**, S40, 1972

46.C   **Trisodium 6 - phosphogluconate dihydrate**
$C_6H_{10}O_{10}P^{3-}$, $3Na^+$, $2H_2O$
For complete entry see 45.5

46.6   **Di - sodium trans - 2 - hydroxycyclohexyl phosphate trihydrate**
$C_6H_{11}O_5P^{2-}$, $2Na^+$, $3H_2O$
P.J.Roberts, O.Kennard *Cryst. Struct. Comm.*, **2**, 233, 1973
Residue 1 also classified in 21

46.C   **Phosphogluconic acid**
$C_6H_{13}O_{10}P$
For complete entry see 45.14

46.C   **Pyridoxal phosphate monohydrate**
$C_8H_{10}NO_6P$, H_2O
For complete entry see 33.35

46.C   **6 - Azauridine - 5′ - phosphoric acid**
$C_8H_{12}N_3O_9P$
For complete entry see 47.2

46.C   **Sodium β - cytidine 2′,3′ - cyclic phosphate dihydrate**
$C_9H_{11}N_3O_7P^-$, Na^+ , $2H_2O$
For complete entry see 47.5

46.C   **Cytidine 2′ - phosphate trihydrate**
Cytidylic acid a trihydrate
$C_9H_{14}N_3O_8P$, $3H_2O$
For complete entry see 47.13

46.C   **8,2′ - S - Cycloadenosine - 3′,5′ - cyclic monophosphate trihydrate**
$C_{10}H_{11}N_5O_5PS$, $3H_2O$
For complete entry see 47.15

46.C   **Sodium 2′ - deoxyadenosine - 5′ - phosphate hexahydrate**
$C_{10}H_{13}N_5O_6P^-$, Na^+ , $6H_2O$
For complete entry see 47.20

46.C   **Adenosine triphosphate disodium salt trihydrate**
$C_{10}H_{14}N_5O_{13}P_3^{2-}$, $2Na^+$, $3H_2O$
For complete entry see 47.24

46.7   **2 - p - Bromophenoxy - 5 - chloromethyl - 5 - methyl - 2 - oxo - 1,3,2 - dioxaphosphorinane**
$C_{11}H_{13}BrClO_4P$
R.Wagner, W.Jensen, W.Wadsworth, Q.Johnson
Cryst. Struct. Comm., **2**, 327, 1973

46.8   **Di - (p - nitrophenyl) phosphate**
$C_{12}H_9N_2O_8P$
C.N.Caughlan, C.K.Wang, A.Fitzgerald   *Acta Cryst. (A)*, **28**, S40, 1972
Also classified in 15

46.9   **Diphenyl phosphate**
$C_{12}H_{11}O_4P$
C.N.Caughlan, C.K.Wang, A.Fitzgerald   *Acta Cryst. (A)*, **28**, S40, 1972

46.10   **trans - Methyl - mesohydrobenzoin phosphate**
$C_{15}H_{13}O_4P$
B.S.Campbell, M.G.Newton
Amer. Cryst. Assoc., Abstr. Papers (Winter Meeting), 85, 1973

46.C   **Tyrosine pyridoxyl phosphate heptahydrate**
$C_{17}H_{21}N_2O_8P$, $7H_2O$
For complete entry see 48.73

46.11   **Triethylammonium tris(o - phenylenedioxy)phosphate**
$C_{18}H_{12}O_6P^-$, $C_6H_{16}N^+$
H.R.Allcock, E.C.Bissell   *J. Amer. Chem. Soc.*, **95**, 3154, 1973
Residue 2 classified in 3

46.C   **Sodium adenylyl - 3',5' - uridine phosphate hydrate**
$C_{19}H_{23}N_7O_{12}P^-$, Na^+ , $6.5H_2O$
For complete entry see 47.33

46.C   **Sodium guanylyl - 3',5' - cytidine nonahydrate**
$C_{19}H_{24}N_8O_{12}P^-$, Na^+ , $9H_2O$
For complete entry see 47.34

46.C   **Sodium thymidylyl - 5',3' - thymidylate - 5' hydrate**
$C_{20}H_{27}N_4O_{15}P_2^-$, Na^+ , $12H_2O$
For complete entry see 47.35

NUCLEOSIDES AND NUCLEOTIDES

47.1 **6 - Azauridine**
$C_8H_{11}N_3O_6$
W.Saenger, D.Suck *Stockholm Symposium,* 90, 1973
Also classified in 44, 45

47.2 **6 - Azauridine - 5' - phosphoric acid**
$C_8H_{12}N_3O_9P$
W.Saenger, D.Suck *Stockholm Symposium,* 90, 1973
Also classified in 44, 45, 46

47.3 **$O^2,2'$ - Cyclouridine**
$C_9H_{10}N_2O_5$
D.Suck, W.Saenger *Acta Cryst. (B),* **29,** 1323, 1973
Also classified in 44, 45

47.4 **5 - Chloro - 2' - deoxyuridine**
$C_9H_{11}ClN_2O_5$
D.W.Young, E.M.Morris *Acta Cryst. (B),* **29,** 1259, 1973
Also classified in 44, 45

47.5 **Sodium β - cytidine 2',3' - cyclic phosphate dihydrate**
$C_9H_{11}N_3O_7P^-$, Na^+, $2H_2O$
C.L.Coulter *J. Amer. Chem. Soc.,* **95,** 570, 1973
Residue 1 also classified in 44, 45, 46

47.6 **5 - Iodo - 2' - deoxycytidine**
$C_9H_{12}IN_3O_4$
G.Ambady, T.Phillips II, G.Kartha
Amer. Cryst. Assoc., Abstr. Papers (Winter Meeting), 60, 1973
Also classified in 44, 45

47.7 **Deoxyuridine**
$C_9H_{12}N_2O_5$
A.Rahman, H.R.Wilson *Acta Cryst. (B),* **28,** 2260, 1972
Also classified in 44, 45

47.8 **Uracil - β - D - arabinofuranoside**
$C_9H_{12}N_2O_6$
P.Tollin, H.R.Wilson, D.W.Young *Nature New Biology,* **242,** 49, 1973
Also classified in 44, 45

47.9 **Cyclocytidine hydrochloride**
2,2' - Anhydro - 1 - β - D - arabinofuranosylcytosine hydrochloride
$C_9H_{12}N_3O_4{}^+$, Cl^-
G.Kartha, T.Phillips *Stockholm Symposium,* 91, 1973
Residue 1 also classified in 44, 45

47.10 **1 - β - D - Arabinofuranosyl - cytosine**
$C_9H_{13}N_3O_5$
O.Lefebvre-Soubeyran, M.P.Tougard
C. R. Acad. Sci., Fr., C, **276,** 403, 1973
Also classified in 44, 45

47.11 **5,6 - Dihydro - 2 - thiouridine**
$C_9H_{14}N_2O_5S$
B.Kojic-Prodic, R.Liminga, M.Sljukic, Z.Ruzic-Toros
Abstr. Ital.-Yug. Congr, 149, 1973
Also classified in 44, 45

47.12 **1 - (β - D - Arabinofuranosyl)cytosine hydrochloride (absolute configuration)**
$C_9H_{14}N_3O_5{}^+$, Cl^-
J.S.Sherfinski, R.E.Marsh *Acta Cryst. (B),* **29,** 192, 1973
Residue 1 also classified in 44, 45

47.13 **Cytidine 2' - phosphate trihydrate**
Cytidylic acid a trihydrate
$C_9H_{14}N_3O_8P$, $3H_2O$
G.Kartha, G.Ambady, M.A.Viswamitra *Science,* **179,** 495, 1973
Residue 1 also classified in 44, 45, 46

47.14 **8,3' - S - Cycloadenosine monohydrate**
$C_{10}H_{11}N_5O_3S$, H_2O
K.Tomita, T.Tanaka, M.Yoneda, T.Fujiwara, M.Ikehara
Acta Cryst. (A), **28,** S45, 1972
Residue 1 also classified in 44, 45, 39

47.15 **8,2' - S - Cycloadenosine - 3',5' - cyclic monophosphate trihydrate**
$C_{10}H_{11}N_5O_5PS$, $3H_2O$
K.Tomita, T.Tanaka, M.Yoneda, T.Fujiwara, M.Ikehara
Acta Cryst. (A), **28,** S45, 1972
Residue 1 also classified in 44, 45, 46, 39

47.16 **Inosine (orthorhombic form)**
$C_{10}H_{12}N_4O_5$
E.Subramanian
Amer. Cryst. Assoc., Abstr. Papers (Winter Meeting), 60, 1973
Also classified in 44, 45

47.17   **Adenosine (absolute configuration)**
$C_{10}H_{13}N_5O_4$
T.F.Lai, R.E.Marsh   *Acta Cryst. (B),* **28,** 1982, 1972
Also classified in 44, 45

47.18   **9 - β - D - Arabinofuranosyl - adenine**
$C_{10}H_{13}N_5O_4$
G.J.Bunick, D.Voet
Amer. Cryst. Assoc., Abstr. Papers (Winter Meeting), 59, 1973
Also classified in 44, 45

47.19   **6 - Thioguanosine monohydrate**
$C_{10}H_{13}N_5O_4S$, H_2O
U.Thewalt, C.E.Bugg   *J. Amer. Chem. Soc.,* **94,** 8892, 1972
Residue 1 also classified in 44, 45

47.20   **Sodium 2' - deoxyadenosine - 5' - phosphate hexahydrate**
$C_{10}H_{13}N_5O_6P^-$, Na^+ , $6H_2O$
B.S.Reddy, M.A.Viswamitra   *Cryst. Struct. Comm.,* **2,** 9, 1973
Residue 1 also classified in 44, 45, 46

47.21   **6 - Methyluridine**
$C_{10}H_{14}N_2O_6$
D.Suck, W.Saenger   *J. Amer. Chem. Soc.,* **94,** 6520, 1972
Also classified in 44, 45

47.22   **Adenosine hydrochloride**
$C_{10}H_{14}N_5O_4^+$, Cl^-
K.Shikata, T.Ueki, T.Mitsui   *Acta Cryst. (B),* **29,** 31, 1973
Residue 1 also classified in 44, 45

47.23   **Thioguanosine monohydrate**
$C_{10}H_{14}N_5O_4S$, H_2O
C.E.Bugg, U.Thewalt, E.Subramanian   *Stockholm Symposium,* 81, 1973
Residue 1 also classified in 44, 45

47.24   **Adenosine triphosphate disodium salt trihydrate**
$C_{10}H_{14}N_5O_{13}P_3^{2-}$, $2Na^+$, $3H_2O$
O.Kennard, N.W.Isaacs, W.D.S.Motherwell, J.C.Coppola, D.L.Wampler,
A.C.Larson, D.G.Watson   *Proc. R. Soc., A,* **325,** 401, 1971
Residue 1 also classified in 44, 45, 46

47.25   **3 - Methyl - cytidine methosulfate**
$C_{10}H_{16}N_3O_5^+$, $CH_3O_4S^-$
E.Shefter, T.F.Brennan, P.Sackman   *Acta Cryst. (A),* **28,** S44, 1972
Residue 1 also classified in 44, 45; residue 2 classified in 11

47.26   **5' - Methylene - adenosine - 3',5' - cyclic monophosphonate monohydrate**
$C_{11}H_{14}N_5O_5P$, H_2O
M.Sundaralingam, J.Abola   *J. Amer. Chem. Soc.,* **94,** 5070, 1972
Residue 1 also classified in 44, 45, 64

47.27   **2,5' - Anhydro - 2',3' - isopropylidene - cyclouridine**
$C_{12}H_{14}N_2O_5$
L.T.J.Delbaere, M.N.G.James   *Acta Cryst. (B),* **29,** 404, 1973
Also classified in 44, 45

47.28   **N_2 - Dimethylguanosine**
$C_{12}H_{17}N_5O_5$
T.Brennan, C.Weeks, E.Shefter, S.T.Rao, M.Sundaralingam
J. Amer. Chem. Soc., **94,** 8548, 1972
Also classified in 44, 45

47.29   **1,7 - Dimethylguanosine iodide**
$C_{12}H_{18}N_5O_5^+$, I^-
E.Shefter, T.F.Brennan, P.Sackman   *Acta Cryst. (A),* **28,** S44, 1972
Residue 1 also classified in 44, 45

47.30   **3,1' - Anhydro - 2 - (4,6 - di - O - acetyl - 2,3 - dideoxy - α - D - ribopyranose) - 3 - hydroxy - 5 - methyl - 2H - 1,2,6 - thiadiazine - 1,1 - dioxide**
$C_{14}H_{18}N_2O_8S$
C.Foces-Foces, P.Smith-Verdier, S.Garcia-Blanco
Eur. Cryst. Meeting, 1973
Also classified in 41, 45

47.31   **2' - O - Tetrahydropyranyl - adenosine**
$C_{15}H_{21}N_5O_5$
O.Kennard, W.D.S.Motherwell, J.C.Coppola, B.E.Griffin, C.B.Reese,
A.C.Larson   *J. Chem. Soc. (B),* 1940, 1971
Also classified in 44, 45

47.32   **N - 5 - L - Phenylalanyl - 5 - amino - uridine**
$C_{18}H_{22}N_4O_7$
H.M.Berman, W.C.Hamilton, R.Rousseau
Amer. Cryst. Assoc., Abstr. Papers (Winter Meeting), 95, 1973
Also classified in 44, 45, 48

47.33   **Sodium adenylyl - 3',5' - uridine phosphate hydrate**
$C_{19}H_{23}N_7O_{12}P^-$, Na^+ , $6.5H_2O$
N.C.Seeman, J.M.Rosenberg, R.O.Day, F.L.Suddath, J.J.Kim, H.Nicholas,
A.Rich   *Stockholm Symposium,* 94, 1973
Residue 1 also classified in 44, 45, 46

47.34   **Sodium guanylyl - 3',5' - cytidine nonahydrate**
$C_{19}H_{24}N_8O_{12}P^-$, Na^+ , $9H_2O$
N.C.Seeman, J.M.Rosenberg, R.O.Day, F.L.Suddath, J.J.Kim, H.Nicholas,
A.Rich   *Stockholm Symposium,* 94, 1973
Residue 1 also classified in 44, 45, 46

47.35   **Sodium thymidylyl - 5',3' - thymidylate - 5' hydrate**
$C_{20}H_{27}N_4O_{15}P_2^-$, Na^+ , $12H_2O$
N.Camerman, K.Fawcett, A.Camerman   *Stockholm Symposium*, 82, 1973
Residue 1 also classified in 44, 45, 46

AMINO-ACIDS AND PEPTIDES

48.1   **Di - glycine hydroiodide**
$C_2H_5NO_2$, $C_2H_6NO_2^+$, I^-
P.Piret, J.Meunier-Piret, J.Verbist, M.van Meerssche
Bull. Soc. Chim. Belges, **81,** 539, 1972

48.2   **Triglycine fluoroberyllate (ferroelectric form, at 22°C)**
$C_2H_5NO_2$, $2C_2H_6NO_2^+$, BeF_4^{2-}
K.Lukaszewicz, F.Warkusz *Acta Cryst. (A),* **28,** S186, 1972

48.3   **Triglycine fluoroberyllate (paraelectric form, at 85°C)**
$C_2H_5NO_2$, $2C_2H_6NO_2^+$, BeF_4^{2-}
K.Lukaszewicz, F.Warkusz *Acta Cryst. (A),* **28,** S186, 1972

48.4   **Triglycine sulfate (ferroelectric form, at 37°C)**
$C_2H_5NO_2$, $2C_2H_6NO_2^+$, O_4S^{2-}
K.Itoh, T.Mitsui *Acta Cryst. (A),* **28,** S184, 1972

48.5   **Triglycine sulfate (at 57°C)**
$C_2H_5NO_2$, $2C_2H_6NO_2^+$, O_4S^{2-}
K.Itoh, T.Mitsui *Acta Cryst. (A),* **28,** S184, 1972

48.6   **Triglycine sulfate (ferroelectric form, at 19°C)**
$C_2H_5NO_2$, $2C_2H_6NO_2^+$, O_4S^{2-}
K.Itoh, T.Mitsui *Ferroelectrics,* **2,** 225, 1971

48.7   **Triglycine sulfate (ferroelectric form, irradiation - field treated)**
$C_2H_5NO_2$, $2C_2H_6NO_2^+$, O_4S^{2-}
S.R.Fletcher, A.C.Skapski, E.T.Keve
J. Phys. Chem. Solids, C, **4,** L255, 1971

48.8   **Diglycine sulfate monohydrate**
$2C_2H_6NO_2^+$, O_4S^{2-} , H_2O
S.Martinez-Carrera, F.H.Cano *Eur. Cryst. Meeting,* 1973

48.9   **DL - Serine (neutron study)**
$C_3H_7NO_3$
M.N.Frey, M.S.Lehmann, T.F.Koetzle, W.C.Hamilton
Acta Cryst. (B), **29,** 876, 1973

48.10   **L - Serine monohydrate (neutron study)**
$C_3H_7NO_3$, H_2O
M.N.Frey, M.S.Lehmann, T.F.Koetzle, W.C.Hamilton
Acta Cryst. (B), **29,** 876, 1973

48.11   **L - Cysteic acid monohydrate (neutron study)**
$C_3H_7NO_5S$, H_2O
M.Ramanadham, S.K.Sikka, R.Chidambaram
Acta Cryst. (B), **29,** 1167, 1973

48.12 γ **- Aminocrotonic acid hydrobromide**
$C_4H_8NO_2{}^+$, Br^-
K.-I.Tomita   *Tetrahedron Letters,* 2587, 1971
Residue 1 also classified in 1, 3

48.13 γ **- Aminocrotonic acid hydrobromide**
$C_4H_8NO_2{}^+$, Br^-
K.Tomita   *Jap. J. Brain Physiol.,* **71,** 9, 1966
Residue 1 also classified in 1, 3

48.14   **L - Asparagine monohydrate (neutron study)**
$C_4H_8N_2O_2$, H_2O
J.J.Verbist, M.S.Lehmann, T.F.Koetzle, W.C.Hamilton
Acta Cryst. (B), **28,** 3006, 1972

48.15   **L - Asparagine monohydrate (neutron study)**
$C_4H_8N_2O_3$
M.Ramanadham, S.K.Sikka, R.Chidambaram
Acta Cryst. (B), **28,** 3000, 1972

48.16   **Disodium N - phosphorylcreatine hydrate (further refinement)**
$C_4H_8N_3O_5P^{2-}$, $2Na^+$, xH_2O
J.C.Hanson, L.H.Jensen, J.R.Herriott   *Acta Cryst. (A),* **28,** S40, 1972
Residue 1 also classified in 8, 64

48.17   **DL -** α **- Amino - n - butyric acid (low temp.form, at $-90°C$)**
$C_4H_9NO_2$
T.Akimoto, Y.Iitaka   *Acta Cryst. (B),* **28,** 3106, 1972

48.18 γ **- Aminobutyric acid**
$C_4H_9NO_2$
K.-I.Tomita   *Tetrahedron Letters,* 2587, 1971
Also classified in 1, 3

48.19 γ **- Amino -** β **- hydroxybutyric acid**
$C_4H_9NO_3$
K.-I.Tomita   *Tetrahedron Letters,* 2587, 1971
Also classified in 1, 3

48.20   **L - Allothreonine**
$C_4H_9NO_3$
R.Srinivasan, K.I.Varughese, P.Swaminathan
Stockholm Symposium, 70, 1973

48.21   **Glycylglycine hydrochloride monohydrate (neutron diffraction)**
$C_4H_9N_2O_3^+$, Cl^-, H_2O
T.F.Koetzle, W.C.Hamilton, R.Parthasarathy
Acta Cryst. (B), **28,** 2083, 1972

48.22   **Glycylglycine nitrate**
$C_4H_9N_2O_3^+$, NO_3^-
S.N.Rao, R.Parthasarathy
Amer. Cryst. Assoc., Abstr. Papers (Winter Meeting), 80, 1973

48.23   **Glycylglycine phosphate monohydrate**
$C_4H_9N_2O_3^+$, $H_2O_4P^-$, H_2O
G.R.Freeman, R.A.Hearn, C.E.Bugg   *Acta Cryst. (B),* **28,** 2906, 1972

48.24   **Creatine monohydrate (further refinement)**
$C_4H_9N_3O_2$, H_2O
J.C.Hanson, L.H.Jensen, J.R.Herriott   *Acta Cryst. (A),* **28,** S40, 1972
Residue 1 also classified in 8

48.25 γ - **Aminobutyric acid hydrochloride**
$C_4H_{10}NO_2^+$, Cl^-
K.Tomita   *Jap. J. Brain Physiol.,* **61,** 1, 1965
Residue 1 also classified in 1

48.26   **DL - Allothreonine hydrobromide**
$C_4H_{10}NO_3^+$, Br^-
R.Srinivasan, K.I.Varughese, P.Swaminathan
Stockholm Symposium, 70, 1973

48.27   **4 - Hydroxy - L - proline (neutron study)**
$C_5H_9NO_3$
T.F.Koetzle, M.S.Lehmann, W.C.Hamilton   *Acta Cryst. (B),* **29,** 231, 1973
Also classified in 32

48.28   **Chondrine**
$C_5H_9NO_3S$
K.J.Palmer, K.S.Lee, R.Y.Wong, J.F.Carson
Acta Cryst. (B), **28,** 2789, 1972
Also classified in 41

48.29   **L - Glutamic acid (β form, neutron study)**
$C_5H_9NO_4$
M.S.Lehmann, T.F.Koetzle, W.C.Hamilton
J. Cryst. Mol. Struct., **2,** 225, 1972

48.30 L - **Glutamic acid hydrochloride (neutron study)**
$C_5H_{10}NO_4^+$, Cl^-
A.Sequeira, H.Rajagopal. R.Chidambaram
Acta Cryst. (B), **28,** 2514, 1972

48.31 **Acetylglycine N - methylamide**
$C_5H_{10}N_2O_2$
F.Iwasaki *Acta Cryst. (A)*, **28,** S13, 1972

48.32 L - **Methionine**
$C_5H_{11}NO_2S$
K.Torii, Y.Iitaka *Acta Cryst. (A)*, **28,** S39, 1972

48.33 L - **Valine hydrochloride (neutron study)**
$C_5H_{12}NO_2^+$, Cl^-
T.F.Koetzle, M.N.Frey, L.Golic, W.C.Hamilton, P.-G.Jonsson, A.Kvick,
M.S.Lehmann, J.J.Verbist *Acta Cryst. (A)*, **28,** S193, 1972

48.34 **Chloroacetyl - glycylglycine**
$C_6H_9ClN_2O_4$
S.T.Rao, M.Mallikarjunan *Cryst. Struct. Comm.*, **2,** 257, 1973

48.35 L - (+) - **Histidine (monoclinic form, data set of Madden,McGandy and Seeman)**
$C_6H_9N_3O_2$
J.J.Madden, E.L.McGandy, N.C.Seeman, M.M.Harding, A.Hoy
Acta Cryst. (B), **28,** 2382, 1972

48.36 L - (+) - **Histidine (monoclinic form, data set of Harding and Hoy)**
$C_6H_9N_3O_2$
J.J.Madden, E.L.McGandy, N.C.Seeman, M.M.Harding, A.Hoy
Acta Cryst. (B), **28,** 2382, 1972

48.37 L - (+) - **Histidine (orthorhombic form)**
$C_6H_9N_3O_2$
J.J.Madden, E.L.McGandy, N.C.Seeman *Acta Cryst. (B)*, **28,** 2377, 1972

48.38 L - **Histidine (orthorhombic form,neutron study)**
$C_6H_9N_3O_2$
M.S.Lehmann, T.F.Koetzle, W.C.Hamilton
Int. J. Peptide Prot. Res., **4,** 229, 1972

48.39 **Cyclo - di - β - alanyl**
$C_6H_{10}N_2O_2$
D.N.J.White, J.D.Dunitz *Israel J. Chem.*, **10,** 249, 1972
Also classified in 34

48.40 γ - **Hydroxy - L - isoleucine - lactone hydrobromide (absolute configuration)**
$C_6H_{12}NO_2^+$, Br^-
W.Hoppe, A.Gieren, P.Narayanan, M.Sturm *Eur. Cryst. Meeting*, 1973
Residue 1 also classified in 38

48.41   **Glycyl - L - threonine dihydrate**
$C_6H_{12}N_2O_4$, $2H_2O$
V.S.Yadava, V.M.Padmanabhan   *Acta Cryst. (B)*, **29**, 854, 1973

48.C ε **- Amino - n - caproic acid**
$C_6H_{13}NO_2$
For complete entry see 1.20

48.42   **DL - Isoleucine**
$C_6H_{13}NO_2$
E.Benedetti, C.Pedone, A.Sirigu   *Acta Cryst. (B)*, **29**, 730, 1973

48.43   **L - Norleucine**
L - α - Amino - n - caproic acid
$C_6H_{13}NO_2$
K.Torii, Y.Iitaka   *Acta Cryst. (A)*, **28**, S39, 1972

48.44   **D - Alloisoleucine hydrochloride monohydrate**
$C_6H_{14}NO_2^+$, Cl^- , H_2O
R.Srinivasan, K.I.Varughese, P.Swaminathan
Stockholm Symposium, 70, 1973

48.45   **L - Isoleucine hydrochloride monohydrate**
$C_6H_{14}NO_2^+$, Cl^- , H_2O
R.Srinivasan, K.I.Varughese, P.Swaminathan
Stockholm Symposium, 70, 1973

48.46   **meso - Lanthionine dihydrochloride**
$C_6H_{14}N_2O_4S^{2+}$, $2Cl^-$
R.E.Rosenfield Junior, R.Parthasarathy   *Acta Cryst. (A)*, **28**, S39, 1972

48.47   **L - Arginine dihydrate (neutron study)**
$C_6H_{14}N_4O_2$, $2H_2O$
M.S.Lehmann, J.J.Verbist, W.C.Hamilton, T.F.Koetzle
J. C. S. Perkin II, 133, 1973

48.48   **L - Lysine monohydrochloride dihydrate (neutron study)**
$C_6H_{15}N_2O_2^+$, Cl^- , $2H_2O$
R.R.Bugayong, A.Sequeira, R.Chidambaram
Acta Cryst. (B), **28**, 3214, 1972

48.49   **L - Lysine monohydrochloride dihydrate (neutron study)**
$C_6H_{15}N_2O_2^+$, Cl^- , $2H_2O$
T.F.Koetzle, M.S.Lehmann, J.J.Verbist, W.C.Hamilton
Acta Cryst. (B), **28**, 3207, 1972

48.50   **L - Arginine phosphate monohydrate**
$C_6H_{15}N_4O_2^+$, $H_2O_4P^-$, H_2O
W.Saenger, K.G.Wagner   *Acta Cryst. (B)*, **28**, 2237, 1972

48.51   **Arginium diethylphosphate**
$C_6H_{15}N_4O_2^+$, $C_4H_{10}O_4P^-$
S.Furberg, J.Solbakk   *Stockholm Symposium,* 83, 1973
Residue 2 classified in 46

48.C   **1 - Amino - 3 - methyl - cyclopentane carboxylic acid hemihydrate**
$C_7H_{13}NO_2$, $0.5H_2O$
For complete entry see 20.9

48.52 β **- Methyl - lanthionine tetrahydrate**
$C_7H_{14}N_2O_4S$, $4H_2O$
J.R.Know, P.Keck
Amer. Cryst. Assoc., Abstr. Papers (Winter Meeting), 81, 1973

48.53   **L - Mimosine**
β - N - (3 - Hydroxy - 4 - pyridone) - α - amino - propionic acid
$C_8H_{10}N_2O_4$
A.Mostad, C.Romming, E.Rosenqvist   *Acta Chem. Scand.,* **27,** 164, 1973
Also classified in 33

48.54   **L - N - Acetylhistidine monohydrate**
$C_8H_{11}N_3O_3$, H_2O
T.J.Kistenmacher, D.J.Hunt, R.E.Marsh   *Acta Cryst. (B),* **28,** 3352, 1972

48.55   **o - (β - D - Xylopyranosyl) - L - serine**
$C_8H_{15}NO_7$
L.T.J.Delbaere, M.Higham, B.Kamenar, P.W.Kent, C.K.Prout
Biochim. Biophys. Acta, **286,** 441, 1972
Also classified in 45

48.56   **Glycyl - L - leucine**
$C_8H_{16}N_2O_3$
V.Pattabhi, K.Venkatesan, S.R.Hall   *Cryst. Struct. Comm.,* **2,** 223, 1973

48.C   **Cytosine - N - benzoylglycine complex monohydrate**
$C_9H_8NO_3^-$, $C_4H_5N_3O^+$, H_2O
For complete entry see 44.4

48.57   **DL - Tyrosine**
$C_9H_{11}NO_3$
A.Mostad, C.Romming   *Acta Chem. Scand.,* **27,** 401, 1973

48.58   **L - Tyrosine**
$C_9H_{11}NO_3$
A.Mostad, H.M.Nissen, C.Romming   *Acta Chem. Scand.,* **26,** 3819, 1972

48.59   **L - Tyrosine (neutron study)**
$C_9H_{11}NO_3$
M.N.Frey, T.F.Koetzle, M.S.Lehmann, W.C.Hamilton
J. Chem. Phys., **58,** 2547, 1973

48.60   **L - Tyrosine hydrochloride (neutron study)**
$C_9H_{12}NO_3^+$, Cl^-
M.N.Frey, T.F.Koetzle, M.S.Lehmann, W.C.Hamilton
J. Chem. Phys., **58,** 2547, 1973

48.C   **Potassium N - (purine - 6 - yl carbamoyl) - L - threonine tetrahydrate**
$C_{10}H_{11}N_6O_4^-$, K^+ , $4H_2O$
For complete entry see 44.15

48.61   **N - Pivalyl - glycyl - isopropylamide**
$C_{10}H_{19}N_2O_2$
A.Aubry, M.Marraud, J.Protas, J.Neel
C. R. Acad. Sci., Fr., C, **276,** 1089, 1973

48.62   **N - Acetyl - DL - pseudo - leucyl - dimethylamide**
$C_{10}H_{20}N_2O_2$
A.Aubry, M.Marraud, J.Protas, J.Neel
C. R. Acad. Sci., Fr., C, **276,** 579, 1973
Also classified in 1

48.63   **DL - Tryptophan formate**
$C_{11}H_{13}N_2O_2^+$, CHO_2^-
E.Bye, A.Mostad, C.Romming   *Acta Chem. Scand.*, **27,** 471, 1973

48.64   **Glycyl - L - tyrosine dihydrate**
$C_{11}H_{14}N_2O_4$, $2H_2O$
M.Cotrait, Y.Barrans   *Cryst. Struct. Comm.*, **1,** 301, 1972

48.65   **Glycyl - L - phenylalanine p - bromobenzenesulfonate**
$C_{11}H_{15}N_2O_3^+$, $C_6H_4BrO_3S^-$
J.M.van der Veen, B.W.Low   *Acta Cryst. (B)*, **28,** 3548, 1972

48.66   **Glycyl - L - phenylalanine p - toluenesulfonate**
$C_{11}H_{15}N_2O_3^+$, $C_7H_7O_3S^-$
J.M.van der Veen, B.W.Low   *Acta Cryst. (B)*, **28,** 3548, 1972

48.67   **Acetyl - L - tyrosine methylamide**
$C_{12}H_{16}N_2O_3$
M.Cotrait, J.-P.Bideau   *Cryst. Struct. Comm.*, **2,** 111, 1973

48.68   **L - Alanyl - L - phenylalanine hydrochloride dihydrate**
$C_{12}H_{17}N_2O_3^+$, Cl^- , $2H_2O$
M.Cotrait, Y.Barrans   *Cryst. Struct. Comm.*, **2,** 213, 1973

48.69   **(Ethylsulfinyl) - L - tryptophan (absolute configuration)**
$C_{13}H_{16}N_2O_3S$
W.Hoppe, A.Gieren, P.Narayanan, M.Sturm   *Eur. Cryst. Meeting,* 1973
Also classified in 11

48.70   N(α) - Tosyl - **arginine methyl ester hydrochloride**
$C_{14}H_{23}N_4O_4S^+$, Cl^-
A.Michel, G.Evrard, F.Durant   *Stockholm Symposium,* 71, 1973

48.C   **3,5 - Di - iodo - L - thyronine - N - methylacetamide complex (absolute configuration)**
$C_{15}H_{13}INO_4$, C_3H_7NO
For complete entry see 60.28

48.71   **Cyclo - (tri - L - prolyl)**
$C_{15}H_{21}N_3O_3$
G.Kartha, G.Ambady   *Acta Cryst. (A),* **28,** S33, 1972
Also classified in 34

48.72   **Cyclo - (L - prolyl - L - prolyl - L - hydroxyprolyl)**
$C_{15}H_{21}N_3O_4$
G.Kartha, G.Ambady   *Acta Cryst. (A),* **28,** S33, 1972
Also classified in 34

48.73   **Tyrosine pyridoxyl phosphate heptahydrate**
$C_{17}H_{21}N_2O_8P$, $7H_2O$
A.Mangia, M.Nardelli, G.Pelizzi   *Eur. Cryst. Meeting,* 1973
Residue 1 also classified in 46, 33

48.74   **N - Benzoyl - L - tyrosine ethyl ester**
$C_{18}H_{19}NO_4$
F.Durant, A.F.Pieret, G.Germain, M.Koch
Cryst. Struct. Comm., **1,** 435, 1972

48.C   **N - 5 - L - Phenylalanyl - 5 - amino - uridine**
$C_{18}H_{22}N_4O_7$
For complete entry see 47.32

48.75   **N - Acetyl - bis(dehydrophenylalanyl) - glycine**
$C_{22}H_{21}N_3O_5$
S.Merlino, G.Montagnoli, O.Pieroni, E.Monti   *Eur. Cryst. Meeting,* 1973

48.76   **Cyclochlorotine**
$C_{24}H_{31}Cl_2N_5O_7$
H.Yoshioka, K.Nakatsu   *Stockholm Symposium,* 45, 1973

48.77   **Cyclo - octasarcosyl tetrahydrate**
$C_{24}H_{40}N_8O_8$, $4H_2O$
K.Titlestad, P.Groth, J.Dale, M.Y.Ali   *J. C. S. Chem. Comm.,* 346, 1973

48.78   **t - Butyloxycarbonyl - tetra - L - proline benzyl ester monohydrate**
$C_{32}H_{44}N_4O_7$, H_2O
T.Matsuzaki   *Acta Cryst. (A),* **28,** S38, 1972

48.C α - **Cyclodextrin** - **iodine**
$C_{36}H_{60}O_{30}$, I_2
For complete entry see 61.6

48.C α - **Cyclodextrin dihydrate**
$C_{36}H_{60}O_{30}$, $2H_2O$
For complete entry see 61.7

48.C α - **Cyclodextrin** - **methanol**
$C_{36}H_{60}O_{30}$, CH_4O
For complete entry see 61.8

48.C α - **Cyclodextrin** - **n** - **propanol**
$C_{36}H_{60}O_{30}$, C_3H_8O
For complete entry see 61.9

48.79 **N** - **Tetracosanoyl** - **phytosphingosine**
$C_{42}H_{85}NO_4$
B.Dahlen, I.Pascher *Acta Cryst. (B),* **28,** 2396, 1972

48.C **Valinomycin**
$C_{54}H_{90}N_6O_{18}$
For complete entry see 50.29

48.C **Antamanide lithium bromide acetonitrile solvate**
$C_{64}H_{70}LiN_{10}O_{10}^+$, Br^- , xC_2H_3N
For complete entry see 50.30

PORPHYRINS AND CORRINS

49.1 **Methylpheophorbide A**
$C_{36}H_{38}N_4O_5$
J.Gassmann, I.Strell, F.Brandl, M.Sturm, W.Hoppe
Tetrahedron Letters, 4609, 1971

49.2 **Nickel(ii) octaethylporphyrin (tetragonal form)**
$C_{36}H_{44}N_4Ni$
E.F.Meyer Junior *Acta Cryst. (B)*, **28**, 2162, 1972

49.3 **Nickel(ii) octaethylporphyrin (triclinic form)**
$C_{36}H_{44}N_4Ni$
D.L.Cullen, E.F.Meyer Junior
Amer. Cryst. Assoc., Abstr. Papers (Winter Meeting), 21, 1973

49.4 **Octaethylporphyrin**
$C_{36}H_{46}N_4$
J.W.Lauher, J.A.Ibers
Amer. Cryst. Assoc., Abstr. Papers (Winter Meeting), 21, 1973

49.5 **bis(Dimethylamine)etio(i)porphinato rhodium(iii) chloride dihydrate**
$C_{36}H_{50}N_6Rh^+$, Cl^-, $2H_2O$
L.K.Hanson, J.C.Hanson *Acta Cryst. (A)*, **28**, S74, 1972

49.6 **bis(Imidazole) - octaethylporphinato - iron(iii) perchlorate chloroform solvate**
$C_{42}H_{52}FeN_8^+$, ClO_4^-, $2CHCl_3$
A.Takenaka, Y.Sasada, E.-I.Watanabe, H.Ogoshi, Z.-I.Yoshida
Chem. Letters, 1235, 1972

49.7 $\alpha,\beta,\gamma,\delta$ - **Tetraphenylporphinato - dichloro - tin(iv)**
$C_{44}H_{28}Cl_2N_4Sn$
D.M.Collins, W.R.Scheidt, J.L.Hoard *J. Amer. Chem. Soc.*, **94**, 6689, 1972

49.8 **Tetraphenylporphine - carbonyl - ruthenium ethanol**
$C_{47}H_{34}N_4O_2Ru$
J.J.Bonnet, S.S.Eaton, G.R.Eaton, R.H.Holm, J.A.Ibers
J. Amer. Chem. Soc., **95**, 2141, 1973

49.9 μ - **(Tetraphenylporphyrin) - bis(tricarbonyl - rhenium(i))**
$C_{50}H_{20}O_6Re_2$
D.L.Cullen, E.F.Meyer *Acta Cryst. (A)*, **28**, S92, 1972

49.10   **meso - Tetraphenylporphinato - bis(tricarbonyl - rhenium(i))**
$C_{50}H_{28}N_4O_6Re_2$
D.Cullen, E.Meyer, T.S.Srivastava, M.Tsutsui
J. Amer. Chem. Soc., **94,** 7603, 1972

49.11   **Stannic phthalocyanine**
$C_{64}H_{32}N_{16}Sn$
W.E.Bennett, D.E.Broberg, N.C.Baenziger   *Inorg. Chem.*, **12,** 930, 1973

ANTIBIOTICS

50.1 αS,5S - α - **Amino - 3 - chloro - 2 - isoxazoline - 5 - acetic acid (absolute configuration)**
$C_5H_7ClN_2O_3$
D.G.Martin, D.J.Duchamp, C.G.Chidester
Tetrahedron Letters, 2549, 1973
Also classified in 40

50.2 **Validamine hydrobromide (absolute configuration)**
1S - 1 - Amino - 5 - hydroxymethyl - 2,3,4 - cyclohexanetriol hydrobromide
$C_7H_{15}NO_4^+$, Br^-
K.Kamiya, Y.Wada, S.Horii, M.Nishikawa
J. Antibiot., A, Jap., **24,** 317, 1971
Residue 1 also classified in 21

50.3 **DL - Methylkasugaminide selenate monohydrate methanol solvate**
$C_7H_{18}N_2O_2^{2+}$, O_4Se^{2-} , H_2O , CH_4O
Y.Suhara, F.Sasaki, G.Koyama, K.Maeda, H.Umezawa, M.Ohno
J. Amer. Chem. Soc., **94,** 6501, 1972
Residue 1 also classified in 38

50.4 **(+) - Aeroplysinin - I (absolute configuration)**
(+) - 3,5 - Dibromo - 1,6 - dihydroxy - 4 - methoxy - cyclohexa - 1,4 - diene acetonitrile
$C_9H_9Br_2NO_2$
L.Mazzarella, R.Puliti   *Gazz. Chim. Ital.,* **102,** 391, 1972
Also classified in 21

50.5 **Pyrazomycin monohydrate**
$C_9H_{13}N_3O_6$, H_2O
N.D.Jones, M.O.Chaney   *Acta Cryst. (A),* **28,** S48, 1972
Residue 1 also classified in 44, 45

50.6 **Phenoxymethyl - anhydropenicillin ethanol solvate (absolute configuration)**
$C_{16}H_{16}N_2O_4S$, C_2H_6O
G.L.Simon, R.B.Morin, L.F.Dahl   *J. Amer. Chem. Soc.,* **94,** 8557, 1972
Residue 1 also classified in 41

50.7 **6 - Methoxy phenoxymethyl anhydropenicillin**
$C_{17}H_{18}N_2O_5S$
M.O.Chaney, N.D.Jones *Cryst. Struct. Comm.*, **2**, 367, 1973
Also classified in 41

50.8 **7 - Chloro - 6 - demethyltetracycline hydrochloride trihydrate**
$C_{21}H_{21}ClN_2O_8{}^+$, Cl^-, $3H_2O$
G.J.Palenik, M.Mathew *Acta Cryst. (A)*, **28**, S47, 1972
Residue 1 also classified in 29

50.9 **$4\alpha,8\alpha,10\beta,14\beta$ - Tetramethyl - $\Delta^{9,11}$ - dodecahydro -**
cyclopenta(a)phenanthrene - 3,17 - dione
$C_{21}H_{30}O_2$
W.G.Dauben, G.Ahlgren, T.J.Leitereg, W.C.Schwarzel, M.Yoshioko
J. Amer. Chem. Soc., **94**, 8593, 1972
Also classified in 51, 29

50.10 **Aphidicolin bis(acetonide)**
$C_{26}H_{42}O_4$
K.M.Brundret, W.Dalziel, B.Hesp, J.A.J.Jarvis, S.Neidle
J. C. S. Chem. Comm., 1027, 1972
Also classified in 31, 38

50.11 **N - Bromoacetyl - daunomycin acetone solvate**
$C_{29}H_{30}BrNO_{11}$, C_3H_6O
R.Angiuli, E.Foresti, L.Riva di Sanseverino, N.W.Isaacs, O.Kennard,
W.D.S.Motherwell, D.L.Wampler, F.Arcamone
Nature New Biology, **234**, 78, 1971
Residue 1 also classified in 29, 38

50.12 **Chaetocin (absolute configuration)**
$C_{30}H_{28}N_6O_6S_4$
H.P.Weber *Acta Cryst. (B)*, **28**, 2945, 1972
Also classified in 41

50.13 **Cesium chlorothircolide methyl ester trihydrate (absolute configuration)**
$C_{30}H_{37}O_8{}^-$, Cs^+, $3H_2O$
M.Brufani, S.Cerrini, W.Fedeli, F.Mazza, R.Muntwyler
Helv. Chim. Acta, **55**, 2094, 1972
Residue 1 also classified in 38

50.14 **Kinamycin C p - bromobenzoate benzene solvate (absolute configuration)**
$C_{31}H_{23}BrN_2O_{11}$, C_6H_6
A.Furusaki, M.Matsui, T.Watanabe, S.Omura, A.Nakagawa, T.Hata
Israel J. Chem., **10**, 173, 1972
Residue 1 also classified in 36

50.15 **5 - Bromo - antibiotic X - 537A hemihydrate**
$C_{34}H_{53}BrO_8$, $0.5H_2O$
E.C.Bissell, I.C.Paul *J. C. S. Chem. Comm.*, 967, 1972
Residue 1 also classified in 38

50.16 **Nonactin**
$C_{40}H_{64}O_{12}$
M.Dobler *Helv. Chim. Acta,* **55,** 1371, 1972
Also classified in 38

50.17 **Tetranactin - potassium thiocyanate (form ii)**
$C_{44}H_{72}KO_{12}^+$, CNS^-
Y.Iitaka, T.Sakamaki, Y.Nawata *Chem. Letters,* 1225, 1972
Residue 1 also classified in 38, 67

50.18 **Tetranactin - potassium thiocyanate (form i)**
$C_{44}H_{72}KO_{12}^+$, CNS^-
Y.Iitaka, T.Sakamaki, Y.Nawata *Chem. Letters,* 1225, 1972
Residue 1 also classified in 38, 67

50.19 **Tetranactin - sodium thiocyanate (form ii)**
$C_{44}H_{72}NaO_{12}^+$, CNS^-
Y.Iitaka, T.Sakamaki, Y.Nawata *Chem. Letters,* 1225, 1972
Residue 1 also classified in 38, 67

50.20 **Tetranactin**
$C_{44}H_{72}O_{12}$
Y.Iitaka, T.Sakamaki, Y.Nawata *Chem. Letters,* 1225, 1972
Also classified in 38

50.21 **Tetranactin - rubidium thiocyanate**
$C_{44}H_{72}O_{12}Rb^+$, CNS^-
Y.Iitaka, T.Sakamaki, Y.Nawata *Acta Cryst. (A),* **28,** S47, 1972
Residue 1 also classified in 67

50.22 **Tetranactin - rubidium thiocyanate (form i)**
$C_{44}H_{72}O_{12}Rb^+$, CNS^-
Y.Iitaka, T.Sakamaki, Y.Nawata *Chem. Letters,* 1225, 1972
Residue 1 also classified in 38, 67

50.23 **Oligomycin B**
$C_{45}H_{72}O_{12}$
R.Norrestam, M.von Glehn *Stockholm Symposium,* 79, 1973
Also classified in 38

50.24 **Oligomycin B hydrate methanol solvate**
$C_{45}H_{72}O_{12}$, xH_2O , yCH_4O
M.von Glehn, R.Norrestam, P.Kierkegaard, L.Maron, L.Ernster
Fed. Eur. Biochem. Soc., **20,** 267, 1972

50.25 **Kidamycin derivative di - iodide methanol solvate monohydrate**
$C_{48}H_{64}N_2O_{13}^{2+}$, $2I^-$, $4CH_4O$, H_2O
M.Furukawa, A.Itai, Y.Iitaka *Tetrahedron Letters,* 1065, 1973
Residue 1 also classified in 38

50.26 **Antibiotic A204A silver salt acetone solvate (absolute configuration)**
$C_{49}H_{83}O_{17}^-$, Ag^+ , C_3H_6O
N.D.Jones, M.O.Chaney, J.W.Chamberlin, R.L.Hamill, S.Chen
J. Amer. Chem. Soc., **95,** 3399, 1973
Residue 1 also classified in 38

50.27 **Antibiotic A204A sodium salt acetone solvate**
$C_{49}H_{83}O_{17}^-$, Na^+ , C_3H_6O
N.D.Jones, M.O.Chaney, J.W.Chamberlin, R.L.Hamill, S.Chen
J. Amer. Chem. Soc., **95,** 3399, 1973

50.28 **Antibiotic A204A**
$C_{49}H_{84}O_{17}$
N.D.Jones, M.O.Chaney, J.W.Chamberlin, R.L.Hamill, S.Chen
Amer. Cryst. Assoc., Abstr. Papers (Winter Meeting), 26, 1973
Also classified in 38

50.29 **Valinomycin**
$C_{54}H_{90}N_6O_{18}$
W.L.Duax, H.Hauptman, C.M.Weeks, D.A.Norton
Science, **176,** 911, 1972
Also classified in 48

50.30 **Antamanide lithium bromide acetonitrile solvate**
$C_{64}H_{70}LiN_{10}O_{10}^+$, Br^- , xC_2H_3N
I.Karle *Stockholm Symposium,* 44, 1973
Residue 1 also classified in 48, 67

STEROIDS

51.1   **10 - Aza - 19 - nor - 5β,9β - androst - 8(14) - en - 3,17 - dione**
$C_{17}H_{23}NO_2$
J.G.H.de Jong, C.J.Dik-Edixhoven, H.Schenk
Cryst. Struct. Comm., **2,** 33, 1973
Also classified in 36

51.2   **Estrone (form ii)**
$C_{18}H_{22}O_2$
B.Busetta, C.Courseille, M.Hospital   *Acta Cryst. (B)*, **29,** 298, 1973

51.3   **Estrone (form i)**
$C_{18}H_{22}O_2$
B.Busetta, C.Courseille, M.Hospital   *Acta Cryst. (B)*, **29,** 298, 1973

51.4   **Estrone (form iii)**
$C_{18}H_{22}O_2$
B.Busetta, C.Courseille, M.Hospital   *Acta Cryst. (B)*, **29,** 298, 1973

51.5   **17β - Hydroxyestr - 5(10) - en - 3 - one**
$C_{18}H_{26}O_2$
R.R.Sobti, S.G.Levine, J.Bordner   *Acta Cryst. (B)*, **28,** 2292, 1972

51.6   **19 - nor - Androstenediol**
$C_{18}H_{28}O_2$
D.J.Duchamp, J.A.Campbell   *Acta Cryst. (A)*, **28,** S50, 1972

51.C   **DL - 3a,6 - Dimethyl - 6 - (trans - 3 - chlorobut - 2 - ene - 1 - yl) -
2,4,5,6,8,9 - hexahydro - 3H - benz(e)indene - 3,7 - (3aH) - dione**
$C_{19}H_{23}ClO_2$
For complete entry see 28.8

51.7   **19 - Hydroxy - 4 - androsten - 3,17 - dione**
$C_{19}H_{26}O_3$
W.L.Duax, Y.Osawa
Amer. Cryst. Assoc., Abstr. Papers (Winter Meeting), 94, 1973

51.8   **5α - Androstan - 3,17 - dione**
$C_{19}H_{28}O_2$
V.M.Coiro, E.Giglio, R.Puliti   *Stockholm Symposium*, 67, 1973

51.9 17α - **Hydroxyandrost - 4 - en - 3 - one**
Epitestosterone
$C_{19}H_{28}O_2$
N.W.Isaacs, W.D.S.Motherwell, J.C.Coppola, O.Kennard
J. C. S. Perkin II, 2335, 1972

51.10   **Testosterone**
$C_{19}H_{28}O_2$
R.C.Pettersen, P.J.Roberts, G.M.Sheldrick, O.Kennard
Amer. Cryst. Assoc., Abstr. Papers (Winter Meeting), 24, 1973

51.11   **Testosterone monohydrate**
$C_{19}H_{28}O_2$, H_2O
B.Busetta, C.Courseille, F.Leroy, M.Hospital
Acta Cryst. (B), **28,** 3293, 1972

51.12 7β - **Methyl - 19 - norandrostenediol monohydrate**
$C_{19}H_{30}O_2$, H_2O
D.J.Duchamp, J.A.Campbell   *Acta Cryst. (A),* **28,** S50, 1972

51.13 $3\beta,17\beta$ - **Dihydroxy - 5α - androstane monohydrate**
$C_{19}H_{32}O_2$, H_2O
G.Precigoux, J.Fornies-Marquina   *Cryst. Struct. Comm.,* **2,** 287, 1973

51.14 17β - **Hydroxyestr - 5(10) - ene iodoacetate**
$C_{20}H_{29}IO_2$
J.Bordner, R.L.Greene, S.G.Levine, R.R.Sobti
Cryst. Struct. Comm., **2,** 55, 1973

51.15   **Steroid Nic - 10**
$C_{21}H_{22}O_4$
M.J.Begley, L.Crombie, P.J.Ham, D.A.Whiting
J. C. S. Chem. Comm., 1250, 1972
Also classified in 38

51.16 9α - **Bromocortisol**
$C_{21}H_{29}BrO_5$
C.M.Weeks, W.L.Duax, D.C.Rohrer, M.E.Wolff
Amer. Cryst. Assoc., Abstr. Papers (Winter Meeting), 25, 1973

51.17 9α - **Fluorocortisol**
$C_{21}H_{29}FO_5$
L.Dupont, O.Dideberg, H.Campsteyn   *Acta Cryst. (B),* **28,** 3023, 1972

51.C $4\alpha,8\alpha,10\beta,14\beta$ - **Tetramethyl - $\Delta^{9,11}$ - dodecahydro -**
cyclopenta(a)phenanthrene - 3,17 - dione
$C_{21}H_{30}O_2$
For complete entry see 50.9

51.18 **Progesterone**
$C_{21}H_{30}O_2$
H.Campsteyn, L.Dupont, O.Dideberg   *Acta Cryst. (B)*, **28**, 3032, 1972

51.19 **11 - Desoxycorticosterone**
$C_{21}H_{30}O_3$
O.Dideberg, H.Campsteyn, L.Dupont   *Acta Cryst. (B)*, **29**, 103, 1973

51.20 **6 - Oxo - 3β,5 - cycloandrostan - 17 - yl acetate**
$C_{21}H_{30}O_3$
R.C.Pettersen, O.Kennard, W.G.Dauben   *J. C. S. Perkin II*, 1929, 1972
Also classified in 30

51.21 **4 - Pregnene - 17α,21 - diol - 3,20 - dione**
$C_{21}H_{30}O_4$
L.Dupont, O.Dideberg, H.Campsteyn   *Acta Cryst. (B)*, **29**, 205, 1973

51.22 **Cortisol methanol solvate**
11β,17α,21 - Trihydroxy - pregn - 4 - ene - 3,20 - dione methanol solvate
$C_{21}H_{30}O_5$, CH_4O
P.J.Roberts, J.C.Coppola, N.W.Isaacs, O.Kennard
J. C. S. Perkin II, 774, 1973

51.23 **Aldosterone 18 - acetal - 20 - hemiketal**
$C_{21}H_{30}O_6$
W.L.Duax, H.Hauptman   *J. Amer. Chem. Soc.*, **94**, 5467, 1972

51.24 **3β - Acetoxy - 17α - iodo - Δ⁵ - androstene**
$C_{21}H_{31}IO_2$
H.-C.Mez, G.Rihs   *Helv. Chim. Acta*, **55**, 375, 1972

51.25 **20(S) - Hydroxypregn - 4 - en - 3 - one**
20(S) - Hydroxyprogesterone
$C_{21}H_{32}O_2$
N.W.Isaacs, W.D.S.Motherwell, J.C.Coppola, O.Kennard
J. C. S. Perkin II, 2331, 1972

51.26 **17 - Ethylenedioxy - 3 - methoxy - 6,7,8 - methylidyne - 1,3,5(10) - estratriene**
$C_{22}H_{26}O_3$
H.P.Weber, E.Galantay   *Helv. Chim. Acta*, **55**, 544, 1972
Also classified in 31, 38

51.27 **17β - Bromoacetoxy - 3 - methoxy - 8α - methyl - 1,3,5(10),6 - estratetraene**
$C_{22}H_{27}BrO_3$
H.P.Weber, E.Galantay   *Helv. Chim. Acta*, **55**, 544, 1972

51.28 9α - Fluoro - 16α - methyl - $11\beta,17\alpha,21\beta$ - trihydroxy - 1,4 - pregnadiene -
3,20 - dione
Dexamethasone
$C_{22}H_{29}FO_5$
D.C.Rohrer, W.L.Duax
Amer. Cryst. Assoc., Abstr. Papers (Winter Meeting), 25, 1973

51.29   19 - Hydroxy - 3,17 - di(acetoxy) - androstan - $14\alpha,15\alpha$ - epoxide
$C_{23}H_{34}O_5$
G.Kruger, G.I.Birnbaum   *Tetrahedron Letters*, 1501, 1973

51.30   19,19 - Dimethoxy - 17β - acetoxy - 4 - androsten - 3 - one irradiation
product
$C_{23}H_{34}O_5$
D.S.Jones, I.L.Karle
Amer. Cryst. Assoc., Abstr. Papers (Winter Meeting), 23, 1973
Also classified in 38

51.31 $3\beta,17\beta$ - Diacetoxy - $14\alpha,15\alpha$ - oxido - 19 - hydroxy - $5\alpha,8\beta$ - androstane
$C_{23}H_{34}O_6$
G.I.Birnbaum
Amer. Cryst. Assoc., Abstr. Papers (Winter Meeting), 23, 1973

51.32 $6\alpha,9\alpha$ - Difluoro - $11\beta,16\alpha,17\alpha,21$ - tetrahydroxy - pregna - 1,4 - diene - 3,20 -
dione 16,17 - acetonide
Synalar
$C_{24}H_{30}F_2O_6$
W.E.Thiessen, A.T.Christensen   *Stockholm Symposium*, 68, 1973

51.33 7α - Acetylthio - 3 - oxo - 17α - pregn - 4 - ene 21,17β - carbolactone
Spironolactone
$C_{24}H_{32}O_4S$
O.Dideberg, L.Dupont   *Acta Cryst. (B)*, **28,** 3014, 1972

51.C   **Deoxycholic acid - p - di - iodobenzene**
$2C_{24}H_{40}O_4$, $C_6H_4I_2$
For complete entry see 60.40

51.C   **Deoxycholic acid - phenanthrene**
$3C_{24}H_{40}O_4$, $C_{14}H_{10}$
For complete entry see 60.41

51.C   **Cholic acid - ethanol complex**
$C_{24}H_{40}O_5$, C_2H_6O
For complete entry see 60.42

51.34 $14\alpha,17\alpha$ - Etheno - 15,16 - di(trifluoromethyl) - 4,15 - pregnadiene - 3,20 -
dione
$C_{25}H_{26}F_6O_2$
G.I.Birnbaum   *Acta Cryst. (B)*, **29,** 54, 1973
Also classified in 31

51.35 2β - Methyl - 19 - nortestosterone p - bromobenzenesulfonate
$C_{25}H_{31}BrO_4S$
V.Cody, W.L.Duax   *Cryst. Struct. Comm.*, **1,** 439, 1972

51.36   Androst - 4 - en - 17β - ol brosylate
$C_{25}H_{33}BrO_3S$
J.Bordner, S.G.Levine, Y.Mazur, L.R.Morrow
Cryst. Struct. Comm., **2,** 59, 1973

51.37   3,20 - bis(Ethylenedioxy) - pregna - 5,7 - diene
$C_{25}H_{36}O_4$
P.B.Braun, J.Hornstra, C.Knobler, E.W.M.Rutten, C.Romers
Acta Cryst. (B), **29,** 463, 1973
Also classified in 38

51.38   3,20 - bis(Ethylenedioxy) - 9,10 - seco - pregna - 5,7,10(19) - triene
$C_{25}H_{36}O_4$
C.Knobler, C.Romers, P.B.Braun, J.Hornstra
Acta Cryst. (B), **28,** 2097, 1972

51.39 9α - Cyano - 19 - nortestosterone p - bromobenzoate
$C_{26}H_{28}BrNO_3$
R.E.Ireland, G.Pfister, D.J.Dawson, D.Dennis, R.H.Stanford
Synthetic Comm., **2,** 175, 1972

51.40 17β - Benzoyloxy - 3 - oxo - 4 - androsten - 19 - al
$C_{26}H_{30}O_4$
W.L.Duax, Y.Osawa
Amer. Cryst. Assoc., Abstr. Papers (Winter Meeting), 94, 1973

51.41   6 - Oxo - $3\alpha,5$ - cycloandrostan - 17 - yl p - bromobenzoate
$C_{26}H_{31}BrO_3$
R.C.Pettersen, O.Kennard, W.G.Dauben   *J. C. S. Perkin II,* 1929, 1972
Also classified in 30

51.42 3α - Hydroxy - 12 - methyl - 18 - nor - $5\beta,17\alpha$ - chola - 8,11,13 - trien - 24 - oic acid methyl ester iodoacetate
$C_{27}H_{37}IO_4$
J.Meney, Y.-H.Kim, R.Stevenson, T.N.Margulis   *Tetrahedron,* **29,** 21, 1973

51.43   (23R) - 23 - Hydroxy - $3\alpha,5\alpha$ - cycloergost - 7 - en - 6 - one
$C_{28}H_{44}O_2$
M.B.Hursthouse, S.Neidle   *J. C. S. Perkin II,* 781, 1973
Also classified in 30

51.C   N - Ethylbaikeine hydrobromide monohydrate
$C_{29}H_{50}NO_3^+$, Br^- , H_2O
For complete entry see 58.43

51.44   **Nic - 3 acetate**
$C_{30}H_{42}O_7$
M.J.Begley, L.Crombie, P.J.Ham, D.A.Whiting
J. C. S. Chem. Comm., 1108, 1972
Also classified in 56

51.45   **17 - Hydroxyprogesterone 17 - (10 - chloro - 9 - ketodecanoate)**
$C_{31}H_{45}ClO_5$
W.H.Watson, K.T.Go, R.H.Purdy   *Acta Cryst. (B)*, **29**, 199, 1973

51.46   **17 - Hydroxyprogesterone 17 - (10 - hydroxy - 9 - ketodecanoate)**
$C_{31}H_{46}O_6$
W.H.Watson, K.T.Go, R.H.Purdy   *Acta Cryst. (B)*, **29**, 199, 1973

51.C   **Datiscoside bis(p - iodobenzoate) dihydrate (absolute configuration)**
$C_{52}H_{60}I_2O_{14}$, $2H_2O$
For complete entry see 59.34

MONOTERPENES

52.1 **(3R,4S,7S) - trans,trans - 3,7 - Dimethyl - 1,8,8 - tribromo - 3,4,7 - trichloro - 1,5 - octadiene (absolute configuration)**
$C_{10}H_{12}Br_3Cl_3$
D.J.Faulkner, M.O.Stallard, J.Fayos, J.Clardy
J. Amer. Chem. Soc., **95**, 3413, 1973

52.2 **Anhydro - bromo - nitro - camphane**
$C_{10}H_{13}BrNO$
G.L.Dwivedi, R.C.Srivastava   *Acta Cryst. (B)*, **28**, 2567, 1972
Also classified in 40

52.3 **$2\alpha,4\alpha$ - Dibromo - 10β - pinan - 3 - one**
$C_{10}H_{14}Br_2O$
P.P.Williams   *Cryst. Struct. Comm.*, **2**, 15, 1973

52.4 **(+) - 3 - Diazocamphor**
$C_{10}H_{14}N_2O$
A.F.Cameron, N.J.Hair, D.G.Morris   *J. C. S. Perkin II*, 1331, 1972

52.C **(+) - α - Naphthyl - phenyl - 1 - menthoxy - methoxy - silane (absolute configuration)**
$C_{27}H_{34}O_2Si$
For complete entry see 63.12

SESQUITERPENES

53.1   **Miscandenin**
$C_{15}H_{14}O_5$
P.J.Cox, G.A.Sim, J.S.Roberts, W.Herz *J. C. S. Chem. Comm.*, 428, 1973

53.2   **Dihydromikanolide**
$C_{15}H_{16}O_6$
P.J.Cox, G.A.Sim, J.S.Roberts, W.Herz *J. C. S. Chem. Comm.*, 428, 1973

53.3   **Chanootin**
$C_{15}H_{18}O_3$
B.Karlsson, A.-M.Pilotti, A.-C.Wiehager *Acta Cryst. (B)*, **29,** 1209, 1973

53.4   **Dihydrofukinolidol sulfite (stereoisomer S - 2, absolute configuration)**
$C_{15}H_{22}O_5S$
A.Furusaki, T.Watanabe *Bull. Chem. Soc. Jap.*, **45,** 2288, 1972

53.5   **Dihydrofukinolidol sulfite (stereoisomer S - 1, absolute configuration)**
$C_{15}H_{22}O_5S$
A.Furusaki, T.Watanabe *Bull. Chem. Soc. Jap.*, **45,** 2288, 1972

53.6   **Breynolide (absolute configuration)**
$C_{15}H_{22}O_7S$
K.Sasaki, Y.Hirata *Tetrahedron Letters,* 2439, 1973

53.7   **7 - Bromo - cyclo(3.15)longifolane (absolute configuration)**
$C_{15}H_{23}Br$
J.C.Thierry, R.Weiss *Acta Cryst. (B)*, **28,** 3228, 1972

53.8   **ω - Bromo - longifolene**
$C_{15}H_{23}Br$
J.C.Thierry, R.Weiss *Acta Cryst. (B)*, **28,** 3249, 1972

53.9   **3α - Bromo - longifolene**
$C_{15}H_{23}Br$
J.C.Thierry, R.Weiss *Acta Cryst. (B)*, **28,** 3249, 1972

53.10   **Humulene diepoxide**
$C_{15}H_{24}O_2$
M.E.Cradwick, P.D.Cradwick, G.A.Sim *J. C. S. Perkin II,* 404, 1973

53.11   3α - Bromo - (7βH)longifolane (absolute configuration)
C$_{15}$H$_{25}$Br
J.C.Thierry, R.Weiss   *Acta Cryst. (B)*, **28**, 3234, 1972

53.12   Pseudoivalin bromoacetate
C$_{17}$H$_{21}$BrO$_4$
G.D.Anderson, R.Gitany, R.S.McEwen, W.Herz
Tetrahedron Letters, 2409, 1973

53.13   Centaurepensin
C$_{19}$H$_{24}$Cl$_2$O$_7$
A.T.Hewson, R.C.Pettersen, O.Kennard
Cryst. Struct. Comm., **1**, 383, 1972

53.14   Melampodin (neutron study)
C$_{21}$H$_{24}$O$_9$
I.Bernal, S.F.Watkins   *Science*, **178**, 1282, 1972

53.15   Elephantol p - bromobenzoate (absolute configuration)
C$_{22}$H$_{19}$BrO$_7$
A.T.McPhail, G.A.Sim   *J. C. S. Perkin II*, 1313, 1972

53.16   15 - 7αH - Longifolyl p - bromobenzoate (absolute configuration)
C$_{22}$H$_{29}$BrO$_2$
J.C.Thierry, R.Weiss   *Acta Cryst. (B)*, **28**, 3241, 1972

53.17   Enhydrin bromohydrin
C$_{23}$H$_{29}$BrO$_{10}$
G.Kartha, K.T.Go, B.S.Joshi   *J. C. S. Chem. Comm.*, 1327, 1972

53.18   Molephantin p - bromobenzenesulfonate (absolute configuration)
C$_{25}$H$_{25}$BrO$_8$S
K.-H.Lee, H.Furukawa, M.Kozuka, H.-C.Huang, P.A.Luhan,
A.T.McPhail   *J. C. S. Chem. Comm.*, 476, 1973

53.19   13β - p - Bromophenylthio - 11α,13 - dihydropulchellin - C diacetate
(absolute configuration)
C$_{25}$H$_{29}$BrO$_6$S
M.Currie, G.A.Sim   *J. C. S. Perkin II*, 400, 1973

DITERPENES

54.1   **Crocetindialdehyde**
2,6,11,15 - Tetramethyl - hexadeca - 2,4,6,8,10,12,14 - heptaene - 1,16 - dial
$C_{20}H_{24}O_2$
J.Hjortas   *Acta Cryst. (B),* **28,** 2252, 1972

54.2   **11 - cis - Retinal**
$C_{20}H_{28}O$
R.D.Gilardi, I.L.Karle, J.Karle   *Acta Cryst. (B),* **28,** 2605, 1972

54.3   **Levopimaric acid**
$C_{20}H_{30}O_2$
I.L.Karle   *Acta Cryst. (B),* **28,** 2000, 1972

54.4   **Leucothol A (absolute configuration)**
$C_{20}H_{30}O_3$
A.Furusaki, N.Hamanaka, H.Miyakoshi, T.Okuno, T.Matsumoto
Chem. Letters, 783, 1972

54.5   **Grayanotoxin XV**
$C_{20}H_{30}O_4$
N.Hamanaka, H.Miyakoshi, A.Furusaki, T.Matsumoto
Chem. Letters, 779, 1972

54.6   **Leucothol B monohydrate**
$C_{20}H_{32}O_5$, H_2O
N.Hamanaka, H.Miyakoshi, A.Furusaki, T.Matsumoto
Chem. Letters, 787, 1972

54.7   **Condinndiol (absolute configuration)**
$C_{20}H_{35}BrO_2$
J.J.Sims, G.H.Y.Lin, R.M.Wing, W.Fenical
J. C. S. Chem. Comm., 470, 1973

54.8   **Nagilactone A diacetate**
$C_{23}H_{28}O_8$
Y.Hayashi, T.Sakan, K.Hirotsu, A.Shimada   *Chem. Letters,* 349, 1972

54.9   **Barbatusin p - bromobenzoyl ester benzene solvate (absolute configuration)**
$C_{31}H_{33}BrO_9$, C_6H_6
A.H.-J.Wang, I.C.Paul, R.Zelnik, K.Mizuta, D.Lavie
J. Amer. Chem. Soc., **95,** 598, 1973

54.10   **Clerodendrin A p - bromobenzoate chlorohydrin ethanol solvate (absolute configuration)**

$C_{38}H_{46}BrClO_{13}$, C_2H_6O

N.Kato, K.Munakata, C.Katayama   *J. C. S. Perkin II,* 69, 1973

SESTERTERPENES

No entries in this volume.

TRITERPENES

56.1   **Hopane derivative H**
$C_{27}H_{46}$
G.W.Smith, E.V.Whitehead   *Stockholm Symposium*, 88, 1973

56.2   **Maitenina hemihydrate**
$C_{28}H_{36}O_3$, $0.5H_2O$
S.Cerrini, W.Fedeli, A.Vaciago   *Abstr. Ital.-Yug. Congr*, 131, 1973

56.3   **Hopane derivative D**
$C_{29}H_{50}$
G.W.Smith, E.V.Whitehead   *Stockholm Symposium*, 88, 1973
Also classified in 30

56.C   **Nic - 3 acetate**
$C_{30}H_{42}O_7$
For complete entry see 51.44

56.4   **Campanulin**
$C_{30}H_{50}O$
J.D.White, J.Fayos, J.Clardy   *J. C. S. Chem. Comm.*, 357, 1973

56.5   **Oleanane derivative E**
$C_{30}H_{52}$
G.W.Smith, E.V.Whitehead   *Stockholm Symposium*, 88, 1973

56.6   **Oleanane derivative F (triclinic form)**
$C_{30}H_{52}$
G.W.Smith, E.V.Whitehead   *Stockholm Symposium*, 88, 1973

56.7   **Oleanane derivative F (orthorhombic form)**
$C_{30}H_{52}$
G.W.Smith, E.V.Whitehead   *Stockholm Symposium*, 8, 1973

56.8   **Retigeranic acid p - bromoanilide (absolute configuration)**
$C_{31}H_{42}BrNO$
M.Kaneda, R.Takahashi, Y.Iitaka, S.Shibata
Tetrahedron Letters, 4609, 1972

56.9 **Abieslactone**
(23R) - 3 - α - Methoxy - $5\alpha,9\beta$ - lanosta - 7,24 - diene - 26,23 - lactone
$C_{31}H_{48}O_3$
F.H.Allen, N.W.Isaacs, O.Kennard, W.D.S.Motherwell
J. C. S. Perkin II, 498, 1973

56.10 **Glabretal iodo - chloro derivative**
$C_{33}H_{48}ClIO_7$
G.Ferguson, P.A.Gunn, W.C.Marsh, R.McCrindle, R.Restivo,
J.D.Connolly, J.W.B.Fulke, M.S.Henderson
J. C. S. Chem. Comm., 159, 1973

56.11 **Glabretal dichloro derivative**
$C_{33}H_{48}Cl_2O_7$
G.Ferguson, P.A.Gunn, W.C.Marsh, R.McCrindle, R.Restivo,
J.D.Connolly, J.W.B.Fulke, M.S.Henderson
J. C. S. Chem. Comm., 159, 1973

56.12 **3 - O - Acetyl - 16 - O - p - bromobenzoylpachysandiol - B (absolute configuration)**
$C_{39}H_{57}BrO_4$
T.Kikuchi, M.Niwa, N.Masaki   *Tetrahedron Letters,* 5249, 1972

56.13 **Pristimerol bis(p - bromobenzoate)**
$C_{44}H_{48}Br_2O_6$
P.J.Ham, D.A.Whiting   *J. C. S. Perkin I,* 330, 1972

TETRATERPENES

57.1   **Vitamin A acid (monoclinic form)**
$C_{20}H_{28}O_2$
C.H.Stam   *Acta Cryst. (B)*, **28,** 2936, 1972

57.2   **Vitamin A acid (triclinic form)**
$C_{20}H_{28}O_2$
C.H.Stam   *Acta Cryst. (B)*, **28,** 2936, 1972

ALKALOIDS

58.1 **Harmidine**
Harmaline
$C_{13}H_{14}N_2O$
A.-ur-Rahman, E.Foresti Serantoni, L.Riva di Sanseverino
Pakistan J. Sci. Ind. Res., **14,** 490, 1971

58.2 **6 - Hydroxy - 2' - (2 - methylpropyl) - 3,3' - spirotetrahydropyrrolidino - oxindole**
$C_{15}H_{20}N_2O_2$
M.N.G.James, G.J.B.Williams   *Canad. J. Chem.*, **50,** 2407, 1972

58.3 **Eserine**
Physostigmine
$C_{15}H_{21}N_3O_2$
T.J.Petcher, P.Pauling   *Nature,* **241,** 277, 1973

58.4 **Porantherine hydrobromide (absolute configuration)**
$C_{15}H_{24}N^+$, Br^-
W.A.Denne, A.McL.Mathieson   *J. Cryst. Mol. Struct.*, **3,** 79, 1973

58.5 **Nupharidine hydrobromide (absolute configuration)**
$C_{15}H_{24}NO_2^+$, Br^-
J.Ohrt, R.Parthasarathy, R.T.LaLonde, C.F.Wong
J. Cryst. Mol. Struct., **3,** 3, 1973

58.6 **Lupanine - N - oxide perchlorate**
$C_{15}H_{25}N_2O_2^+$, ClO_4^-
Z.Kaluski, A.I.Gusiev, Yu.T.Struchkov, J.Skolik, P.Baranowski,
M.Wiewiorowski   *Bull. Acad. Polon. Sci., Ser. Sci. Chim.*, **20,** 1, 1972

58.7 **Porantheridine hydrobromide (absolute configuration)**
$C_{15}H_{28}NO^+$, Br^-
W.A.Denne, A.McL.Mathieson   *J. Cryst. Mol. Struct.*, **3,** 87, 1973

58.8 **anti - Desethyl - eburnamonine**
$C_{17}H_{18}N_2O$
M.Cesario, C.Pascard-Billy   *Acta Cryst. (B),* **29,** 529, 1973

58.9 **syn - Desethyl - eburnamonine**
$C_{17}H_{18}N_2O$
M.Cesario, C.Pascard-Billy   *Acta Cryst. (B),* **29,** 529, 1973

58.10 **Anthramycin methyl ester monohydrate**
$C_{17}H_{19}N_3O_4$, H_2O
J.S.Sherfinski, R.E.Marsh, R.M.Sweet, L.F.Dahl
Amer. Cryst. Assoc., Abstr. Papers (Winter Meeting), 26, 1973

58.11 **6 - Desoxy - 6 - azido - dihydro - isomorphine**
$C_{17}H_{20}N_4O_2$
K.Sasvari, K.Simon *Stockholm Symposium*, 47, 1973

58.12 **Mesembranol**
$C_{17}H_{24}NO_3$
P.A.Luhan, A.T.McPhail *J. C. S. Perkin II*, 51, 1973

58.13 **1 - Hyoscyamine hydrobromide**
$C_{17}H_{24}NO_3^+$, Br^-
E.Kussather, J.Haase *Acta Cryst. (B)*, **28,** 2896, 1972

58.14 **Stemonamine hydrochloride dihydrate**
$C_{18}H_{24}NO_4^+$, Cl^- , $2H_2O$
H.Iizuka, H.Irie, N.Masaki, K.Osaki, S.Uyeo
J. C. S. Chem. Comm., 125, 1973

58.15 **Lycorenine methiodide**
$C_{19}H_{26}NO_4^+$, I^-
J.Clardy, J.A.Chan, W.C.Wildman *J. Org. Chem.*, **37,** 49, 1972

58.16 **Swazine methiodide**
$C_{19}H_{26}NO_6^+$, I^-
M.Laing, P.Sommerville *Tetrahedron Letters*, 5183, 1972

58.17 **Dihydroisohistrionicotoxin hydrochloride**
7 - (cis - 1 - Buten - 3 - ynyl) - 8 - hydroxy - 2 - (3,4 - pentadienyl) - 1 -
azaspiro(5.5)undecane hydrochloride
$C_{19}H_{28}NO^+$, Cl^-
I.L.Karle *Amer. Cryst. Assoc., Abstr. Papers (Winter Meeting)*, 22, 1973

58.18 **6,7 - Dimethoxy - 1 - veratryl - isoquinoline**
Papaverine
$C_{20}H_{21}NO_4$
R.F.Baggio, S.Baggio *Cryst. Struct. Comm.*, **2,** 251, 1973

58.19 **Cryptostyline I methiodide (absolute configuration)**
$C_{20}H_{24}NO_4^+$, I^-
L.Westin *Acta Chem. Scand.*, **26,** 2305, 1972

58.20 **Sceletium alkaloid A_4**
$C_{20}H_{24}N_2O_2$
P.A.Luhan, A.T.McPhail *J. C. S. Perkin II*, 2006, 1972

ALKALOIDS

58.21 P - Bromobenzoyl dihydroluciduline
$C_{20}H_{26}BrNO_2$
D.Hall, N.Masaki   *Cryst. Struct. Comm.*, **2**, 271, 1973

58.22 Unnatural cryptostyline II hydrobromide (absolute configuration)
$C_{20}H_{26}NO_4^+$, Br^-
J.F.Blount, V.Toome, S.Teitel, A.Brossi   *Tetrahedron*, **29**, 31, 1973

58.23 Lysergic acid diethylamide o - iodobenzoate monohydrate
$C_{20}H_{26}N_3O^+$, $C_7H_4IO_2^-$, H_2O
R.W.Baker, C.Chothia, P.Pauling, H.P.Weber   *Science*, **178**, 614, 1972

58.24 Podopetaline hydrobromide (absolute configuration)
$C_{20}H_{34}N_3^+$, Br^-
N.K.Hart, S.R.Johns, J.A.Lamberton, M.F.Mackay, A.McL.Mathieson, L.Satzke   *Tetrahedron Letters*, 5333, 1972

58.25 5 - Bromo - 12S - tetrahydroaustamide
$C_{21}H_{24}BrN_3O_3$
J.Coetzer, P.S.Steyn   *Acta Cryst. (B)*, **29**, 685, 1973

58.26 N - Desacetyl - N - methyl - colchicine
$C_{21}H_{25}NO_5$
T.N.Margulis
Amer. Cryst. Assoc., Abstr. Papers (Winter Meeting), 95, 1973

58.27 Cularine methiodide (absolute configuration)
$C_{21}H_{26}NO_4^+$, I^-
T.Kametani, T.Honda, H.Shimanouchi, Y.Sasada
J. C. S. Chem. Comm., 1072, 1972

58.28 Clivorine monohydrate (at −160°C)
$C_{21}H_{27}NO_7$, H_2O
K.B.Birnbaum   *Acta Cryst. (B)*, **28**, 2825, 1972

58.29 Yohimbine hydrochloride (absolute configuration)
$C_{21}H_{27}N_2O_3^+$, Cl^-
G.Ambady, G.Kartha   *J. Cryst. Mol. Struct.*, **3**, 37, 1973

58.30 Narciclasine tetra - acetate
$C_{22}H_{21}NO_{11}$
A.Immirzi   *Cryst. Struct. Comm.*, **2**, 359, 1973

58.31 Daphnilactone B benzene solvate
$C_{22}H_{31}NO_2$, $0.5C_6H_6$
K.Sasaki, Y.Hirata   *Acta Cryst. (B)*, **29**, 547, 1973

58.32 Oxotuberostemonine
$C_{22}H_{31}NO_5$
C.S.Huber   *Acta Cryst. (B)*, **28**, 2015, 1972

58.33 **Miyaconitine hydrobromide dihydrate (absolute configuration)**
$C_{23}H_{30}NO_6^+$, Br^- , $2H_2O$
H.Shimanouchi, Y.Sasada, T.Takeda
Bull. Chem. Soc. Jap., **44,** 3020, 1971

58.34 **Protostemonine methanol solvate**
$C_{23}H_{31}NO_6$, CH_4O
K.Ishizuka, K.Ohno, T.Taga, N.Masaki   *Acta Cryst. (A)*, **28,** S20, 1972

58.35 **Daphnilactone A**
$C_{23}H_{35}NO_2$
K.Sasaki, Y.Hirata   *J. C. S. Perkin II*, 1411, 1972

58.36 **Alkaloid CC - 2 methiodide (from colchicum cornigerum,absolute configuration)**
$C_{24}H_{32}NO_6^+$, I^-
A.F.Cameron, C.Hannaway   *J. C. S. Perkin II*, 1002, 1973

58.37 **Protostemonine methiodide**
$C_{24}H_{34}NO_6^+$, I^-
K.Ishizuka, K.Ohno, T.Taga, N.Masaki   *Acta Cryst. (A)*, **28,** S20, 1972

58.38 **Cephalotaxine p - bromobenzoate (absolute configuration)**
$C_{25}H_{24}BrNO_5$
S.K.Arora, R.B.Bates, R.A.Grady, R.G.Powell
Stockholm Symposium, 90, 1973

58.39 **Demissidine hydroiodide ethanol solvate**
5α - Solanidan - 3β - ol hydroiodide ethanol solvate
$C_{27}H_{46}NO^+$, I^- , $0.5C_2H_6O$
E.Hohne   *J. Prakt. Chem.*, **314,** 371, 1972

58.40 **Tetrahydroveralkamine - 16 - one hydroiodide**
$C_{27}H_{48}NO_2^+$, I^-
E.Hohne, I.Seidel, G.Adam, K.Schreiber, J.Tomko
Tetrahedron, **28,** 4019, 1972

58.41 **Corynoline p - bromobenzoate**
$C_{28}H_{18}BrNO_6$
T.Kametani, T.Honda, M.Ihara, H.Shimanouchi, Y.Sasada
Tetrahedron Letters, 3729, 1972

58.42 **Bromolythranine hydrobromide ethanol solvate (absolute configuration)**
$C_{28}H_{37}BrNO_5^+$, Br^- , C_2H_6O
R.J.McClure, G.A.Sim   *J. C. S. Perkin II*, 2073, 1972

58.43 **N - Ethylbaikeine hydrobromide monohydrate**
$C_{29}H_{50}NO_3^+$, Br^- , H_2O
S.Ito, M.Miyashita, Y.Fukazawa, A.Mori, I.Iwai, M.Yoshimura
Tetrahedron Letters, 2961, 1972
Residue 1 also classified in 51

58.44 **Isocinchophyllamine methanol solvate**
$C_{31}H_{36}N_4O_2$, $2CH_4O$
J.Guilhem *Acta Cryst. (A)*, **28**, S19, 1972

58.45 **Zygacine acetonide hydroiodide acetone solvate (absolute configuration)**
$C_{32}H_{50}NO_8^+$, I^- , $2C_3H_6O$
R.F.Bryan, R.J.Restivo, S.M.Kupchan *J. C. S. Perkin II*, 386, 1973

58.46 **D - Tubocurarine dichloride pentahydrate**
$C_{37}H_{42}N_2O_6^{2+}$, $2Cl^-$, $5H_2O$
P.W.Codding, M.N.G.James *Acta Cryst. (B)*, **29**, 935, 1973

58.47 **Maytansine (3 - bromopropyl) ether (absolute configuration)**
$C_{37}H_{51}BrClN_3O_{10}$
R.F.Bryan, C.J.Gilmore, R.C.Haltiwanger *J. C. S. Perkin II*, 897, 1973

58.48 **O,O' - Trimethyl - D - tubocurarine di - iodide**
$C_{40}H_{48}N_2O_6^{2+}$, $2I^-$
H.M.Sobell, T.D.Sakore, S.S.Tavale, F.G.Canepa, P.Pauling, T.J.Petcher
Proc. Nation. Acad. Sci. U. S. A., **69**, 2212, 1972

58.49 **Dibromo - vobtusine**
$C_{43}H_{46}BrN_4O_6$
O.Lefebvre-Soubeyran *Acta Cryst. (A)*, **28**, S12, 1972

MISCELLANEOUS NATURAL PRODUCTS

59.1   **Calcium phosphoryl choline chloride**
$C_5H_{13}NO_4P^-$, Ca^{2+} , Cl^-
C.N.Caughlan, C.K.Wang, A.Fitzgerald *Acta Cryst. (A),* **28,** S40, 1972
Residue 1 also classified in 3, 46

59.C   **Calcium(+) - allo - hydroxycitrate lactone tetrahydrate**
Hibiscus acid calcium salt tetrahydrate
$C_6H_4O_7^{2-}$, Ca^{2+} , $4H_2O$
For complete entry see 2.26

59.2   **Carbamoyl - choline chloride**
$C_6H_{15}N_2O_2^+$, Cl^-
B.Jensen *Eur. Cryst. Meeting,* 1973
Residue 1 also classified in 3

59.3   **Carbamoyl - choline iodide**
$C_6H_{15}N_2O_2^+$, I^-
B.Jensen *Eur. Cryst. Meeting,* 1973
Residue 1 also classified in 3

59.4   **5 - Hydroxydopamine hydrochloride**
3,4,5 - Trihydroxyphenyl - ethylamine hydrochloride
$C_8H_{12}NO_3^+$, Cl^-
A.M.Andersen, A.Mostad, C.Romming
Acta Chem. Scand., **26,** 2670, 1972
Residue 1 also classified in 3, 17

59.5   **Sceleratinic acid**
$C_{10}H_{13}ClO_4$
J.Coetzer, A.Wiechers *Acta Cryst. (B),* **29,** 917, 1973
Also classified in 38

59.6   **Xanthotoxin**
8 - Methoxy - 3,'2' - 6,7 - furocoumarin
$C_{12}H_8O_4$
N.R.Stemple, W.H.Watson *Acta Cryst. (B),* **28,** 2485, 1972
Also classified in 38

59.7   **Phenyl β - methylcholine ether bromide**
$C_{12}H_{20}NO^+$, Br^-
A.J.Geddes, B.Sheldrick   *Eur. Cryst. Meeting*, 1973
Residue 1 also classified in 17, 3

59.8   **o - Methylphenyl - β - methylcholine ether bromide**
$C_{13}H_{22}NO^+$, Br^-
A.J.Geddes, B.Sheldrick   *Eur. Cryst. Meeting*, 1973
Residue 1 also classified in 17, 3

59.9   **4,9 - Dimethoxy - 7 - methyl - 5H - furo(3,2 - g)benzopyran - 5 - one**
Khellin
$C_{14}H_{12}O_5$
J.P.Beale   *Cryst. Struct. Comm.*, **2**, 123, 1973
Also classified in 38

59.10   **Succinyl - choline picrate**
$C_{14}H_{30}N_2O_4{}^{2+}$, $2C_6H_2N_3O_7{}^-$
B.Jensen   *Eur. Cryst. Meeting*, 1973
Residue 1 also classified in 3; residue 2 classified in 15, 17

59.11   **Butein monohydrate**
$C_{15}H_{12}O_5$, H_2O
N.Saito, K.Ueno, Y.Sasada   *Bull. Chem. Soc. Jap.*, **45**, 2274, 1972
Residue 1 also classified in 17

59.12   **Anthrotaxin monohydrate**
$C_{17}H_{16}O_6$, H_2O
A.-M.Pilotti   *Acta Cryst. (B)*, **28**, 2123, 1972
Residue 1 also classified in 38

59.13   **Axivalin monohydrate**
$C_{17}H_{22}O_5$, H_2O
G.D.Anderson, R.S.McEwen, W.Herz   *Tetrahedron Letters*, 4423, 1972
Residue 1 also classified in 38

59.14   **Usnic acid (at $-110°C$)**
$C_{18}H_{16}O_7$
R.Norrestam, M.von Glehn   *Stockholm Symposium*, 79, 1973
Also classified in 38

59.15   **Mesylmethynolide**
$C_{18}H_{30}O_7S$
R.E.Hughes, C.-C.Tsai   *Stockholm Symposium*, 92, 1973
Also classified in 38

59.16   **Strigol**
$C_{19}H_{22}O_6$
P.Coggon, P.A.Luhan, A.T.McPhail   *J. C. S. Perkin II*, 465, 1973
Also classified in 38

59.17   **Averufin.**
$C_{20}H_{16}O_7$
Y.Katsube, T.Tsukihara, N.Tanaka, K.Ando, T.Hamasaki, Y.Hatsuda
Bull. Chem. Soc. Jap., **45,** 2091, 1972
Also classified in 38

59.18   **Triptolide (absolute configuration)**
$C_{20}H_{24}O_6$
C.J.Gilmore, R.F.Bryan   *J. C. S. Perkin II,* 816, 1973
Also classified in 38

59.19   **Tripdiolide**
$C_{20}H_{24}O_7$
C.J.Gilmore, R.F.Bryan   *J. C. S. Perkin II,* 816, 1973
Also classified in 38

59.20   **Monobromo - dehydro - bispulegone**
$C_{20}H_{29}BrO$
S.Martinez-Carrera, J.M.Franco   *Acta Cryst. (A),* **28,** S20, 1972
Also classified in 28

59.21   **Stemodinone**
$C_{20}H_{32}O_2$
P.S.Manchand, J.D.White, H.Wright, J.Clardy
J. Amer. Chem. Soc., **95,** 2705, 1973
Also classified in 31

59.22   **Prostaglandin A$_1$**
$C_{20}H_{32}O_4$
J.W.Edmonds, W.L.Duax
Amer. Cryst. Assoc., Abstr. Papers (Winter Meeting), 22, 1973

59.23   **7 - Methylthio - 7 - phenylacetamido - deacetoxycephalosporanic acid t - butyl ester**
$C_{21}H_{26}N_2O_4S_2$
W.A.Slusarchyk, H.E.Applegate, P.Funke, W.Koster, M.S.Puar, M.Young,
J.E.Dolfini   *J. Org. Chem.*, **38,** 943, 1973
Also classified in 41

59.24   **2' - Bromopodophyllotoxin ethyl acetate solvate (absolute configuration)**
$C_{22}H_{21}BrO_8$, $0.5C_4H_8O_2$
T.J.Petcher, H.P.Weber, M.Kuhn, A.von Wartburg
J. C. S. Perkin II, 288, 1973
Residue 1 also classified in 17, 38

59.25   **5' - Demethoxy - β - peltatin A methyl ether**
$C_{22}H_{22}O_7$
R.B.Bates, J.B.Wood III   *J. Org. Chem.*, **37,** 562, 1972
Also classified in 38

59.26 **Episteganol (absolute configuration)**
$C_{22}H_{22}O_8$
S.M.Kupchan, R.W.Britton, M.F.Ziegler, C.J.Gilmore, R.J.Restivo,
R.F.Bryan *J. Amer. Chem. Soc.*, **95,** 1335, 1973
Also classified in 38

59.27 **Wortmannin**
$C_{23}H_{24}O_8$
T.J.Petcher, H.-P.Weber, Z.Kis *J. C. S. Chem. Comm.*, 1061, 1972
Also classified in 38

59.28 **Leuco - thelephoric acid hexamethyl ether**
(2,3,6,8,9,12 - Hexamethoxybenzo - bis(1,2 - b.4,5 - b')benzofuran)
$C_{24}H_{22}O_8$
J.V.Silverton *Acta Cryst. (B)*, **29,** 293, 1973
Also classified in 17, 38

59.29 **(+) - Allethronyl (+) - trans - chrysanthemate 6 - bromo - 2,4 - dinitrophenyl - hydrazone (absolute configuration)**
$C_{25}H_{29}BrN_4O_6$
M.J.Begley, L.Crombie, D.J.Simmonds, D.A.Whiting
J. C. S. Chem. Comm., 1276, 1972
Also classified in 20

59.30 **Rearrangement product of a Beyer - 15(16) - en - 12 - one system**
$C_{27}H_{29}BrO_4$
M.Laing, P.Sommerville, D.Hanouskova *Acta Cryst. (A)*, **28,** S20, 1972
Also classified in 29, 31

59.31 **Dothistromin bromoethyl ether tetra - acetate (absolute configuration)**
$C_{28}H_{23}BrO_{13}$
C.A.Bear, J.M.Waters, T.N.Waters *J. C. S. Perkin II*, 2375, 1972
Also classified in 38

59.32 **Wortmannin p - bromobenzoate (absolute configuration)**
$C_{30}H_{26}BrO_9$
T.J.Petcher, H.-P.Weber, Z.Kis *J. C. S. Chem. Comm.*, 1061, 1972
Also classified in 38

59.33 **Bromo - heptamethyl - poriolide acetone solvate**
$C_{36}H_{41}BrO_{11}$, xC_3H_6O
A.Ogiso, A.Sato, S.Sato, C.Tamura *Tetrahedron Letters,* 3071, 1972

59.34 **Datiscoside bis(p - iodobenzoate) dihydrate (absolute configuration)**
$C_{52}H_{60}I_2O_{14}$, $2H_2O$
R.J.Restivo, R.F.Bryan, S.M.Kupchan *J. C. S. Perkin II*, 892, 1973
Residue 1 also classified in 51, 45

MOLECULAR COMPLEXES

60.1 **Oxalic acid - acetamide complex**
$C_2H_2O_4$, C_2H_5NO
L.Leiserowitz, F.Nader *Angew. Chem.*, **84,** 536, 1972
Residue 1 also classified in 1; residue 2 classified in 60, 1

60.C **Furamide - oxalic acid complex**
$C_2H_2O_4$, $2C_5H_5NO_2$
For complete entry see 60.3

60.C **Oxalic acid - acetamide complex**
C_2H_5NO , $C_2H_2O_4$
For complete entry see 60.1

60.C **Cholic acid - ethanol complex**
C_2H_6O , $C_{24}H_{40}O_5$
For complete entry see 60.42

60.C **3,5 - Di - iodo - L - thyronine - N - methylacetamide complex (absolute configuration)**
C_3H_7NO , $C_{15}H_{13}INO_4$
For complete entry see 60.28

60.C **Benzamide - succinic acid complex**
$C_4H_6O_4$, $2C_7H_7NO$
For complete entry see 60.9

60.C **3,5,3' - Tri - iodothyroacetic acid - N - diethanolamine**
$C_4H_{11}NO_2$, $C_{14}H_9I_3O_4$
For complete entry see 60.26

60.2 **4 - Nitropyridine - N - oxide - hydroquinone complex**
$C_5H_4N_2O_3$, $C_6H_6O_2$
M.Shiro, T.Kubota *Chem. Letters,* 1151, 1972
Residue 1 also classified in 33, 10; residue 2 classified in 60, 17

60.3 **Furamide - oxalic acid complex**
$2C_5H_5NO_2$, $C_2H_2O_4$
C.-M.Huang, L.Leiserowitz, G.M.J.Schmidt *J. C. S. Perkin II*, 503, 1973
Residue 1 also classified in 38; residue 2 classified in 60, 1

60.4   **Thymine N,N - diethylmelamine monohydrate**
$C_5H_6N_2O_2$, $C_7H_{14}N_6$, H_2O
C.Tamura, T.Hata, S.Sato *Acta Cryst. (A),* **28,** S46, 1972
Residue 1 also classified in 44; residue 2 classified in 60, 33

60.C   **Acenaphthene - tetrachloro - p - benzoquinone complex**
$C_6Cl_4O_2$, $C_{12}H_{10}$
For complete entry see 60.19

60.C   **9 - Methylanthracene - tetrachloro - p - benzoquinone complex**
$C_6Cl_4O_2$, $C_{15}H_{12}$
For complete entry see 60.27

60.C   **Hexamethylbenzene - hexafluorobenzene complex (trigonal form, at 5°C)**
C_6F_6 , $C_{12}H_{18}$
For complete entry see 60.24

60.C   **Hexamethylbenzene - hexafluorobenzene complex (triclinic form, at −40°C)**
C_6F_6 , $C_{12}H_{18}$
For complete entry see 60.25

60.C   **5,7 - (12) - Paracyclophadiyne - tetracyanoethylene complex**
C_6N_4 , $2C_{18}H_{20}$
For complete entry see 60.35

60.5   **Trinitrobenzene - benzothiophene complex**
$C_6H_3N_3O_6$, C_8H_6S
C.Pascard-Billy, R.Pascard *Eur. Cryst. Meeting,* 1973
Residue 1 also classified in 15; residue 2 classified in 60, 39

60.C   **Pyrene - 1,3,5 - trinitrobenzene complex**
$C_6H_3N_3O_6$, $C_{16}H_{10}$
For complete entry see 60.29

60.C   **Deoxycholic acid - p - di - iodobenzene**
$C_6H_4I_2$, $2C_{24}H_{40}O_4$
For complete entry see 60.40

60.6   **2,2' - Bi - 1,3 - dithiole 7,7,8,8 - tetracyanoquinodimethanide**
$C_6H_4S_4^+$, $C_{12}H_4N_4^-$
T.E.Phillips, T.J.Kistenmacher, J.P.Ferraris, D.O.Cowan
J. C. S. Chem. Comm., 471, 1973
Residue 1 also classified in 39; residue 2 classified in 60, 7

60.C   **4 - Nitropyridine - N - oxide - hydroquinone complex**
$C_6H_6O_2$, $C_5H_4N_2O_3$
For complete entry see 60.2

60.7   **p - Phenylenediamine - 1,2,4,5 - tetracyanobenzene complex**
$C_6H_8N_2$, $C_{10}H_2N_4$
H.Tsuchiya, F.Marumo, Y.Saito   *Acta Cryst. (B),* **29,** 659, 1973
Residue 1 also classified in 16; residue 2 classified in 60, 7

60.8   **N - Ethyl - 2 - methyl - thiazoline - bis(7,7,8,8 - tetracyanoquinodimethane)**
$C_6H_{12}NS^+$, $C_{12}H_4N_4^-$, $C_{12}H_4N_4$
R.P.Shibaeva, L.O.Atovmyan, V.F.Kaminskii   *Eur. Cryst. Meeting,* 1973
Residue 1 also classified in 41; residue 2 classified in 60, 7

60.C   **bis(1,1,1,5,5,5 - Hexafluoropentaen - 2,4 - dionato) copper(ii) - 1,4 - diazabicyclo(2.2.2)octane complex**
$C_6H_{12}N_2$, $C_{10}H_2CuF_{12}O_4$
For complete entry see 60.12

60.9   **Benzamide - succinic acid complex**
$2C_7H_7NO$, $C_4H_6O_4$
C.-M.Huang, L.Leiserowitz, G.M.J.Schmidt   *J. C. S. Perkin II,* 503, 1973
Residue 1 also classified in 13; residue 2 classified in 60, 1

60.C   **Amobarbital - salicylamide complex**
$C_7H_7NO_2$, $C_{11}H_{18}N_2O_3$
For complete entry see 60.13

60.10   **9 - Ethyladenine - 5,5 - diethylbarbituric acid complex**
$C_7H_9N_5$, $C_8H_{12}N_2O_3$
D.Voet   *J. Amer. Chem. Soc.,* **94,** 8213, 1972
Residue 1 also classified in 44; residue 2 classified in 60, 43

60.11   **9 - Ethyladenine - 5 - isopropyl - 5 - bromoallylbarbituric acid complex**
$C_7H_9N_5$, $C_{10}H_{15}BrN_2O$
D.Voet, A.Rich   *J. Amer. Chem. Soc.,* **94,** 5888, 1972
Residue 1 also classified in 44; residue 2 classified in 60, 43

60.C   **Thymine N,N - diethylmelamine monohydrate**
$C_7H_{14}N_6$, $C_5H_6N_2O_2$, H_2O
For complete entry see 60.4

60.C   **Trinitrobenzene - benzothiophene complex**
C_8H_6S , $C_6H_3N_3O_6$
For complete entry see 60.5

60.C   **9 - Ethyladenine - 5,5 - diethylbarbituric acid complex**
$C_8H_{12}N_2O_3$, $C_7H_9N_5$
For complete entry see 60.10

60.12   **bis(1,1,1,5,5,5 - Hexafluoropentaen - 2,4 - dionato) copper(ii) - 1,4 - diazabicyclo(2.2.2)octane complex**
$C_{10}H_2CuF_{12}O_4$, $C_6H_{12}N_2$
R.C.E.Belford, D.E.Fenton, M.R.Truter   *J. C. S. Dalton,* 2208, 1972
Residue 1 also classified in 77; residue 2 classified in 60, 37

60.C   **p - Phenylenediamine - 1,2,4,5 - tetracyanobenzene complex**
$C_{10}H_2N_4$, $C_6H_8N_2$
For complete entry see 60.7

60.C   **Pyrene - 1,2,4,5 - tetracyanobenzene complex (at 290°K)**
$C_{10}H_2N_4$, $C_{16}H_{10}$
For complete entry see 60.30

60.C   **Pyrene - 1,2,4,5 - tetracyanobenzene complex (at 178°K)**
$C_{10}H_2N_4$, $C_{16}H_{10}$
For complete entry see 60.31

60.C   **9 - Ethyladenine - 5 - isopropyl - 5 - bromoallylbarbituric acid complex**
$C_{10}H_{15}BrN_2O$, $C_7H_9N_5$
For complete entry see 60.11

60.13   **Amobarbital - salicylamide complex**
$C_{11}H_{18}N_2O_3$, $C_7H_7NO_2$
I.-N.Hsu, B.M.Craven
Amer. Cryst. Assoc., Abstr. Papers (Winter Meeting), 86, 1973
Residue 1 also classified in 43; residue 2 classified in 60, 13, 17

60.14   **7,7,8,8 - Tetracyanoquinodimethane - 1,10 - phenanthroline complex**
$C_{12}H_4N_4$, $C_{12}H_8N_2$
I.Goldberg, U.Shmueli   *Cryst. Struct. Comm.*, **2,** 175, 1973
Residue 1 also classified in 7; residue 2 classified in 60, 36

60.15   **7,7,8,8 - Tetracyanoquinodimethane - phenazine**
$C_{12}H_4N_4$, $C_{12}H_8N_2$
I.Goldberg, U.Shmueli   *Acta Cryst. (B)*, **29,** 440, 1973
Residue 1 also classified in 7; residue 2 classified in 60, 36

60.16   **7,7,8,8 - Tetracyanoquinodimethane - dibenzo - p - dioxin**
$C_{12}H_4N_4$, $C_{12}H_8O_2$
I.Goldberg, U.Shmueli   *Acta Cryst. (B)*, **29,** 432, 1973
Residue 1 also classified in 7; residue 2 classified in 60, 38

60.C   **Acenaphthene - 7,7,8,8 - tetracyanoquinodimethane complex**
$C_{12}H_4N_4$, $C_{12}H_{10}$
For complete entry see 60.20

60.C   **Benzidine - 7,7,8,8 - tetracyano - p - quinodimethane complex**
$C_{12}H_4N_4$, $C_{12}H_{12}N_2$
For complete entry see 60.21

60.C   **Benzidine - 7,7,8,8 - tetracyano - p - quinodimethane complex dichloromethane solvate**
$C_{12}H_4N_4$, $C_{12}H_{12}N_2$, $1.8CH_2Cl_2$
For complete entry see 60.22

60.C **Benzidine - 7,7,8,8 - tetracyano - p - quinodimethane complex benzene solvate**
$C_{12}H_4N_4$, $C_{12}H_{12}N_2$, $1.35C_6H_6$
For complete entry see 60.23

60.17 **7,7,8,8 - Tetracyanoquinodimethane - N,N' - dimethyldihydrophenazine (photographic data)**
$C_{12}H_4N_4$, $C_{14}H_{14}N_2$
I.Goldberg, U.Shmueli *Acta Cryst. (B)*, **29,** 421, 1973
Residue 1 also classified in 7; residue 2 classified in 60, 36

60.18 **7,7,8,8 - Tetracyanoquinodimethane - N,N' - dimethyldihydrophenazine (diffractometer data)**
$C_{12}H_4N_4$, $C_{14}H_{14}N_2$
I.Goldberg, U.Shmueli *Acta Cryst. (B)*, **29,** 421, 1973
Residue 1 also classified in 7; residue 2 classified in 60, 36

60.C **Pyrene - 7,7,8,8 - tetracyanoquinodimethane complex (data of Prout and Tickle)**
$C_{12}H_4N_4$, $C_{16}H_{10}$
For complete entry see 60.32

60.C **Pyrene - 7,7,8,8 - tetracyanoquinodimethane complex (data of Wright)**
$C_{12}H_4N_4$, $C_{16}H_{10}$
For complete entry see 60.33

60.C **Perylene - 7,7,8,8 - tetracyanoquinodimethane complex**
$C_{12}H_4N_4$, $C_{20}H_{12}$
For complete entry see 60.38

60.C **2,2' - Bi - 1,3 - dithiole 7,7,8,8 - tetracyanoquinodimethanide**
$C_{12}H_4N_4^-$, $C_6H_4S_4^+$
For complete entry see 60.6

60.C **N - Ethyl - 2 - methyl - thiazoline - bis(7,7,8,8 - tetracyanoquinodimethane)**
$C_{12}H_4N_4^-$, $C_6H_{12}NS^+$, $C_{12}H_4N_4$
For complete entry see 60.8

60.C **Dimethyl - thiacyanine bis(7,7,8,8 - tetracyanoquinodimethane)**
$C_{12}H_4N_4^-$, $C_{17}H_{15}N_2S_2^+$, $C_{12}H_4N_4$
For complete entry see 60.34

60.C **Dimethyl - thiacarbocyanine bis(7,7,8,8 - tetracyanoquinodimethane)**
$C_{12}H_4N_4^-$, $C_{19}H_{17}N_2S_2^+$, $C_{12}H_4N_4$
For complete entry see 60.36

60.C **Diethyl - thiacyanine bis(7,7,8,8 - tetracyanoquinodimethane)**
$C_{12}H_4N_4^-$, $C_{19}H_{19}N_2S_2^+$, $C_{12}H_4N_4$
For complete entry see 60.37

60.C   **Diethyl - thiacarbocyanine bis(7,7,8,8 - tetracyanoquinodimethane) (triclinic form)**

$C_{12}H_4N_4^-$, $C_{21}H_{21}N_2S_2^+$, $C_{12}H_4N_4$

For complete entry see 60.39

60.C   **7,7,8,8 - Tetracyanoquinodimethane - 1,10 - phenanthroline complex**

$C_{12}H_8N_2$, $C_{12}H_4N_4$

For complete entry see 60.14

60.C   **7,7,8,8 - Tetracyanoquinodimethane - phenazine**

$C_{12}H_8N_2$, $C_{12}H_4N_4$

For complete entry see 60.15

60.C   **7,7,8,8 - Tetracyanoquinodimethane - dibenzo - p - dioxin**

$C_{12}H_8O_2$, $C_{12}H_4N_4$

For complete entry see 60.16

60.19   **Acenaphthene - tetrachloro - p - benzoquinone complex**

$C_{12}H_{10}$, $C_6Cl_4O_2$

I.J.Tickle, C.K.Prout *J. C. S. Perkin II,* 724, 1973

Residue 1 also classified in 28; residue 2 classified in 60, 18

60.20   **Acenaphthene - 7,7,8,8 - tetracyanoquinodimethane complex**

$C_{12}H_{10}$, $C_{12}H_4N_4$

I.J.Tickle, C.K.Prout *J. C. S. Perkin II,* 727, 1973

Residue 1 also classified in 28; residue 2 classified in 60, 7

60.21   **Benzidine - 7,7,8,8 - tetracyano - p - quinodimethane complex**

$C_{12}H_{12}N_2$, $C_{12}H_4N_4$

H.Kuroda, I.Ikemoto, K.Yakushi, K.Chikaishi

Acta Cryst. (A), **28,** S15, 1972

Residue 1 also classified in 16; residue 2 classified in 60, 7

60.22   **Benzidine - 7,7,8,8 - tetracyano - p - quinodimethane complex dichloromethane solvate**

$C_{12}H_{12}N_2$, $C_{12}H_4N_4$, $1.8CH_2Cl_2$

I.Ikemoto, K.Chikaishi, K.Yakushi, H.Kuroda

Acta Cryst. (B), **28,** 3502, 1972

Residue 1 also classified in 16; residue 2 classified in 60, 7

60.23   **Benzidine - 7,7,8,8 - tetracyano - p - quinodimethane complex benzene solvate**

$C_{12}H_{12}N_2$, $C_{12}H_4N_4$, $1.35C_6H_6$

H.Kuroda, I.Ikemoto, K.Yakushi, K.Chikaishi

Acta Cryst. (A), **28,** S15, 1972

Residue 1 also classified in 16; residue 2 classified in 60, 7

60.24   **Hexamethylbenzene - hexafluorobenzene complex (trigonal form, at 5°C)**

$C_{12}H_{18}$, C_6F_6

T.Dahl *Acta Chem. Scand.,* **26,** 1569, 1972

Residue 1 also classified in 19; residue 2 classified in 60, 19

60.25 **Hexamethylbenzene - hexafluorobenzene complex (triclinic form, at −40°C)**
$C_{12}H_{18}$, C_6F_6
T.Dahl *Acta Chem. Scand.*, **27**, 995, 1973
Residue 1 also classified in 19; residue 2 classified in 60, 19

60.26 **3,5,3' - Tri - iodothyroacetic acid - N - diethanolamine**
$C_{14}H_9I_3O_4$, $C_4H_{11}NO_2$
V.Cody *Stockholm Symposium*, 48, 1973
Residue 1 also classified in 17; residue 2 classified in 60, 3

60.C **Deoxycholic acid - phenanthrene**
$C_{14}H_{10}$, $3C_{24}H_{40}O_4$
For complete entry see 60.41

60.C **7,7,8,8 - Tetracyanoquinodimethane - N,N' - dimethyldihydrophenazine (photographic data)**
$C_{14}H_{14}N_2$, $C_{12}H_4N_4$
For complete entry see 60.17

60.C **7,7,8,8 - Tetracyanoquinodimethane - N,N' - dimethyldihydrophenazine (diffractometer data)**
$C_{14}H_{14}N_2$, $C_{12}H_4N_4$
For complete entry see 60.18

60.27 **9 - Methylanthracene - tetrachloro - p - benzoquinone complex**
$C_{15}H_{12}$, $C_6Cl_4O_2$
I.J.Tickle, C.K.Prout *J. C. S. Perkin II*, 731, 1973
Residue 1 also classified in 26; residue 2 classified in 60, 18

60.28 **3,5 - Di - iodo - L - thyronine - N - methylacetamide complex (absolute configuration)**
$C_{15}H_{13}INO_4$, C_3H_7NO
V.Cody, W.L.Duax, D.A.Norton *Acta Cryst. (B)*, **28**, 2244, 1972
Residue 1 also classified in 48; residue 2 classified in 60, 1

60.29 **Pyrene - 1,3,5 - trinitrobenzene complex**
$C_{16}H_{10}$, $C_6H_3N_3O_6$
C.K.Prout, I.J.Tickle *J. C. S. Perkin II*, 734, 1973
Residue 1 also classified in 29; residue 2 classified in 60, 15

60.30 **Pyrene - 1,2,4,5 - tetracyanobenzene complex (at 290°K)**
$C_{16}H_{10}$, $C_{10}H_2N_4$
C.K.Prout, T.Morley, I.J.Tickle, J.D.Wright *J. C. S. Perkin II*, 523, 1973
Residue 1 also classified in 29; residue 2 classified in 60, 7

60.31 **Pyrene - 1,2,4,5 - tetracyanobenzene complex (at 178°K)**
$C_{16}H_{10}$, $C_{10}H_2N_4$
C.K.Prout, T.Morley, I.J.Tickle, J.D.Wright *J. C. S. Perkin II*, 523, 1973
Residue 1 also classified in 29; residue 2 classified in 60, 7

60.32 **Pyrene - 7,7,8,8 - tetracyanoquinodimethane complex (data of Prout and Tickle)**
$C_{16}H_{10}$, $C_{12}H_4N_4$
C.K.Prout, I.J.Tickle, J.D.Wright *J. C. S. Perkin II*, 528, 1973
Residue 1 also classified in 29; residue 2 classified in 60, 7

60.33 **Pyrene - 7,7,8,8 - tetracyanoquinodimethane complex (data of Wright)**
$C_{16}H_{10}$, $C_{12}H_4N_4$
C.K.Prout, I.J.Tickle, J.D.Wright *J. C. S. Perkin II*, 528, 1973
Residue 1 also classified in 29; residue 2 classified in 60, 7

60.34 **Dimethyl - thiacyanine bis(7,7,8,8 - tetracyanoquinodimethane)**
$C_{17}H_{15}N_2S_2^+$, $C_{12}H_4N_4^-$, $C_{12}H_4N_4$
R.P.Shibaeva, L.O.Atovmyan, V.F.Kaminskii *Eur. Cryst. Meeting*, 1973
Residue 1 also classified in 41; residue 2 classified in 60, 7

60.35 **5,7 - (12) - Paracyclophadiyne - tetracyanoethylene complex**
$2C_{18}H_{20}$, C_6N_4
K.Harata, T.Aono, K.Sakabe, N.Sakabe, J.Tanaka
Acta Cryst. (A), **28**, S14, 1972
Residue 1 also classified in 31; residue 2 classified in 60, 7

60.36 **Dimethyl - thiacarbocyanine bis(7,7,8,8 - tetracyanoquinodimethane)**
$C_{19}H_{17}N_2S_2^+$, $C_{12}H_4N_4^-$, $C_{12}H_4N_4$
R.P.Shibaeva, L.O.Atovmyan, V.F.Kaminskii *Eur. Cryst. Meeting*, 1973
Residue 1 also classified in 41; residue 2 classified in 60, 7

60.37 **Diethyl - thiacyanine bis(7,7,8,8 - tetracyanoquinodimethane)**
$C_{19}H_{19}N_2S_2^+$, $C_{12}H_4N_4^-$, $C_{12}H_4N_4$
R.P.Shibaeva, L.O.Atovmyan, V.F.Kaminskii *Eur. Cryst. Meeting*, 1973
Residue 1 also classified in 41; residue 2 classified in 60, 7

60.38 **Perylene - 7,7,8,8 - tetracyanoquinodimethane complex**
$C_{20}H_{12}$, $C_{12}H_4N_4$
I.J.Tickle, C.K.Prout *J. C. S. Perkin II*, 720, 1973
Residue 1 also classified in 30; residue 2 classified in 60, 7

60.39 **Diethyl - thiacarbocyanine bis(7,7,8,8 - tetracyanoquinodimethane) (triclinic form)**
$C_{21}H_{21}N_2S_2^+$, $C_{12}H_4N_4^-$, $C_{12}H_4N_4$
R.P.Shibaeva, L.O.Atovmyan, V.F.Kaminskii *Eur. Cryst. Meeting*, 1973
Residue 1 also classified in 41; residue 2 classified in 60, 7

60.40 **Deoxycholic acid - p - di - iodobenzene**
$2C_{24}H_{40}O_4$, $C_6H_4I_2$
S.C.de Sanctis, E.Giglio, V.Pavel, C.Quagliata
Acta Cryst. (B), **28**, 3656, 1972
Residue 1 also classified in 51; residue 2 classified in 60, 19

60.41 **Deoxycholic acid - phenanthrene**
$3C_{24}H_{40}O_4$, $C_{14}H_{10}$
S.C.de Sanctis, E.Giglio, V.Pavel, C.Quagliata
Acta Cryst. (B), **28,** 3656, 1972
Residue 1 also classified in 51; residue 2 classified in 60, 28

60.42 **Cholic acid - ethanol complex**
$C_{24}H_{40}O_5$, C_2H_6O
P.L.Johnson, J.P.Schaefer *Acta Cryst. (B),* **28,** 3083, 1972
Residue 1 also classified in 51; residue 2 classified in 60, 5

CLATHRATES

61.1   Urea - 1,4 - dichlorobutane
CH_4N_2O , $C_4H_8Cl_2$
J.Otto *Acta Cryst. (B)*, **28**, 543, 1972
Residue 1 also classified in 8; residue 2 classified in 61, 5

61.C α - Cyclodextrin - methanol
CH_4O , $C_{36}H_{60}O_{30}$
For complete entry see 61.8

61.C α - Cyclodextrin - n - propanol
C_3H_8O , $C_{36}H_{60}O_{30}$
For complete entry see 61.9

61.2   catena - μ - Ethylenediamine - cadmium(ii) tetracyanonickelate(ii) dibenzene
$(C_4H_8CdN_4Ni)_n$, $2nC_6H_6$
T.Miyoshi, T.Iwamoto, Y.Sasaki *Inorg. Chim. Acta*, **6**, 59, 1972
Residue 1 also classified in 76; residue 2 classified in 61, 19

61.C   Urea - 1,4 - dichlorobutane
$C_4H_8Cl_2$, CH_4N_2O
For complete entry see 61.1

61.C   Hydrogen tetrakis(benzoylacetonato) europate diethylamine complex
$xC_4H_{11}N$, $C_{40}H_{37}EuO_8$
For complete entry see 61.10

61.3   Cadmium diammine tetracyanomercury(ii) benzene
$2C_6H_6$, $C_4H_6CdHgN_6$
R.Kuroda *Inorg. Nucl. Chem. Letters*, **9**, 13, 1973
Residue 1 also classified in 19

61.C   catena - μ - Ethylenediamine - cadmium(ii) tetracyanonickelate(ii) dibenzene
$2nC_6H_6$, $(C_4H_8CdN_4Ni)_n$
For complete entry see 61.2

61.C   Iron 1,8 - bis(fluoroboro) - 2,7,9,14,15,20 - hexaoxa - 3,6,10,13,16,19 -
hexa - aza - 4,5,11,12,17,18 - hexamethylbicyclo(6.6.6)eicosa -
3,5,10,12,16,18 - hexaene cyclohexane
$0.5C_6H_{12}$, $C_{12}H_{18}B_2F_2FeN_6O_6$
For complete entry see 61.5

61.4   **Tetra - n - butylammonium bis(tetramethylene)trihydrodiborane**
$C_8H_{19}B_2^-$, $C_{16}H_{36}N^+$
W.R.Clayton, D.J.Saturnino, P.W.R.Corfield, S.G.Shore
J. C. S. Chem. Comm., 377, 1973
Residue 1 also classified in 62; residue 2 classified in 3

61.5   **Iron 1,8 - bis(fluoroboro) - 2,7,9,14,15,20 - hexaoxa - 3,6,10,13,16,19 -
hexa - aza - 4,5,11,12,17,18 - hexamethylbicyclo(6.6.6)eicosa -
3,5,10,12,16,18 - hexaene cyclohexane**
$C_{12}H_{18}B_2F_2FeN_6O_6$, $0.5C_6H_{12}$
E.C.Lingafelter, M.Dunaj-Jurco *Acta Cryst. (A)*, **28,** S79, 1972
Residue 1 also classified in 83; residue 2 classified in 61, 21

61.6 α - **Cyclodextrin - iodine**
$C_{36}H_{60}O_{30}$, I_2
W.Saenger, P.C.Manor *Eur. Cryst. Meeting*, 1973
Residue 1 also classified in 48

61.7 α - **Cyclodextrin dihydrate**
$C_{36}H_{60}O_{30}$, $2H_2O$
W.Saenger, P.C.Manor *Eur. Cryst. Meeting*, 1973
Residue 1 also classified in 48

61.8 α - **Cyclodextrin - methanol**
$C_{36}H_{60}O_{30}$, CH_4O
W.Saenger, P.C.Manor *Eur. Cryst. Meeting*, 1973
Residue 1 also classified in 48; residue 2 classified in 61, 5

61.9 α - **Cyclodextrin - n - propanol**
$C_{36}H_{60}O_{30}$, C_3H_8O
W.Saenger, P.C.Manor *Eur. Cryst. Meeting*, 1973
Residue 1 also classified in 48; residue 2 classified in 61, 5

61.10   **Hydrogen tetrakis(benzoylacetonato) europate diethylamine complex**
$C_{40}H_{37}EuO_8$, $xC_4H_{11}N$
A.L.Il'inskii, M.A.Porai-Koshits, L.A.Aslanov, P.I.Lazarev
Zh. Strukt. Khim., **13,** 277, 1972
Residue 1 also classified in 77; residue 2 classified in 61, 3

BORON COMPOUNDS

62.1 **Triethanolamine borate**
triptych - Boroxazolidine
$C_6H_{12}BNO_3$
R.Mattes, D.Fenske, K.-F.Tebbe *Chem. Ber.*, **105,** 2089, 1972

62.2 **tris(2,2 - Dimethylhydrazino)borane**
$C_6H_{21}BN_6$
H.Noth, R.Ullmann, H.Vahrenkamp *Chem. Ber.*, **106,** 1165, 1973
Also classified in 9

62.C **1 - (π - Cyclopentadienyl)dodecahydro - 1 - cobalta - 2,4 - dicarba - closo - tridecaborane**
$C_7H_{17}B_{10}Co$
For complete entry see 73.3

62.3 **Compound II**
$C_7H_{20}B_9MnO_4$
J.W.Lott, D.F.Gaines, H.Shenhav, R.Schaeffer
J. Amer. Chem. Soc., **95,** 3042, 1973

62.4 **Potassium boromalate monohydrate**
$C_8H_8BO_{10}^-$, K^+ , H_2O
R.A.Mariezcurrena, S.E.Rasmussen *Acta Cryst. (B)*, **29,** 1035, 1973

62.5 **Tetra - acetyl diborate**
$C_8H_{12}B_2O_2$
A.D.Negro, L.Ungaretti, A.Perotti *J. C. S. Dalton,* 1639, 1972

62.C **Tetra - n - butylammonium bis(tetramethylene)trihydrodiborane**
$C_8H_{19}B_2^-$, $C_{16}H_{36}N^+$
For complete entry see 61.4

62.6 **Tetraethylammonium 2,2' - cobaltobiscarborane**
$C_8H_{40}B_{40}Co^-$, $C_8H_{20}N^+$
R.A.Love, R.Bau *J. Amer. Chem. Soc.*, **94,** 8274, 1972
Residue 2 classified in 3

62.7 **Tri - n - propylamine borate trihydrate**
$C_9H_{18}BNO_3$, $3H_2O$
Z.Taira, K.Osaki *Inorg. Nucl. Chem. Letters,* **9,** 207, 1973

62.8 Tetra - ethylammonium (3,11') - commo - (undecahydro - 1,2 - dicarba - 3 - cobalta - closo - dodecaborato)(decahydro - 9' - pyridyl - 7',8' - dicarba - 11' - nido - undecaborate)

$C_9H_{26}B_{17}CoN^-$, $C_8H_{20}N^+$

M.R.Churchill, K.Gold *Inorg. Chem.*, **12,** 1157, 1973

Residue 2 classified in 3

62.C Cyclopentadienyl - carbonyl - iron - bis(methylisocyanide) borohydride adduct

$C_{10}H_{15}BFeN_2O$

For complete entry see 71.11

62.C 5,6 - μ - Diphenylphosphino - decaborane(14)

$C_{12}H_{23}B_{10}P$

For complete entry see 64.16

62.9 Compound III

$C_{13}H_{17}B_3Co_2$

D.C.Beer, V.R.Miller, L.G.Sneddon, R.N.Grimes, M.Mathew, G.J.Palenik
J. Amer. Chem. Soc., **95,** 3046, 1973

Also classified in 75, 73

62.10 Compound X

$C_{13}H_{35}BMnNO_4$

D.F.Gaines, J.W.Lott, J.C.Calabrese *J. C. S. Chem. Comm.*, 295, 1973

Also classified in 3

62.11 bis(Tropolonato) boron bromide monohydrate

$C_{14}H_{10}BO_4^+$, Br^- , H_2O

A.Karipides, J.Graf *Inorg. Nucl. Chem. Letters*, **8,** 161, 1972

62.12 Boron compound

$C_{14}H_{21}B_{10}^-$, $C_4H_{12}N^+$

E.I.Tolpin. W.N.Lipscomb *J. C. S. Chem. Comm.*, 257, 1973

Residue 2 classified in 3

62.C (Dihydro - bis(3,5 - dimethyl - 1 - pyrazolyl)borate) - (trihapto - cycloheptatrienyl)dicarbonyl - molybdenum

$C_{19}H_{23}BMoN_4O_2$

For complete entry see 75.31

62.C Ethylene - (dicyclohexyl - (3 - (ethyl - vinyl - boro)propyl) - phosphine) nickel

$C_{21}H_{40}BNiP$

For complete entry see 72.30

62.C Chugaev's red salt (form ii)

$C_{24}H_{20}B^-$, $C_9H_7N_6Pt^+$

For complete entry see 71.8

62.C   **Isothiocyanato - (bis(2 - (diethylamino)ethyl) - 2 - (diphenylarsino)ethylamine) nickel(ii) tetraphenylborate**
$C_{24}H_{20}B^-$, $C_{27}H_{42}AsN_4NiS^+$
For complete entry see 76.51

62.C   **Dibromo - tetrakis(1,8 - naphthyridine) dinickel tetraphenylborate**
$C_{24}H_{20}B^-$, $C_{32}H_{24}Br_2N_8Ni^+$
For complete entry see 83.112

62.C   **Sodium - bis(tetrahydrofuran) - bis(N,N' - ethylenebis(salicylideneiminato) cobalt(ii)) tetraphenylborate**
$C_{24}H_{20}B^-$, $C_{40}H_{44}Co_2N_4NaO_6^+$
For complete entry see 67.22

62.C   **Nitrosyl - bis(1,2 - bis(diphenylphosphino)ethane) ruthenium tetraphenylborate acetone solvate**
$C_{24}H_{20}B^-$, $C_{52}H_{48}NOP_4Ru^+$, C_3H_6O
For complete entry see 86.75

62.C   **Di - μ - cyanato - bis(2,2',2'' - triaminotriethylamine) dinickel(ii) tetraphenylborate (form I)**
$2C_{24}H_{20}B^-$, $C_{14}H_{36}N_{10}Ni_2O_2^{2+}$
For complete entry see 76.46

62.C   **hexakis(tris - p - Tolylphosphine - gold) tetraphenylborate**
$2C_{24}H_{30}B^-$, $C_{126}H_{126}Au_6P_6^{2+}$
For complete entry see 86.86

62.13   **(Diphenylmethyleneamino) - dimesityl - borane**
$C_{31}H_{32}BN$
G.J.Bullen   *J. C. S. Dalton,* 858, 1973

62.C   **1,1,3,3,5,5 - Hexaphenyl - cyclotriborata - phosphoniane (α form)**
$C_{36}H_{36}B_3P_3$
For complete entry see 64.45

SILICON COMPOUNDS

63.1   Silicon(iv) acetate
$C_8H_{12}O_8Si$
B.Kamenar, M.Bruvo *Abstr. Ital.-Yug. Congr*, 73, 1973

63.2   1,3,5,7 - Tetramethyl - tetrasila - adamantane
$C_{10}H_{24}Si_4$
E.W.Krahe, R.Mattes, K.-F.Tebbe, H.G.von Schnering, G.Fritz
Z. Anorg. Allg. Chem., **393**, 74, 1972

63.3   (tris(Trimethylsilyl)silyl)pentacarbonyl - manganese
$C_{14}H_{27}MnO_5Si_4$
B.K.Nicholson, J.Simpson, W.T.Robinson
J. Organometal. Chem., **47**, 403, 1973

63.4   Di - μ - (dimethylsilylene) - bis(tricarbonyl(trimethylsilyl) ruthenium(iii))
$C_{16}H_{30}O_6Ru_2Si_4$
M.M.Crozat, S.F.Watkins *J. C. S. Dalton*, 2512, 1972

63.C   tetrakis(Trimethylsilylmethyl - copper(i)) (at −40°C)
$C_{16}H_{44}Cu_4Si_4$
For complete entry see 71.38

63.5   Hexadecamethyl - bicyclo(3.3.1)nonasilane
$C_{16}H_{48}Si_9$
W.Stallings, J.Donohue
Amer. Cryst. Assoc., Abstr. Papers (Winter Meeting), 29, 1973

63.6   tris(Hexamethyldisilylamido) iron(iii)
$C_{18}H_{54}FeN_3Si_6$
M.B.Hursthouse, P.F.Rodesiler *J. C. S. Dalton*, 2100, 1972
Also classified in 83

63.7   2 - (α - Naphthyl) - 2 - fluoro - 2 - sila - 1,2,3,4 - tetrahydronaphthalene
$C_{19}H_{17}FSi$
J.-P.Vidal *Eur. Cryst. Meeting*, 1973

63.8   2,2 - Diphenyl - 2 - sila - indane
$C_{20}H_{18}Si$
J.P.Vidal, J.Falgueirettes *Acta Cryst. (B)*, **29**, 263, 1973

63.9   **2,2 - Diphenyl - 2 - sila - 3 - bromo - Δ^3 - 1 - tetralone**
$C_{21}H_{15}BrOSi$
J.P.Vidal, J.L.Galigne, J.Falgueirettes *Acta Cryst. (B)*, **28**, 3130, 1972

63.10   **2,2 - Diphenyl - 2 - sila - 1,3,4 - trihydronaphthalene**
$C_{21}H_{20}Si$
J.P.Vidal, J.Lapasset, J.Falgueirettes *Acta Cryst. (B)*, **28**, 3137, 1972

63.C   **1 - Tricarbonyl - 2,3,4,5 - tetra(trimethylsiloxy) - 1 - ferracyclopentadiene tricarbonyl - iron**
$C_{22}H_{36}Fe_2O_{10}Si_4$
For complete entry see 71.54

63.11   **Phenyl(triphenylsilyl)diazomethane**
$C_{25}H_{20}N_2Si$
C.Glidewell, G.M.Sheldrick *J. C. S. Dalton*, 2409, 1972
Also classified in 9

63.C   **π - Cyclopentadienyl - (π - 1,3 - diphenyl - bis(trimethylsilyl) cyclobutadiene) - cobalt**
$C_{27}H_{33}CoSi_2$
For complete entry see 75.40

63.C   **π - Cyclopentadienyl - (π - 1,3 - diphenyl - 2,4 - bis(trimethylsilyl) cyclobutadiene) cobalt**
$C_{27}H_{33}CoSi_2$
For complete entry see 75.41

63.C   **π - Cyclopentadienyl - (π - 1,2 - diphenyl - 3,4 - bis(trimethylsilyl) cyclobutadiene) cobalt**
$C_{27}H_{33}CoSi_2$
For complete entry see 75.42

63.12   **(+) - α - Naphthyl - phenyl - 1 - menthoxy - methoxy - silane (absolute configuration)**
$C_{27}H_{34}O_2Si$
J.A.Kanters, A.M.van Veen *Cryst. Struct. Comm.*, **2**, 261, 1973
Also classified in 52

63.C   **bis(Trimethylsilylmethyl) - bis(bipyridyl) chromium(iii) iodide**
$C_{28}H_{27}CrN_4Si_2^+$, I^-
For complete entry see 83.108

63.13   **bis(Trimethylsilyl - methyl) - bis(2,2' - bipyridyl) chromium(iii) iodide**
$C_{28}H_{38}CrN_4Si_2^+$, I^-
J.J.Daly, F.Sanz, R.P.A.Sneeden, H.H.Zeiss
Helv. Chim. Acta, **56**, 503, 1973
Residue 1 also classified in 71, 83

63.14 **Di - μ - diphenylsilyl - bis(tetracarbonyl manganese)**
$C_{32}H_{20}Mn_2O_8Si_2$
S.L.Simon, L.F.Dahl   *J. Amer. Chem. Soc.*, **95**, 783, 1973

63.15 **Hexaphenyl - cyclotrisiloxane**
$C_{36}H_{30}O_3Si_3$
N.G.Bokii, G.N.Zakharova, Yu.T.Struchkov
Zh. Strukt. Khim., **13**, 291, 1972

PHOSPHORUS COMPOUNDS

64.1 **Ethane - 1 - hydroxy - 1,1 - diphosphonic acid monohydrate**
$C_2H_8O_7P_2$, H_2O
V.A.Uchtman, R.A.Gloss *J. Phys. Chem.*, **76**, 1298, 1972

64.C **Calcium dihydrogen ethane - 1 - hydroxy - 1,1 - diphosphonate dihydrate**
$(C_2H_{10}CaO_9P_2)_n$
For complete entry see 67.2

64.C **N - Ammonium - phosphoryl - taurocyamine**
$C_3H_8N_3O_6PS^{2-}$, $2H_4N^+$
For complete entry see 8.10

64.C **Disodium N - phosphorylcreatine hydrate (further refinement)**
$C_4H_8N_3O_5P^{2-}$, $2Na^+$, xH_2O
For complete entry see 48.16

64.2 **Calcium 1,4 - dimethyl - 3,6 - dioxo - 2,5 - dioxa - 3,6 - diphosphacyclohexane - 1,4 - di(hydrogenphosphonate) - 3,6 - diolate decahydrate**
$C_4H_8O_{12}P_4^{4-}$, $2Ca^{2+}$, $10H_2O$
E.Philippot, G.Brun, J.-C.Jumas, M.Maurin
Rev. Chim. Miner., Fr., **9**, 591, 1972

64.C **2 - Methyl - 5 - tetrafluorophosphoranyl - pyrrole**
$C_5H_6F_4NP$
For complete entry see 32.5

64.3 **Dichlorophosphinyl(dichlorophosphinothioyl)aniline**
$C_6H_5Cl_4NOP_2S$
K.M.Ghouse, R.Keat, H.H.Mills, J.M.Robertson, T.S.Cameron,
K.D.Howlett, C.K.Prout *Phosphorus, **2**, 47, 1972
Also classified in 16

64.4 **Phenyl phosphorodiamidate**
$C_6H_9N_2O_2P$
G.J.Bullen, P.E.Dann *Acta Cryst. (B)*, **29**, 331, 1973

64.5 **3 - α - Oxo - 3β - hydrido - 7β - hydroxy - 2,4 - dioxa - 3 - phosphabicyclo(3.3.1)nonane**
$C_6H_{11}O_4P$
D.M.Nimrod, D.R.Fitzwater, J.G.Verkade *Inorg. Chim. Acta*, **2**, 149, 1968

64.6   **2,4,6 - Trichloro - 2,4,6 - tris(dimethylamino)cyclotriphosphazatriene**
$C_6H_{18}Cl_3N_6P_3$
F.R.Ahmed, D.R.Pollard   *Acta Cryst. (B)*, **28**, 3530, 1972

64.7   **tris(Dimethylhydrazino) - bis(phosphine oxide)**
$C_6H_{18}N_6O_2P_2$
J.W.Gilje, K.Seff   *Inorg. Chem.*, **11**, 1643, 1972

64.8   **3,7 - Dicyano - 3,5,7 - triaza - 1 - phosphabicyclo(3.3.1)nonane**
$C_7H_{10}N_5P$
L.M.Trefonas, J.N.Brown   *J. Heterocycl. Chem.*, **9**, 1295, 1972
Also classified in 42

64.9   **N,N - bis(2' - Chloroethyl) - N',O - trimethylene - phosphodiamide**
$C_7H_{14}Cl_2N_2O_2P$
J.C.Clardy, J.A.Mosbo, J.G.Verkade   *J. C. S. Chem. Comm.*, 1163, 1972
Also classified in 42

64.10   **N,N - bis(2 - Chloroethyl) - N' - O - propylene - phosphoric ester diamide**
monohydrate (at −10°C)
$C_7H_{15}Cl_2N_2O_2P$, H_2O
S.Garcia-Blanco, A.Perales   *Acta Cryst. (B)*, **28**, 2647, 1972

64.11   **Phosphonitrilic isothiocyanate tetramer**
$C_8N_{12}P_4S_8$
J.B.Faught   *Canad. J. Chem.*, **50**, 1315, 1972

64.12   **2,cis - 4,trans - 6,trans - 8 - Tetrachloro - 2,4,6,8 - tetrakis(dimethylamino) -**
cyclotetraphosphazatetraene
$C_8H_{24}Cl_4N_8P_4$
G.J.Bullen, P.A.Tucker   *J. C. S. Dalton*, 2437, 1972

64.13   **1,cis - 3,trans - 4,trans - 7 - tetrakis(Dimethylamino) - 1,3,5,7 - tetrafluoro -**
tetraphosphonitrile
$C_8H_{24}F_4N_8P_4$
D.Millington, T.J.King, D.B.Sowerby   *J. C. S. Dalton*, 396, 1973

64.C   **Pyridoxol 5' - methylphosphonate**
$C_9H_{14}NO_4P$
For complete entry see 33.37

64.14   **Tetramethylformamidinium phosphonic anhydride**
$C_{10}H_{24}N_4O_5P_2$
F.Sanz, J.J.Daly   *J. C. S. Dalton*, 2267, 1972

64.C   **5' - Methylene - adenosine - 3',5' - cyclic monophosphonate monohydrate**
$C_{11}H_{14}N_5O_5P$, H_2O
For complete entry see 47.26

64.15 **Diphenylphosphinic acid**
$C_{12}H_{11}O_2P$
D.Fenske, R.Mattes, J.Lons, K.-F.Tebbe *Chem. Ber.*, **106**, 1139, 1973

64.16 **5,6 - μ - Diphenylphosphino - decaborane(14)**
$C_{12}H_{23}B_{10}P$
L.B.Friedman, S.L.Perry *Inorg. Chem.*, **12**, 288, 1973
Also classified in 62

64.17 **hexakis(Dimethylamino)cyclotriphosphazadienium isopolymolybdate**
$2C_{12}H_{36}N_9P_3{}^+$, $Mo_6O_{19}{}^{2-}$
H.R.Allcock, E.C.Bisell, E.T.Shawl *J. Amer. Chem. Soc.*, **94**, 8603, 1972

64.18 **2 - Methyl - 2,2′ - spirobi(1,3,2 - benzodithiaphosphole)**
$C_{13}H_{11}PS_4$
M.Eisenhut, R.Schmutzler, W.S.Sheldrick
J. C. S. Chem. Comm., 144, 1973

64.19 **Potassium O,O - dibenzylphosphorodithioate**
$C_{14}H_{14}O_2PS_2{}^-$, K^+
J.P.Hazel, R.L.Collin *Acta Cryst. (B)*, **28**, 2279, 1972

64.20 **Diphenyl - ethyl - phosphine - oxide tetrakis(μ_3 - hydroxo - tricarbonyl - hydrido - tungsten)**
$4C_{14}H_{15}OP$, $C_{12}H_8O_{16}W_3$
V.G.Albano, G.Ciani, M.Manassero, M.Sansoni
J. Organometal. Chem., **34**, 353, 1972

64.21 **trans - 1,3 - Dimethyl - 2,4 - diphenyl - 2,4 - dithiocyclodiphosphazane**
$C_{14}H_{16}N_2P_2S_2$
E.H.M.Ibrahim, R.A.Shaw, B.C.Smith, C.P.Thakur, M.Woods, G.J.Bullen, J.S.Rutherford, P.A.Tucker, T.S.Cameron, K.D.Howlett, C.K.Prout
Phosphorus, **1**, 153, 1971

64.22 **cis - 1,3 - Diethyl - 2,4 - diphenyl - 2,4 - dithiocyclodiphosphazane**
$C_{16}H_{20}N_2P_2S_2$
E.H.M.Ibrahim, R.A.Shaw, B.C.Smith, C.P.Thakur, M.Woods, G.J.Bullen, J.S.Rutherford, P.A.Tucker, T.S.Cameron, K.D.Howlett, C.K.Prout
Phosphorus, **1**, 153, 1971

64.23 **trans - 1,3 - Diethyl - 2,4 - diphenyl - 2,4 - dithiocyclodiphosphazane**
$C_{16}H_{20}N_2P_2S_2$
E.H.M.Ibrahim, R.A.Shaw, B.C.Smith, C.P.Thakur, M.Woods, G.J.Bullen, J.S.Rutherford, P.A.Tucker, T.S.Cameron, K.D.Howlett, C.K.Prout
Phosphorus, **1**, 153, 1971

64.24 **1,1,2,3,3,4,4 - Heptamethyl - spiro(4,5) - 6,9 - diene - 1 - phosphetanium bromide**
$C_{16}H_{28}P^+$, Br^-
J.N.Brown, L.M.Trefonas, R.L.R.Towns
J. Heterocycl. Chem., **9**, 463, 1972

64.25 **p - Bromophenyl - diphenylphosphine sulfide**
$C_{18}H_{14}BrPS$
W.Dreissig, K.Plieth, P.Zaske *Acta Cryst. (B)*, **28**, 3473, 1972

64.26 **p - Chlorophenyl - diphenylphosphine sulfide**
$C_{18}H_{14}ClPS$
W.Dreissig, K.Plieth *Acta Cryst. (B)*, **28**, 3478, 1972

64.27 **Phenylthionophosphine oxide**
Phenyl - monothio - phosphonic anhydride
$C_{18}H_{15}O_3P_3S_3$
J.J.Daly, L.Maier, F.Sanz *Helv. Chim. Acta*, **55**, 1991, 1972

64.28 **3,5 - Di - (t - butyl) - 1 - phenylphosphoryl - 4 - aza - 2,5 - cyclohexadiene**
$C_{18}H_{26}NOP$
L.Hungerford, L.M.Trefonas *J. Heterocycl. Chem.*, **9**, 347, 1972
Also classified in 42

64.29 **2 - Cyclohexyl - 3 - phenyl - 2 - aza - 3 - phospha - bicyclo(2.2.2)octane - 3 - thione**
$C_{18}H_{26}NPS$
J.D.Healy, E.H.M.Ibrahim, R.A.Shaw, T.S.Cameron, K.D.Howlett,
C.K.Prout *Phosphorus*, **1**, 157, 1971
Also classified in 42

64.30 **Caged adamantanoid oxyphosphorane compound**
$C_{19}H_{19}N_2O_4P$
J.S.Ricci, W.C.Hamilton
Amer. Cryst. Assoc., Abstr. Papers (Winter Meeting), 92, 1973
Also classified in 42

64.31 **(p - Dimethylamino - phenyl)diphenylphosphine**
$C_{20}H_{20}NP$
W.Dreissig, K.Plieth *Z. Kristallogr.*, **135**, 294, 1972
Also classified in 16

64.32 **Triphenylphosphine - carbomethoxymethylene**
$C_{21}H_{19}O_2P$
V.D.Cherepinskii-Malov, G.G.Aleksandrov, A.I.Gusev, Yu.T.Struchkov
Zh. Strukt. Khim., **13**, 298, 1972

64.C **Triphenylphosphine - acetylmethylene trimethyltin chloride**
$C_{23}H_{24}ClO_2PSn$
For complete entry see 69.22

64.33 **10,10' - (5H,5'H) - Spirobiphenophosphazinium chloride methanol solvate**
$C_{24}H_{18}N_2P^+$, Cl^-, CH_4O
R.N.Jenkins, L.D.Freedman, J.Bordner *J. Cryst. Mol. Struct.*, **3**, 103, 1973

64.34 2,cis - 4,cis - 6,cis - 8 - Tetrachloro - 2,4,6,8 - tetraphenylcyclotetraphosphazatetraene (α form)
$C_{24}H_{20}Cl_4N_4P_4$
G.J.Bullen, P.A.Tucker   *J. C. S. Dalton*, 1651, 1972

64.35 1,2,3,4 - Tetraphenyl - 2,4 - dithiocyclodiphosphazane
$C_{24}H_{20}N_2P_2S_2$
M.B.Peterson, A.J.Wagner   *J. C. S. Dalton*, 106, 1973

64.36 Diphenylphosphino(diphenylphosphinothioyl) methylamine
$C_{25}H_{23}NP_2S$
K.M.Ghouse, R.Keat, H.H.Mills, J.M.Robertson, T.S.Cameron,
K.D.Howlett, C.K.Prout   *Phosphorus*, **2**, 47, 1972
Also classified in 3

64.37 Triphenyl - (tropone - 2 - yl) - phosphonium chloride
$C_{26}H_{22}OP^+$, Cl^-
I.Kawamoto, T.Hata, S.Sato, Y.Kishida, C.Tamura
Acta Cryst. (A), **28**, S12, 1972

64.38 4,4,6 - Trimethyl - 1 - oxa - 3 - azacyclohex - 2 - enylcarbenyl phosphonium chloride
$C_{26}H_{29}NOP^+$, Cl^-
L.M.Trefonas, J.N.Brown   *J. Heterocycl. Chem.*, **9**, 985, 1972
Residue 1 also classified in 40

64.39 1 - Ethoxy - 1,2 - diphenyl - 3,3,5 - tricarbethoxy - 1,2 - diphosphocyclopenten - 5 - one
$C_{26}H_{30}O_8P_2$
W.Saenger   *J. Org. Chem.*, **38**, 253, 1973

64.40 1,1,4,4 - Tetraphenyl - 1,4 - diphosphoniacyclohexane dibromide monohydrate
$C_{28}H_{28}P_2^{2+}$, $2Br^-$, H_2O
J.N.Brown, L.M.Trefonas   *J. Heterocycl. Chem.*, **9**, 35, 1972

64.41 (3,7 - Dimethyl - 2,6 - octadienyl)triphenyl - phosphonium bromide
$C_{28}H_{32}P^+$, Br^-
J.Hjortas   *Acta Cryst. (B)*, **29**, 767, 1973

64.42 N^2 - (p - Chlorophenyl) - N^1 - cyano - N^2 - (1,2 - bis(methoxycarbonyl) - 2 - (triphenyl - phosphonio)vinyl)hydrazide
$C_{31}H_{25}ClN_3O_4P$
R.D.Gilardi, I.L.Karle   *Acta Cryst. (B)*, **28**, 3420, 1972

64.43 1,1,4,4 - Tetraphenyl - 2,5 - di(t - butyl) - 1,4 - diphosphoniacyclohexa - 2,5 - diene dichloride hemihydrate
$C_{34}H_{40}P_2^{2+}$, $2Cl^-$, $0.5H_2O$
L.D.Cheung, L.M.Trefonas   *J. Heterocycl. Chem.*, **9**, 991, 1972

64.44 bis(Triphenylphosphine)imine iron - tetracarbonyl - hydride
$C_{36}H_{30}NP_2^+$, $C_4HFeO_4^-$
M.B.Smith, R.Bau *J. Amer. Chem. Soc.*, **95,** 2388, 1973

64.45 1,1,3,3,5,5 - Hexaphenyl - cyclotriborata - phosphoniane (α form)
$C_{36}H_{36}B_3P_3$
G.J.Bullen, P.R.Mallinson *J. C. S. Dalton,* 1295, 1973
Also classified in 62

64.46 1,1,2,4,4,5 - Hexaphenyl - 1,4 - diphosphoniacyclohexa - 2,5 - diene - dibromide monohydrate
$C_{40}H_{32}P_2^{2+}$, $2Br^-$, H_2O
J.N.Brown, L.M.Trefonas *J. Heterocycl. Chem.*, **9,** 187, 1972

64.47 Ethylene - 1 - (diphenylphosphino) - 1 - (triphenylphosphonium) - 2 - (diphenylamino) - 2 - (phenylamide)
$C_{50}H_{40}N_2P_2$
F.K.Ross, L.Manojlovic-Muir, W.C.Hamilton, F.Ramirez, J.F.Pilot
J. Amer. Chem. Soc., **94,** 8738, 1972

ARSENIC COMPOUNDS

65.C $(-)_{589}$ - **Acetylacetonato** - **bis(trimethylenediamine) cobalt(iii) diarsenic** -
ditartrate monohydrate (absolute configuration)
$C_8H_8As_2O_{12}{}^{2-}$, $C_{11}H_{27}CoN_4O_2{}^{2+}$, H_2O
For complete entry see 77.3

65.1   **10 - Phenoxarsine chloride**
$C_{12}H_8AsClO$
J.E.Stuckey, A.W.Cordes, L.B.Handy, R.W.Perry, C.K.Fair
Inorg. Chem., **11,** 1846, 1972

65.2   **10 - Chlorophenoxarsine oxide antimony pentachloride**
$C_{12}H_8AsCl_6O_2Sb$
R.J.Holliday, R.W.Broach, L.B.Handy, A.W.Cordes, L.Thomas
Inorg. Chem., **11,** 1849, 1972

65.3   **Potassium**$(-)_{589}$ - **tris(1,2 - benzenediolato) arsenate(v) sesquihydrate**
(absolute configuration)
$C_{18}H_{12}AsO_6{}^-$, K^+ , $1.5H_2O$
A.Kobayashi, T.Ito, F.Marumo, Y.Saito   *Acta Cryst. (B)*, **28,** 3446, 1972

65.4   **2,3,6 - Triphenylarsenin**
$C_{23}H_{17}As$
F.Sanz, J.J.Daly   *J. C. S. Dalton,* 511, 1973

65.5   **10 - Phenoxarsine sulfide**
$C_{24}H_{16}As_2O_2S$
W.K.Grindstaff, A.W.Cordes, C.K.Fair, R.W.Perry, L.B.Handy
Inorg. Chem., **11,** 1852, 1972

65.6   **Tetraphenylarsonium tri - iodide (at 20°C)**
$C_{24}H_{20}As^+$, $I_3{}^-$
J.Runsink, S.Swan-Walstra, T.Migchelsen   *Acta Cryst. (B)*, **28,** 1331, 1972

65.7   **Tetraphenylarsonium diaquohydrogen dichloride**
$C_{24}H_{20}As^+$, $H_5O_2{}^+$, $2Cl^-$
B.D.Faithful, S.C.Wallwork   *Acta Cryst. (B)*, **28,** 2301, 1972

65.8   **bis(Triphenylarsinehydroxy)bromine tribromide**
$C_{36}H_{32}As_2BrO_2{}^+$, $Br_3{}^-$
M.Calleri, G.Ferguson   *Cryst. Struct. Comm.*, **1,** 331, 1972

ANTIMONY AND BISMUTH COMPOUNDS

66.1   Methyl - trichloro(acetylacetonato) antimony(v)
$C_6H_{10}Cl_3O_2Sb$
N.Kanehisa, Y.Kai, N.Kasai   *Inorg. Nucl. Chem. Letters,* **8,** 375, 1972

66.2   Benzoyl chloride - antimony pentachloride
$C_7H_5Cl_6OSb$
B.Chevrier, J.-M.Le Carpentier, R.Weiss   *Acta Cryst. (B),* **28,** 2667, 1972

66.3   m - Toluoyl chloride - antimony pentachloride
$C_8H_7Cl_6OSb$
B.Chevrier, J.-M.Le Carpentier, R.Weiss   *Acta Cryst. (B),* **28,** 2667, 1972

66.4   p - Toluoyl chloride - antimony pentachloride complex
$C_8H_7Cl_6OSb$
B.Chevrier, J.-M.Le Carpentier, R.Weiss
J. Amer. Chem. Soc., **94,** 5718, 1972

66.5   trans - Diphenylantimony - dichloride - acetylacetonate
$C_{17}H_{17}Cl_2O_2Sb$
K.Onuma, Y.Kai, N.Kasai   *Inorg. Nucl. Chem. Letters,* **8,** 143, 1972

66.6   trans - Diphenylantimony - dichloride - acetylacetonate
$C_{17}H_{17}Cl_2O_2Sb$
J.Kroon, J.B.Hulscher, A.F.Peerdeman
J. Organometal. Chem., **37,** 297, 1972

GROUPS IA AND IIA COMPOUNDS

67.1   **Urea - calcium nitrate trihydrate**
$CH_{10}CaN_6O_{11}$
L.Lebioda   *Rocz. Chem.*, **46,** 373, 1972
Also classified in 8

67.2   **Calcium dihydrogen ethane - 1 - hydroxy - 1,1 - diphosphonate dihydrate**
$(C_2H_{10}CaO_9P_2)_n$
V.A.Uchtman   *J. Phys. Chem.*, **76,** 1304, 1972
Also classified in 64

67.3   **Calcium oxydiacetate hexahydrate**
$C_4H_{14}CaO_{10}$, H_2O
V.A.Uchtman, R.P.Oertel   *J. Amer. Chem. Soc.*, **95,** 1802, 1973
Residue 1 also classified in 2

67.C   **myo - Inositol calcium bromide pentahydrate**
$C_6H_{12}CaO_6^{2+}$, $2Br^-$, $5H_2O$
For complete entry see 45.7

67.C   **α - Galactose calcium bromide trihydrate**
$C_6H_{12}CaO_6^{2+}$, $2Br^-$, $3H_2O$
For complete entry see 45.13

67.C   **Potassium 2,2' - di(methoxy)diethylether bis(cyclo - octatetraenyl) cerium(iii)**
$C_6H_{14}KO_3^+$, $C_{16}H_{16}Ce^-$
For complete entry see 75.26

67.C   **β - D - Mannofuranose - calcium chloride tetrahydrate**
$(C_6H_{18}CaO_9^{2+})_n$, $2nCl^-$, nH_2O
For complete entry see 45.16

67.4   **Strontium violurate tetrahydrate**
$C_8H_4N_6O_8Sr$, $4H_2O$
M.Hamelin   *Acta Cryst. (A)*, **28,** S91, 1972
Residue 1 also classified in 43

67.5   **Triaquo - calcium terephthalate**
$(C_8H_{10}CaO_7)_n$
T.Matsuzaki, Y.Iitaka   *Acta Cryst. (B)*, **28,** 1977, 1972
Also classified in 14

67.C **bis(Tetrahydrofuran) sodium (9,10 - dihydro - 9,10 - anthrylene)dimethyl - aluminate**
$2C_8H_{16}NaO_2{}^+$, $C_{32}H_{32}Al_2{}^{2-}$
For complete entry see 68.31

67.6 **Hexa - aquo - magnesium ethylenediaminetetra - acetato - aquomagnesium dihydrate**
$C_{10}H_{14}MgN_2O_9{}^{2-}$, $H_{12}MgO_6{}^{2+}$, $2H_2O$
A.I.Pzhidaev, T.N.Polynova, M.A.Porai-Koshits
Acta Cryst. (A), **28,** S76, 1972
Residue 1 also classified in 3

67.7 **Calcium mellitate hydrate**
$C_{12}H_2Ca_2O_{12}$, $9H_2O$
R.J.Jandacek, V.A.Uchtman
Amer. Cryst. Assoc., Abstr. Papers (Winter Meeting), 69, 1973
Residue 1 also classified in 14

67.8 **Diaquo - magnesium picolinate**
$C_{12}H_{12}MgN_2O_6$
J.-P.Deloume, H.Loiseleur, G.Thomas   *Acta Cryst. (B)*, **29,** 668, 1973
Also classified in 33

67.C **Calcium ascorbate dihydrate**
$C_{12}H_{18}CaO_{14}$, $2H_2O$
For complete entry see 45.24

67.9 **hexakis(Methylcyanide) magnesium(ii) tetrachloroaluminate(iii)**
$C_{12}H_{18}MgN_6{}^{2+}$, $2AlCl_4{}^-$
B.A.Stork-Blaisse, G.C.Verschoor, C.Romers
Acta Cryst. (B), **28,** 2445, 1972
Residue 1 also classified in 7

67.C **Lactose calcium bromide heptahydrate**
$C_{12}H_{22}CaO_{12}{}^{2+}$, $2Br^-$, $7H_2O$
For complete entry see 45.27

67.C **α - Trehalose calcium bromide dihydrate**
$C_{12}H_{22}O_{12}$, Ca^{2+} , $2Br^-$, $2H_2O$
For complete entry see 45.39

67.10 **Lithium(4,7,13,19 - tetra - oxa - 1,10 - diaza - bicyclo(8.5.5)eicosane) iodide**
$C_{14}H_{28}LiN_2O_4{}^+$, I^-
D.Moras, R.Weiss   *Acta Cryst. (B)*, **29,** 400, 1973
Residue 1 also classified in 40

67.11 **Bromo - (tetraethyl - ethylenediamine) - cyclopentadienyl - magnesium**
$C_{15}H_{29}BrMgN_2$
C.Johnson, J.Toney, G.D.Stucky   *J. Organometal. Chem.*, **40,** C11, 1972
Also classified in 3, 20

67.C   **tetrakis(Tetrahydrofuran) lithium tetrakis(2,6 - dimethylphenyl) lutetiate**
$C_{16}H_{32}LiO_4^+$, $C_{32}H_{36}Lu^-$
For complete entry see 71.79

67.12   **Cesium(4,7,13,16,21,24 - hexa - oxa - 1,10 - diaza - bicyclo(8.8.8)hexaeicosane) thiocyanate monohydrate**
$C_{18}H_{36}CsN_2O_6^+$, CNS^- , H_2O
B.Metz, R.Weiss *Acta Cryst. (B),* **29,** 388, 1973
Residue 1 also classified in 40

67.13   **Potassium(4,7,13,16,21,24 - hexa - oxa - 1,10 - diaza - bicyclo(8.8.8)hexaeicosane)iodide**
$C_{18}H_{36}KN_2O_6^+$, I^-
D.Moras, B.Metz, R.Weiss *Acta Cryst. (B),* **29,** 383, 1973
Residue 1 also classified in 40

67.14   **Sodium(4,7,13,16,21,24 - hexa - oxa - 1,10 - diaza - bicyclo(8.8.8)hexaeicosane)iodide**
$C_{18}H_{36}N_2NaO_6^+$, I^-
D.Moras, R.Weiss *Acta Cryst. (B),* **29,** 396, 1973
Residue 1 also classified in 40

67.15   **Rubidium(4,7,13,16,21,24 - tetra - oxa - 1,10 - diaza - bicyclo(8.8.8)hexaeicosane) thiocyanate monohydrate**
$C_{18}H_{36}N_2O_6Rb^+$, CNS^- , H_2O
D.Moras, B.Metz, R.Weiss *Acta Cryst. (B),* **29,** 388, 1973
Residue 1 also classified in 40

67.C   **2,3.11,12 - Dicyclohexyl - 1,4,7,10,13,16 - hexaoxacyclo - octane sodium bromide dihydrate**
$C_{20}H_{36}NaO_6^+$, Br^- , $2H_2O$
For complete entry see 38.25

67.16 μ - **1,4 - Dioxane - bis(di(tetrahydrofuran) lithium) di(trichloro - cyclopentadienyl - chromium)**
$C_{20}H_{40}Li_2O_6^{2+}$, $2C_5H_5Cl_3Cr^-$
B.Muller, J.Krausse *J. Organometal. Chem.,* **44,** 141, 1972
Residue 1 also classified in 38; residue 2 classified in 73

67.17   **Bicyclo(1,1,0)butan - 1 - yl - lithium tetramethylethylenediamine dimer**
$C_{20}H_{42}Li_2N_4$
R.P.Zerger, G.D.Stucky *J. C. S. Chem. Comm.,* 44, 1973
Also classified in 3, 27

67.18   **Naphthalene - bis(lithium tetramethylethylenediamine)**
$C_{22}H_{40}Li_2N_4$
J.J.Brooks, W.Rhine, G.D.Stucky *J. Amer. Chem. Soc.,* **94,** 7346, 1972
Also classified in 24, 3

67.19 **Triphenylmethyl - lithium tetramethylethylenediamine**
$C_{25}H_{31}LiN_2$
J.J.Brooks, G.D.Stucky *J. Amer. Chem. Soc.*, **94**, 7333, 1972
Also classified in 3, 19

67.20 **Fluorenyl - lithium bis(quinuclidine)**
$C_{27}H_{35}LiN_2$
J.J.Brooks, W.Rhine, G.D.Stucky *J. Amer. Chem. Soc.*, **94**, 7339, 1972
Also classified in 28, 37

67.C **Potassium iodide bis(2,3,5,6,8,9,11,12 - octahydro - 1,4,7,10,13 - benzo -**
pentaoxacyclopentadecine)
Potassium iodide bis(benzo - 15 - crown - 5)
$C_{28}H_{40}KO_{10}^+$, I^-
For complete entry see 38.31

67.21 **bis(Dimethylformamido) - bis(1,3 - diphenyl - 1,3 - propanedionato)**
magnesium
$C_{36}H_{34}MgN_2O_6$
F.J.Hollander, D.H.Templeton, A.Zalkin *Acta Cryst. (B)*, **29**, 1289, 1973

67.22 **Sodium - bis(tetrahydrofuran) - bis(N,N' - ethylenebis(salicylideneiminato)**
cobalt(ii)) tetraphenylborate
$C_{40}H_{44}Co_2N_4NaO_6^+$, $C_{24}H_{20}B^-$
C.Floriani, F.Calderazzo, L.Randaccio *J. C. S. Chem. Comm.*, 384, 1973
Residue 1 also classified in 78, 76, 38; residue 2 classified in 62

67.C **Tetranactin - potassium thiocyanate (form ii)**
$C_{44}H_{72}KO_{12}^+$, CNS^-
For complete entry see 50.17

67.C **Tetranactin - potassium thiocyanate (form i)**
$C_{44}H_{72}KO_{12}^+$, CNS^-
For complete entry see 50.18

67.C **Tetranactin - sodium thiocyanate (form ii)**
$C_{44}H_{72}NaO_{12}^+$, CNS^-
For complete entry see 50.19

67.C **Tetranactin - rubidium thiocyanate**
$C_{44}H_{72}O_{12}Rb^+$, CNS^-
For complete entry see 50.21

67.C **Tetranactin - rubidium thiocyanate (form i)**
$C_{44}H_{72}O_{12}Rb^+$, CNS^-
For complete entry see 50.22

67.23 **bis(6,7,9,10,17,18,20,21 - Octahydro - 7R,9R,18R,20R - tetramethyl -**
dibenzo(b,k)(1,4,7,10,13,16)hexaoxacyclo - octadecin) cesium thiocyanate
$C_{48}H_{64}CsO_{12}^+$, CNS^-
P.R.Mallinson, D.G.Parsons, M.R.Truter *Eur. Cryst. Meeting,* 1973

67.24 **bis(Thiocyanato - cesium - (6,7,9,10,17,18,20,21 - octahydro - 7R,9R,18S,20S - tetramethyldibenzo(b,k)(1,4,7,10,13,16)hexa - oxacyclo - octadecin))**

$C_{50}H_{64}Cs_2N_2O_{12}S_2$

P.R.Mallinson, D.G.Parsons, M.R.Truter *Eur. Cryst. Meeting,* 1973

Also classified in 38

67.C **Antamanide lithium bromide acetonitrile solvate**

$C_{64}H_{70}LiN_{10}O_{10}{}^+$, Br^- , xC_2H_3N

For complete entry see 50.30

67.25 **bis(1,3 - Diphenyl - 1,3 - propanedionato) calcium hemiethanolate**

$C_{124}H_{100}Ca_4O_{18}$

F.J.Hollander, D.H.Templeton, A.Zalkin *Acta Cryst. (B),* **29,** 1295, 1973

67.26 **bis(1,3 - Diphenyl - 1,3 - propanedionato) strontium hemiacetonate**

$C_{126}H_{100}O_{18}Sr_4$

F.J.Hollander, D.H.Templeton, A.Zalkin *Acta Cryst. (B),* **29,** 1303, 1973

GROUP III COMPOUNDS

68.1   **Potassium cyano - trimethyl - aluminate**
$C_4H_9AlN^-$, K^+
J.L.Atwood, R.E.Cannon   *J. Organometal. Chem.*, **47**, 321, 1973

68.2   **Acetato(dimethyl) indium(iii)**
$C_4H_9InO_2$
F.W.B.Einstein, M.M.Gilbert, D.G.Tuck   *J. C. S. Dalton*, 248, 1973

68.3   **Cesium tetramethylindate**
$C_4H_{12}In^-$, Cs^+
K.Hoffmann, E.Weiss   *J. Organometal. Chem.*, **50**, 17, 1973

68.4   **Potassium tetramethylindate**
$C_4H_{12}In^-$, K^+
K.Hoffmann, E.Weiss   *J. Organometal. Chem.*, **50**, 17, 1973

68.5   **Pyrazolyl - dideuterio - gallane dimer**
$C_6H_6D_4Ga_2N_4$
D.F.Rendle, A.Storr, J.Trotter   *J. C. S. Chem. Comm.*, 189, 1973

68.6   **Aziridinylgallane trimer**
$C_6H_{12}Ga_3N_3$
W.Harrison, A.Storr, J.Trotter   *J. C. S. Dalton*, 1554, 1972

68.7   **Diethyl - indium thioacetate**
$(C_6H_{13}InOS)_n$
H.-D.Hausen   *Z. Naturforsch., B*, **27**, 82, 1972

68.8   **Diethyl - indium - acetate**
$C_6H_{13}InO_2$
H.-D.Hausen   *J. Organometal. Chem.*, **39**, C37, 1972

68.9   **Potassium bis(trimethylaluminium)azide**
$C_6H_{18}Al_2N_3^-$, K^+
J.L.Atwood, W.R.Newberry III   *J. Organometal. Chem.*, **42**, C77, 1972

68.10   **Dimethylaminoalane**
$C_6H_{18}Al_3N_3$
K.N.Semeneko, E.B.Lobkovskii, A.L.Dorosinskii
Zh. Strukt. Khim., **13**, 743, 1972

68.11   **m - Toluyl chloride - aluminium chloride**
$C_8H_7AlCl_4O$
B.Chevrier, J.-M.Le Carpentier, R.Weiss   *Acta Cryst. (B)*, **28**, 2659, 1972

68.12   **o - Toluyl chloride - aluminium chloride**
$C_8H_7AlCl_4O$
B.Chevrier, J.-M.Le Carpentier, R.Weiss   *Acta Cryst. (B)*, **28**, 2659, 1972

68.13   **p - Toluyl chloride - aluminium chloride**
$C_8H_7AlCl_4O$
B.Chevrier, J.-M.Le Carpentier, R.Weiss   *Acta Cryst. (B)*, **28**, 2659, 1972

68.14   **Cyclodi - μ - dimethylamido - bis(dimethylaluminium)**
$C_8H_{24}Al_2N_2$
G.M.McLaughlin, G.A.Sim, D.J.Smith   *J. C. S. Dalton*, 2197, 1972

68.15   **cis - Cyclotri - μ - methylamido - tris(dimethylaluminium)**
$C_9H_{27}Al_3N_3$
G.M.McLaughlin, G.A.Sim, J.D.Smith   *J. C. S. Dalton*, 2197, 1972

68.16   **trans - Cyclotri - μ - methylamido - tris(dimethylaluminium)**
$C_9H_{27}Al_3N_3$
G.M.McLaughlin, G.A.Sim, J.D.Smith   *J. C. S. Dalton*, 2197, 1972

68.17   **1,10 - Phenanthroline - dimethyl - thallium(iii) perchlorate**
$C_{14}H_{14}N_2Tl^+$, ClO_4^-
T.L.Blundell, H.M.Powell   *Proc. R. Soc., A*, **331**, 161, 1972

68.18   **Di - μ - di - isopropyldithiocarbamato - dithallium(i)**
$C_{14}H_{28}N_2S_4Tl_2$
P.Jennische, A.Olin, R.Hesse   *Acta Chem. Scand.*, **26**, 2799, 1972

68.19   **tris(Cyclopentadienyl) indium(iii) (at $-100°C$)**
$(C_{15}H_{15}In)_n$
F.W.B.Einstein, M.M.Gilbert, D.G.Tuck   *Inorg. Chem.*, **11**, 2832, 1972

68.20   **Aluminium acetylacetonate (γ form)**
$C_{15}H_{21}AlO_6$
B.W.McClelland
Amer. Cryst. Assoc., Abstr. Papers (Winter Meeting), 68, 1973

68.21   **Di - μ - phenyl - bis(dimethylaluminium)**
$C_{16}H_{22}Al_2$
J.F.Malone, W.S.McDonald   *J. C. S. Dalton*, 2649, 1972

68.22   **Dimethyl - (N - phenylbenzimidato)(trimethylamine - N - oxide) aluminium**
$C_{18}H_{25}AlN_2O_2$
Y.Kai, N.Yasuoka, N.Kasai, M.Kakudo
Bull. Chem. Soc. Jap., **45**, 3388, 1972

68.23 **tris(Pentamethylenedithiocarbamato) indium(iii)**
$C_{18}H_{30}InN_3S_6$
P.J.Hauser, J.Bordner, A.F.Schreiner *Inorg. Chem.*, **12**, 1347, 1973

68.24 **Thallium(i) cryptate formate monohydrate**
$C_{18}H_{36}N_2O_6Tl^+$, CHO_2^- , H_2O
D.Moras, R.Weiss *Acta Cryst. (B)*, **29**, 1059, 1973
Residue 2 classified in 2

68.25 **Trimethylaluminium - (N - phenyl - N - (ethan - 1 - olato)benzamide) - dimethylaluminium**
$C_{20}H_{29}Al_2NO_2$
Y.Kai, N.Yasuoka, N.Kasai, M.Kakudo
Bull. Chem. Soc. Jap., **45**, 3403, 1972

68.26 **tris(Tropolonato) aluminium(iii)**
$C_{21}H_{15}AlO_6$
E.L.Muetterties, L.J.Guggenberger *J. Amer. Chem. Soc.*, **94**, 8046, 1972

68.27 **tris(Cyclopentadienyl - tricarbonyl - molybdenum) thallium(iii)**
$C_{24}H_{15}Mo_3O_9Tl$
J.Rajaram, J.A.Ibers *Inorg. Chem.*, **12**, 1313, 1973
Also classified in 73

68.28 **Sodium tetraphenylindate**
$C_{24}H_{20}In^-$, Na^+
K.Hoffmann, E.Weiss *J. Organometal. Chem.*, **50**, 25, 1973

68.29 **Dimethylaluminium - dihydride - bis - μ - (π - cyclopentadienyl molybdenum) - (methylaluminium - di - μ - (1,1 - cyclopentadienyl) - dimethylaluminium)**
$C_{25}H_{35}Al_3Mo_2$
R.A.Forder, M.L.H.Green, R.E.MacKenzie, J.S.Poland, K.Prout
J. C. S. Chem. Comm., 426, 1973
Also classified in 73

68.30 **Di - μ - hydroxo - bis(chloro(methyl - (6 - methyl - 2 - pyridylmethyl) - (2 - pyridylmethyl)amine)gallium(iii)) chloride monohydrate**
$C_{28}H_{36}Cl_2Ga_2N_6O_2^{2+}$, $2Cl^-$, H_2O
K.Dymock, G.J.Palenik, A.J.Carty *J. C. S. Chem. Comm.*, 1218, 1972

68.31 **bis(Tetrahydrofuran) sodium (9,10 - dihydro - 9,10 - anthrylene)dimethyl - aluminate**
$C_{32}H_{32}Al_2^{2-}$, $2C_8H_{16}NaO_2^+$
D.J.Brauer, G.D.Stucky *J. Organometal. Chem.*, **37**, 217, 1972
Residue 2 classified in 67

68.32 **N - Phenyl - N - (ethan - 1 - olato)benzamide - dimethylaluminium dimer**
$C_{34}H_{40}Al_2N_2O_4$
Y.Kai, N.Yasuoka, N.Kasai, M.Kakudo
Bull. Chem. Soc. Jap., **45**, 3397, 1972

68.33 **Di - μ - phenyl - bis(diphenylaluminium)**
Triphenylaluminium dimer
$C_{36}H_{30}Al_2$
J.F.Malone, W.S.McDonald *J. C. S. Dalton*, 2646, 1972

68.34 **Di - μ - (π - cyclopentadienyl - molybdenum) - bis(methylaluminium - di - μ - (1,1 - cyclopentadienyl) - dimethyl - aluminium)**
$C_{36}H_{44}Al_4Mo_2$
R.A.Forder, M.L.H.Green, R.E.MacKenzie, J.S.Poland, K.Prout
J. C. S. Chem. Comm., 426, 1973

GERMANIUM, TIN, LEAD COMPOUNDS

69.1   **Lead(ii) thiourea dihydrate**
CH_4BrN_2PbS , $2H_2O$
I.Goldberg, F.H.Herbstein, M.Kaftory, M.Kapon
Acta Cryst. (A), **28**, S85, 1972

69.2   **Lead(ii) thiourea bromide**
$CH_4Br_2N_2PbS$
I.Goldberg, F.H.Herbstein, M.Kaftory, M.Kapon
Acta Cryst. (A), **28**, S85, 1972

69.3   **bis(Chloromethyl)dichloro - stannane**
$C_2H_4Cl_4Sn$
N.G.Bokii, Yu.T.Struchkov, A.K.Prokof'ev
Zh. Strukt. Khim., **13**, 665, 1972

69.4   **Dimethyl - dinitrato - tin(iv)**
$C_2H_6N_2O_6Sn$
J.Hilton, E.K.Nunn, S.C.Wallwork   *J. C. S. Dalton*, 173, 1973

69.5   **Lead(ii) bromide di - thiourea (form i)**
$C_2H_8Br_2N_4PbS_2$
I.Goldberg, F.H.Herbstein, M.Kaftory, M.Kapon
Acta Cryst. (A), **28**, S85, 1972

69.6   **Lead(ii) formate tetra - thiourea monohydrate**
$C_4H_{10}N_4O_2PbS_2$, H_2O
I.Goldberg, F.H.Herbstein, M.Kaftory, M.Kapon
Acta Cryst. (A), **28**, S85, 1972

69.7   **Tetra(methyl - germanium) hexasulfide**
$C_4H_{12}Ge_4S_6$
R.H.Benno, C.J.Fritchie Junior   *J. C. S. Dalton*, 543, 1973

69.8   **Aquo - bis(thiourea) lead(ii) formate**
$C_4H_{12}N_4O_5PbS_2$
I.Goldberg, F.H.Herbstein   *Acta Cryst. (B)*, **29**, 246, 1973

69.9   **Tetra(methyltin) hexasulfide**
$C_4H_{12}S_6Sn_4$
D.Kobelt, E.F.Paulus, H.Scherer   *Acta Cryst. (B)*, **28**, 2323, 1972

69.10   **Trimethyltin prop - 1 - yn - 3 - yl sulfinate**
$(C_6H_{12}O_2SSn)_n$
D.Ginderow, M.Huber   *Acta Cryst. (B),* **29,** 560, 1973

69.11   **Trimethyltin isocyanate - hydroxide**
$(C_7H_{19}O_2Sn_2)_n$
J.B.Hall, D.Britton   *Acta Cryst. (B),* **28,** 2133, 1972

69.12   **Dimethyltin - bis(N,N - dimethyldithiocarbamate)**
$C_8H_{18}N_2S_4Sn$
T.Kimura, N.Yasuoka, N.Kasai, M.Kakudo
Bull. Chem. Soc. Jap., **45,** 1649, 1972

69.13   **Dihydrogen ethylenediaminetetra - acetato - stannate(ii)**
$C_{10}H_{14}N_2O_8Sn$
K.G.Shields, R.C.Seccombe, C.H.L.Kennard   *J. C. S. Dalton,* 741, 1973

69.14   **bis(Isonicotinato) lead(ii) hydrate**
$(C_{12}H_8N_2O_4Pb)_n$, nH_2O
M.B.Cingi, A.G.Manfredotti, C.Guastini, M.Nardelli
Gazz. Chim. Ital., **102,** 1034, 1972

69.15   **Diphenyl - dichloro - lead**
$(C_{12}H_{10}Cl_2Pb)_n$
M.Mammi, V.Busetti, A.Del Pra   *Inorg. Chim. Acta,* **1,** 419, 1967

69.16   **bis(Acetylacetonato)dimethyl - tin(iv)**
$C_{12}H_{20}O_4Sn$
G.A.Miller, E.O.Schlemper   *Inorg. Chem.,* **12,** 677, 1973

69.17   **bis(Triethanolaminato) tin(iv)**
$C_{12}H_{26}N_2O_6Sn$
H.Follner   *Monatsh. Chem.,* **103,** 1438, 1972

69.18   **Dithiocyanato - (1,7,10,16 - tetraoxa - 4,13 - diazacyclo - octadecane) lead(ii)**
$C_{14}H_{26}N_4O_4PbS_2$
B.Metz, R.Weiss   *Acta Cryst. (B),* **29,** 1088, 1973

69.C   **Di - μ - dimethylstannylene - bis(carbonyl - π - cyclopentadienyl - cobalt)**
$C_{16}H_{22}Co_2O_2Sn_2$
For complete entry see 73.29

69.19   **N,N' - Ethylene - bis(salicylideneiminato) - dimethyl - tin(iv)**
$C_{18}H_{20}N_2O_2Sn$
M.Calligaris, G.Nardin, L.Randaccio   *J. C. S. Dalton,* 2003, 1972

69.20   **Di(t - butyl) - pyridino - stannyl pentacarbonyl - chromium**
$C_{18}H_{23}CrNO_5Sn$
M.D.Brice, F.A.Cotton
Amer. Cryst. Assoc., Abstr. Papers (Winter Meeting), 45, 1973

69.21 **Dimethyl - di - isothiocyanato - terpyridyl - tin(iv)**
$C_{19}H_{17}N_5S_2Sn$
D.V.Naik, W.R.Scheidt *Inorg. Chem.*, **12,** 272, 1973

69.22 **Triphenylphosphine - acetylmethylene trimethyltin chloride**
$C_{23}H_{24}ClO_2PSn$
J.Buckle, P.G.Harrison, T.J.King, J.A.Richards
J. C. S. Chem. Comm., 1104, 1972
Also classified in 64

69.23 **tetrakis(Pentafluorophenyl) germanium**
$C_{24}F_{20}Ge$
A.Karipides, R.H.P.Thomas *Cryst. Struct. Comm.*, **2,** 275, 1973

69.24 **Tetraphenyl - germanium**
$C_{24}H_{20}Ge$
A.Karipides, D.A.Haller *Acta Cryst. (B)*, **28,** 2889, 1972

69.25 **Tetraphenyl - lead**
$C_{24}H_{20}Pb$
V.Busetti, M.Mammi, A.Signor, A.Del Pra *Inorg. Chim. Acta*, **1,** 424, 1967

69.26 **5,5 - Diphenyl - 10,11 - dihydro - 5H - dibenzo(b,f)germepin**
$C_{26}H_{22}Ge$
J.Y.Corey, E.R.Corey, M.D.Glick, J.S.Dueber
J. Heterocycl. Chem., **9,** 1379, 1972

69.27 **Hexaphenyl - distannane**
$C_{36}H_{30}Sn_2$
H.Pruet, H.-J.Haupt, F.Huber *Z. Anorg. Allg. Chem.*, **396,** 81, 1973

69.C **π - Cyclopentadienyl - π - diphenylacetylene - monocarbonyl - triphenyl - stannyl - iron**
$C_{38}H_{30}FeOSn$
For complete entry see 72.41

69.28 **cis - Triphenylphosphine - phenyl - platinum - bis(triphenyl - lead)**
$C_{60}H_{50}PPb_2Pt$
B.Crociani, M.Nicolini, D.A.Clemente, G.Bandoli
J. Organometal. Chem., **49,** 249, 1973
Also classified in 71, 86

TELLURIUM COMPOUNDS

70.1   **Tellurium di(methylxanthate)**
$C_4H_6O_2S_4Te$
N.J.Brondmo, S.Esperas, H.Graver, S.Husebye
Acta Chem. Scand., **27**, 713, 1973

70.2   **1 - Thia - 4 - telluracyclohexane 4,4 - dibromide**
$C_4H_8Br_2STe$
C.Knobler, J.D.McCullough *Inorg. Chem.*, **11**, 3026, 1972

70.3   **Di - μ - bromo - μ - 1,2 - cyclohexylene - tetrabromoditellurium**
$C_6H_{10}Br_6Te_2$
A.C.Hazell *Acta Chem. Scand.*, **26**, 1510, 1972

70.4   **Tellurium di(methane - thiosulfonate) bis(ethylenethiourea) (triclinic form)**
$C_8H_{18}N_4O_4S_6Te$
O.Foss, N.Lyssandtrae, K.Maartmann-Moe, M.Tysseland
Acta Chem. Scand., **27**, 218, 1973

70.5   **Tellurium di(methane - thiosulfonate) bis(ethylenethiourea) (monoclinic form)**
$C_8H_{18}N_4O_4S_6Te$
O.Foss, N.Lyssandtrae, K.Maartmann-Moe, M.Tysseland
Acta Chem. Scand., **27**, 218, 1973

70.6   **trans - Tetrachloro - bis(tetramethylthiourea) tellurium (monoclinic form)**
$C_{10}H_{24}Cl_4N_4S_2Te$
S.Esperas, J.W.George, S.Husebye, O.Mikalsen
Acta Chem. Scand., **27**, 1089, 1973

70.7   **Diphenyl - ditelluride**
$C_{12}H_{10}Te_2$
G.Llabres, O.Dideberg, L.Dupont *Acta Cryst. (B)*, **28**, 2438, 1972

70.8   **Di - μ - bromo - bis(di(ethylenethiourea) tellurium(ii)) dibromide (monoclinic form)**
$C_{12}H_{24}Br_2N_8S_4Te_2^{2+}$, $2Br^-$
P.Herland, M.Lundeland, K.Maroy *Acta Chem. Scand.*, **26**, 2567, 1972

70.9   **Di - μ - bromo - bis(di(ethylenethiourea) tellurium(ii)) dibromide (orthorhombic form)**
$C_{12}H_{24}Br_2N_8S_4Te_2{}^{2+}$, $2Br^-$
P.Herland, M.Lundeland, K.Maroy   *Acta Chem. Scand.*, **26,** 2567, 1972

70.10   **tetrakis(4 - Morpholine carbodithioato) tellurium(iv) benzene solvate**
$C_{20}H_{32}N_4O_4S_8Te$, $3C_6H_6$
S.Esperas, S.Husebye   *Acta Chem. Scand.*, **27,** 706, 1973

70.11   **tetrakis(Diethyldithiocarbamato) tellurium(iv)**
$C_{20}H_{40}N_4S_8Te$
S.Husebye, S.E.Svaeren   *Acta Chem. Scand.*, **27,** 763, 1973

70.12   **tris(Diethyldithiocarbamato)phenyl - tellurium(iv)**
$C_{21}H_{35}N_3S_6Te$
S.Esperas, S.Husebye   *Acta Chem. Scand.*, **26,** 3293, 1972

TRANSITION METAL-C COMPOUNDS

71.1 **Ethyl zinc iodide**
$(C_2H_5IZn)_n$
P.T.Moseley, H.M.M.Shearer *J. C. S. Dalton*, 64, 1973

71.2 **N - (2 - (Chloromercuri)ethyl)diethylamine**
$C_6H_{14}ClHgN$
K.Toman, G.G.Hess *J. Organometal. Chem.*, **49**, 133, 1973

71.3 **Chloro - (ethylenedioxy - carbene) - tetracarbonyl - manganese(i)**
$C_7H_4ClMnO_6$
M.Green, J.R.Moss, I.W.Nowell, F.G.A.Stone
J. C. S. Chem. Comm., 1339, 1972

71.4 **Bromo - tricarbonyl - bis(methylisocyanide) manganese**
$C_7H_6BrMnN_2O_3$
A.C.Sarapu, R.F.Fenske *Inorg. Chem.*, **11**, 3021, 1972

71.5 **cis - (N - Methylcarboxamido) - (methylamine) - tetracarbonyl - manganese(i) (monoclinic form)**
$C_7H_9MnN_2O_5$
D.M.Chipman, R.A.Jacobson *Inorg. Chim. Acta*, **1**, 393, 1967
Also classified in 83

71.6 **Phenylmercury(ii) acetate**
$C_8H_8HgO_2$
B.Kamenar, M.Penavic *Inorg. Chim. Acta*, **6**, 191, 1972
Also classified in 81

71.7 **cis - β - Benzoylvinyl - mercury(ii) chloride**
C_9H_7ClHgO
L.G.Kuz'mina, N.G.Bokii, M.I.Rybinskaya, Yu.T.Struchkov, T.V.Popova
Zh. Strukt. Khim., **12**, 1026, 1971

71.8 **Chugaev's red salt (form ii)**
$C_9H_7N_6Pt^+$, $C_{24}H_{20}B^-$
W.M.Butler, J.H.Enemark, J.Parks, A.L.Balch *Inorg. Chem.*, **12**, 451, 1973
Residue 2 classified in 62

71.9 **Tetracarbonyl(1,3 - dimethylimidazolinylidene) iron(0)**
$C_9H_8FeN_2O_4$
G.Huttner, W.Gartzke *Chem. Ber.*, **105**, 2714, 1972

71.10   (3 - Acetyl - 4,5 - dihydrofuran - 2 - yl)tetracarbonyl - manganese(i)
$C_{10}H_7MnO_6$
C.P.Casey, R.A.Boggs, D.F.Marten, J.C.Calabrese
J. C. S. Chem. Comm., 243, 1973

71.11   Cyclopentadienyl - carbonyl - iron - bis(methylisocyanide) borohydride adduct
$C_{10}H_{15}BFeN_2O$
W.M.Butler, J.H.Enemark   *J. Organometal. Chem.*, **49,** 233, 1973
Also classified in 73, 62

71.12   trans - Di - iodo - bis(t - butylisonitrile) - palladium(ii)
$C_{10}H_{18}I_2N_2Pd$
N.A.Bailey, N.W.Walker, J.A.W.Williams
J. Organometal. Chem., **37,** C49, 1972

71.13   Chloro - ethyl - (dimethylglyoximato)(dimethylglyoxime) cobalt(iii) monohydrate
$C_{10}H_{20}ClCoN_4O_2$, H_2O
A.L.Crumbliss, J.T.Bowman, P.I.Gaus, A.T.McPhail
J. C. S. Chem. Comm., 415, 1973
Residue 1 also classified in 83

71.14   trans - bis(Methyl isocyanide) - bis(methylamino(thioethoxy)carbene) platinum(ii) hexafluorophosphate
$C_{10}H_{24}N_4PtS_2{}^{2+}$, $2F_6P^-$
W.M.Butler, J.H.Enemark   *Inorg. Chem.*, **12,** 540, 1973

71.15   Dicobalt acetylene nonacarbonyl (triclinic form)
μ - Carbonyl - μ - (5 - oxo - dihydrofuran - 1,1 - diyl) - bis(tricarbonyl cobalt)
$C_{11}H_2Co_2O_9$
O.S.Mills, G.Robinson   *Inorg. Chim. Acta,* **1,** 61, 1967

71.16   trans - Chloromethyl - bis(trimethylarsine) - platinum(ii) hexafluorobut - 2 - yne
$C_{11}H_{21}As_2ClF_6Pt$
B.W.Davies, R.J.Puddephatt, N.C.Payne   *Canad. J. Chem.*, **50,** 2276, 1972
Also classified in 72, 86

71.C   bis(Triphenylcarbenium)di - μ - chloro - bis((di - μ - chloro) - (μ - o - phenylene)bis(dichloroplatinum)) dichloromethane solvate
$C_{12}H_8Cl_{14}Pt_4{}^{2-}$, $2C_{19}H_{15}{}^+$, $2CH_2Cl_2$
For complete entry see 12.10

71.17   Iron carbonyl derivative of barbalone (triclinic form)
$C_{12}H_8FeO_4$
F.A.Cotton, J.M.Troup   *J. Amer. Chem. Soc.*, **95,** 3798, 1973
Also classified in 75

71.18 Tetracyano - tetrakis(methyl isocyanide) molybdenum(iv)
$C_{12}H_{12}MoN_8$
M.Novotny, D.F.Lewis, S.J.Lippard *J. Amer. Chem. Soc.*, **94,** 6961, 1972

71.19 Hexa(methylisocyanide)iron(ii) tetrachloroferrate(iii)
$C_{12}H_{18}FeN_6^{2+}$, $2Cl_4Fe^-$
G.Constant, J.-C.Daran, Y.Jeannin *Eur. Cryst. Meeting,* 1973

71.20 tris(Methyl - vinyl - ketone) tungsten
$C_{12}H_{18}O_3W$
R.E.Moriarty, R.D.Ernst, R.Bau *J. C. S. Chem. Comm.*, 1242, 1972
Also classified in 72, 84

71.21 bis - (μ - N,N - Dimethylthiocarbamoyl) - bis(chloro(trimethylphosphite) - palladium(ii))
$C_{12}H_{21}Cl_2N_2O_6P_2Pd_2S_2$
S.K.Porter, H.White, C.R.Green, R.J.Angelici, J.Clardy
J. C. S. Chem. Comm., 493, 1973
Also classified in 85, 86

71.22 1,2 - bis(Trifluoromethyl) - 1 - (trans - dimethyl - trimethylphosphino - gold(iii)) - 2 - (trimethylphosphino - gold(i)) - ethylene
$C_{12}H_{24}Au_2F_6P_2$
J.A.J.Jarvis, A.Johnson, R.J.Puddephatt *J. C. S. Chem. Comm.*, 373, 1973
Also classified in 86

71.23 1 - (trans - (Dichloro(diethylamine) - platinio)) - 2 - diethylammoniumethane
$C_{12}H_{30}Cl_2N_2Pt$
E.Benedetti, A.de Renzi, G.Paiaro, A.Panunzi, C.Pedone
Gazz. Chim. Ital., **102,** 744, 1972
Also classified in 83

71.24 μ - (1 - 3,6 - η.1,4 - 6 - η - 1,3,6 - tris(Trifluoromethyl)hexa - 1,3,5 - trien - 1,6 - diyl) - bis(dicarbonyl - cobalt)
$C_{13}H_3Co_2F_9O_4$
R.S.Dickson, P.J.Fraser, B.M.Gatehouse *J. C. S. Dalton,* 2278, 1972

71.25 Trichloromethyl - mercury chloride - 1,10 - phenanthroline
$C_{13}H_8Cl_4HgN_2$
A.D.Redhouse *J. C. S. Chem. Comm.*, 1119, 1972
Also classified in 83

71.26 Tetracarbonyl - 2 - (dimethylaminomethyl) phenyl - manganese
$C_{13}H_{12}MnNO_4$
R.G.Little, R.J.Doedens *Inorg. Chem.*, **12,** 844, 1973
Also classified in 83

71.27   **Cyclopentadienyl - (2 - hydroxycyclo - octa - 5 - enyl) - palladium(ii)**
$C_{13}H_{18}OPd$
A.C.Villa, A.G.Manfredotti, C.Guastini
Cryst. Struct. Comm., **2**, 181, 1973
Also classified in 73, 75

71.28   **Dipivaloylmethyl - mercury(ii) acetate**
$C_{13}H_{22}HgO_4$
R.Allmann   *Eur. Cryst. Meeting,* 1973
Also classified in 81

71.29   **trihapto - 1,1′,2′ - ((5′ - Oxocyclopenten - 1 - yl) - 1 - ethyl)(π -
cyclopentadienyl) - dicarbonylmolybdenum**
$C_{14}H_{13}MoO_2$
P.Herpin   *Eur. Cryst. Meeting,* 1973
Also classified in 72, 73

71.30   **Di - (μ - methylisonitrile) di - (h^5 - cyclopentadienyl) dinickel**
$C_{14}H_{16}N_2Ni_2$
R.D.Adams, F.A.Cotton, G.A.Rusholme   *J. Coord. Chem.*, **1**, 275, 1971
Also classified in 73

71.31   **N,N′ - Ethylene - bis(acetylacetoniminato) vinyl - aquo - cobalt(iii)**
$C_{14}H_{23}CoN_2O_3$
S.Bruckner, M.Calligaris, G.Nardin, L.Randaccio
Inorg. Chim. Acta, **2**, 416, 1968
Also classified in 76, 77

71.32   **Hydrido - enneacarbonyl - μ - (3,3 - dimethyl - 1 - butynyl) - triangulo -
triruthenium**
$C_{15}H_{10}O_9Ru_3$
G.Gervasio, G.Ferraris   *Abstr. Ital.-Yug. Congr,* 175, 1973
Also classified in 72

71.33   **Tricyclopentadienyl - scandium**
$(C_{15}H_{15}Sc)_n$
J.L.Atwood, K.D.Smith   *J. Amer. Chem. Soc.*, **95**, 1488, 1973
Also classified in 73

71.34   **Iron(iii) acetylacetonate silver perchlorate**
$C_{15}H_{21}AgFeO_6^+$, ClO_4^-
L.R.Nassimbeni, M.M.Thackeray   *Inorg. Nucl. Chem. Letters,* **9**, 539, 1973
Residue 1 also classified in 77

71.35   **bis(Benzyl - trimethyl - ammonium) carbido - pentadecacarbonyl -
hexarhodate**
$C_{16}O_{15}Rh^{2-}$, $2C_{10}H_{16}N^+$
V.G.Albano, M.Sansoni, P.Chini, S.Martinengo   *J. C. S. Dalton,* 651, 1973
Residue 2 classified in 3

71.36   **(Bullvalene - 2,8 - diyl - iron tricarbonyl) iron tricarbonyl**
$C_{16}H_{10}Fe_2O_6$
G.Huttner, D.Regler   *Chem. Ber.*, **105**, 3936, 1972
Also classified in 75

71.37   **5 - (Tricarbonyl - iron - methyl) - 6 - allyl - cyclohexa - 1,3 - diene iron tricarbonyl**
$C_{16}H_{12}Fe_2O_6$
J.M.Troup, B.A.Frenz, F.A.Cotton
Amer. Cryst. Assoc., Abstr. Papers (Winter Meeting), 41, 1973
Also classified in 72, 75

71.38   **tetrakis(Trimethylsilylmethyl - copper(i)) (at −40°C)**
$C_{16}H_{44}Cu_4Si_4$
J.A.J.Jarvis, B.T.Kilbourn, R.Pearce, M.F.Lappert
J. C. S. Chem. Comm., 475, 1973
Also classified in 63

71.39   **Nonacarbonyl - (phenyl(methoxy)carbene) dimanganese(0)**
$C_{17}H_8Mn_2O_{10}$
G.Huttner, D.Regler   *Chem. Ber.*, **105**, 1230, 1972

71.40   **Tetracarbonyl - 2 - (N - phenylformimidoyl)phenyl - manganese**
$C_{17}H_{10}MnNO_4$
R.G.Little, R.J.Doedens   *Inorg. Chem.*, **12**, 840, 1973
Also classified in 83

71.41   **1,1 - bis(Trimethylphosphite) - 2,3,4,5,6,7 - hexa(trifluoromethyl) - 1 - nickela - cyclohepta - cis,trans,cis - triene**
$C_{18}H_{18}F_{18}NiO_6P_2$
J.Browning, M.Green, B.R.Penfold, J.L.Spencer, F.G.A.Stone
J. C. S. Chem. Comm., 31, 1973
Also classified in 86

71.42   **N,N' - Ethylene - bis(acetylacetoniminato) - methyl - pyridino - cobalt(iii)**
$C_{18}H_{26}CoN_3O_2$
M.Calligaris, G.Nardin, L.Randaccio
Inorg. Nucl. Chem. Letters, **8**, 477, 1972
Also classified in 76, 83

71.43   **(Di - isopropyl - phenyl - phosphine) - π - pentenyl - σ - methyl - nickel**
$C_{18}H_{31}NiP$
B.L.Barnett, C.Kruger   *Acta Cryst. (A)*, **28**, S80, 1972
Also classified in 72, 86

71.44   **Complex III**
$C_{19}H_9CoF_6N_2O_3$
M.I.Bruce, B.L.Goodall, A.D.Redhouse, F.G.A.Stone
J. C. S. Chem. Comm., 1228, 1972
Also classified in 72

71.45 π - Cyclopentadienyl - azotoluene - 2 - yl - nickel
$C_{19}H_{18}N_2Ni$
V.A.Semion, I.V.Barinov, Yu.A.Ustynyuk, Yu.T.Struchkov
Zh. Strukt. Khim., **13**, 543, 1972
Also classified in 73, 83

71.46   Dibenzosemibullvalene - iron tetracarbonyl
$C_{20}H_{12}FeO_4$
R.M.Moriarty, K.-N.Chen, C.-L.Yeh, J.L.Flippen, J.Karle
J. Amer. Chem. Soc., **94**, 8944, 1972

71.47   Tetracyclopentadienyl - hafnium (at 0°C)
$C_{20}H_{20}Hf$
V.I.Kulishov, E.M.Brainina, N.G.Bokiy, Yu.T.Struchkov
J. Organometal. Chem., **36**, 333, 1972
Also classified in 73

71.48   Niobocene dimer
$C_{20}H_{20}Nb_2$
L.J.Guggenberger *Inorg. Chem.*, **12**, 294, 1973
Also classified in 73

71.49   1,1 - bis(Phenyl - dimethyl - arsine) - 1 - nickela - octafluoro - cyclopentane
$C_{20}H_{22}As_2F_8Ni$
R.Countryman, B.R.Penfold *Acta Cryst. (A)*, **28**, S83, 1972
Also classified in 86

71.50   trans - 1,3 - Diphenyl - 2 - (dichloro - triethylphosphine - platinum) - imidazolidine
$C_{21}H_{29}Cl_2N_2PPt$
D.J.Cardin, B.Cetinkaya, E.Cetinkaya, M.F.Lappert, L.J.Manojlovic-Muir, K.W.Muir *J. Organometal. Chem.*, **44**, C59, 1972
Also classified in 86

71.51   Decacarbonyl(cyclododecatrienyl) - tetrahedro - tetraruthenium
$C_{22}H_{16}O_{10}Ru_4$
R.Belford, H.P.Taylor, P.Woodward *J. C. S. Dalton*, 2425, 1972
Also classified in 75

71.52   Phenylmethinyl - tricobalt hexacarbonyl - mesitylene
$C_{22}H_{17}Co_3O_6$
R.J.Dellaca, B.R.Penfold *Inorg. Chem.*, **11**, 1855, 1972
Also classified in 74

71.53   bis(Dipivaloylmethyl) mercury
$C_{22}H_{30}HgO_4$
R.Allmann, K.Flatau, H.Musso *Chem. Ber.*, **105**, 3067, 1972

71.54   **1 - Tricarbonyl - 2,3,4,5 - tetra(trimethylsiloxy) - 1 - ferracyclopentadiene tricarbonyl - iron**
$C_{22}H_{36}Fe_2O_{10}Si_4$
M.J.Bennett, W.A.G.Graham, R.A.Smith, R.P.Stewart Junior
J. Amer. Chem. Soc., **95,** 1684, 1973
Also classified in 75, 63

71.55   **Methyl - tetracarbonyl - triphenylphosphine - manganese(i)**
$C_{23}H_{18}MnO_4P$
A.Mawby, G.E.Pringle   *J. Inorg. Nucl. Chem.*, **34,** 877, 1972
Also classified in 86

71.56   **Tricyclopentadienyl - phenylethynyl - uranium(iv)**
$C_{23}H_{20}U$
J.L.Atwood, C.F.Hains Junior, M.Tsutsui, A.E.Gebala
J. C. S. Chem. Comm., 452, 1973
Also classified in 73

71.57   **N,N' - Ethylene - bis(salicylideneiminato) - pyridine - vinyl - cobalt(iii)**
$C_{23}H_{22}CoN_3O_2$
M.Calligaris, G.Nardin, L.Randaccio   *J. C. S. Dalton,* 1433, 1972
Also classified in 78, 83

71.58   **Compound II**
$C_{23}H_{25}ClO_{14}Pd$, $0.66CHCl_3$
D.M.Roe, C.Calvo, N.Krishnamachari, K.Moseley, P.M.Maitlis
J. C. S. Chem. Comm., 436, 1973
Residue 1 also classified in 77

71.59   **Tri(cyclohexyl)phosphine - ethylidene - nickel tricarbonyl**
$C_{23}H_{37}NiO_3P$
B.L.Barnett, C.Kruger   *J. Cryst. Mol. Struct.*, **2,** 271, 1972

71.60   **(Pentafluorophenyl)(triphenylphosphine) gold(i)**
$C_{24}H_{15}AuF_5P$
R.W.Baker, P.J.Pauling   *J. C. S. Dalton*, 2264, 1972
Also classified in 86

71.61   **o - (1 - Manganese - tetracarbonyl - ethyl)phenyl - diphenylphosphine**
$C_{24}H_{18}MnO_4P$
G.B.Robertson, P.O.Whimp   *J. Organometal. Chem.*, **49,** C27, 1973
Also classified in 86

71.C   **bis(t - Butyl - isocyanide)(diphenylacetylene) nickel(0)**
$C_{24}H_{28}N_2Ni$
For complete entry see 72.33

71.62   **(Tricyclohexylphosphine)methyl - nickel(ii) 2,4 - pentanedionate**
$C_{24}H_{43}NiO_2P$
B.L.Barnett, C.Kruger   *J. Organometal. Chem.*, **42,** 169, 1972
Also classified in 77, 86

71.63 Tricarbonyl - π - (1,1,1 - tricarbonyl - 2 - methyl - 3 - diphenylmethylene - 6 - methoxy - ferra - 2 - oxacyclohexenyl) iron
$C_{25}H_{18}Fe_2O_8$
J.A.D.Jeffreys, C.M.Willis, I.C.Robertson, G.Ferguson, J.G.Sime
J. C. S. Dalton, 749, 1973
Also classified in 75

71.64 Tricarbonyl - π - (1,1,1 - tricarbonyl - 2,3 - dimethoxy - 5 - (diphenylmethyl) ferracyclopentadiene) iron
$C_{25}H_{18}Fe_2O_8$
J.A.D.Jeffreys, C.M.Willis *J. C. S. Dalton,* 2169, 1972
Also classified in 75

71.65 π - Diphenylacetylene - π - cyclopentadienyl - phenyl - oxo - tungsten
$C_{25}H_{20}OW$
N.G.Bokii, A.I.Gusev, V.E.Shklover, Yu.T.Struchkov
Acta Cryst. (A), **28,** S84, 1972
Also classified in 72, 73

71.66 bis(Dimethylglyoximato)(tri - n - butylphosphine)(4 - pyridyl) cobalt(iii)
$C_{25}H_{45}CoN_5O_4P$
W.W.Adams, P.G.Lenhert
Amer. Cryst. Assoc., Abstr. Papers (Winter Meeting), 44, 1973
Also classified in 86, 83

71.67 Di - μ - sulfur dioxide - pentakis(t - butyl isocyanide) - triangulo - tripalladium benzene solvate
$C_{25}H_{45}N_5O_4Pd_3S_2$, $2C_6H_6$
S.Otsuka, Y.Tatsuno, M.Miki, T.Aoki, M.Matsumoto, H.Yoshioka, K.Nakatsu *J. C. S. Chem. Comm.,* 445, 1973

71.68 Di - μ - chloro - dicarbonyl - rhodium(i) - bis(phenylazophenyl - 2C,N') rhodium(iii)
$C_{26}H_{18}Cl_2N_4O_2Rh_2$
R.J.Hoare, O.S.Mills *J. C. S. Dalton,* 2141, 1972
Also classified in 83

71.69 Acetato - bis(phenylazophenyl - 2C,N') rhodium(ii)
$C_{26}H_{21}N_4O_2Rh$
R.J.Hoare, O.S.Mills *J. C. S. Dalton,* 2138, 1972
Also classified in 81, 83

71.70 Norbornadiene - (binorbornenyl - diyl) iridium acetylacetonate
$C_{26}H_{31}IrO_2$
A.R.Fraser, P.H.Bird, S.A.Bezman, J.R.Shapley, R.White, J.A.Osborn
J. Amer. Chem. Soc., **95,** 597, 1973
Also classified in 75, 77

TRANSITION METAL-C COMPOUNDS

71.71   Di - μ - acetato - bis((2 - methallyl - 3 - norbornyl) palladium(ii))
$C_{26}H_{40}O_4Pd_2$
M.Zocchi, G.Tieghi, A.Albinati   *J. C. S. Dalton,* 883, 1973
Also classified in 72, 81

71.72   Nonacarbonyl - μ - dimethylarsino - μ - (2 -
(diphenylphosphino)tetrafluorocyclobut - 1 - enyl) - tri - iron
$C_{27}H_{16}AsF_4Fe_3O_9P$
F.W.B.Einstein, R.D.G.Jones   *J. C. S. Dalton,* 2563, 1972
Also classified in 75, 86

71.73   (Dimenthenyl - methyl - phosphine) - π - pentenyl - σ - methyl - nickel
$C_{27}H_{49}NiP$
B.L.Barnett, C.Kruger   *Acta Cryst. (A),* **28,** S80, 1972
Also classified in 72, 86

71.C   bis(Trimethylsilyl - methyl) - bis(2,2' - bipyridyl) chromium(iii) iodide
$C_{28}H_{38}CrN_4Si_2^+$, I^-
For complete entry see 63.13

71.74   Di - μ - chloro - bis(dicarbonyl - cyclopentadienyl - iron - phenylacetylide)
$C_{30}H_{20}Cl_2Cu_2Fe_2O_4$
M.I.Bruce, R.Clark, J.Howard, P.Woodward
J. Organometal. Chem., **42,** C107, 1972
Also classified in 72, 73

71.75   Sodium pentaphenylchromate tri - diethylether tetrahydrofuran solvate
$C_{30}H_{25}Cr^{2-}$, $2Na^+$, $3C_4H_{10}O$, C_4H_8O
E.Muller, J.Krause, K.Schmiedeknecht
J. Organometal. Chem., **44,** 127, 1972

71.76   1,2,3,4 - Tetra(trifluoromethyl)butadienyl - (cyclopentadienyl) - triphenyl -
phosphine - ruthenium
$C_{31}H_{21}F_{12}PRu$
R.E.Davis, A.Garza, N.V.Raghavan
Amer. Cryst. Assoc., Abstr. Papers (Winter Meeting), 91, 1973
Also classified in 72, 73, 86

71.77   cis - Diphenyl - bis(2,2' - bipyridyl) chromium(iii) iodide
$C_{32}H_{26}CrN_4^+$, I^-
J.J.Daly, F.Sanz, R.P.A.Sneeden, H.H.Zeiss   *J. C. S. Dalton,* 73, 1973
Residue 1 also classified in 83

71.78   Triphenylphosphine - (1 - (di(trifluoromethyl) - hydroxymethyl) -
cyclopentadienyl) - (1,2 - di(carboxymethyl)ethylene - 1 - yl) - ruthenium
$C_{32}H_{27}F_6O_5PRu$
R.E.Davis, A.Garza, N.V.Raghavan
Amer. Cryst. Assoc., Abstr. Papers (Winter Meeting), 91, 1973
Also classified in 73, 84, 86

71.79 **tetrakis(Tetrahydrofuran) lithium tetrakis(2,6 - dimethylphenyl) lutetiate**
$C_{32}H_{36}Lu^-$, $C_{16}H_{32}LiO_4^+$
S.A.Cotton, F.A.Hart, M.B.Hursthouse, A.J.Welch
J. C. S. Chem. Comm., 1225, 1972
Residue 2 classified in 67

71.80 **Dibromo - tetra(2 - dimethylaminophenyl) - hexacopper(i) benzene solvate**
$C_{32}H_{40}Br_2Cu_6N_4$, $1.5C_6H_6$
J.M.Guss, R.Mason, K.M.Thomas, G.van Koten, J.G.Noltes
J. Organometal. Chem., **40**, C79, 1972
Residue 1 also classified in 83

71.81 **(bis(Biphenylylidene)butatriene)hexacarbonyl - di - iron**
$C_{34}H_{16}Fe_2O_6$
D.Bright, O.S.Mills *J. C. S. Dalton,* 2465, 1972
Also classified in 72

71.82 **Structure III**
$C_{34}H_{25}FeNiO_3P$
B.L.Barnett, C.Kruger *Cryst. Struct. Comm.*, **2**, 347, 1973
Also classified in 73, 86

71.83 **cis - bis(- 2 - Methoxyphenyl) - bis(2,2' - bipyridyl) chromium(iii) iodide monohydrate**
$C_{34}H_{30}CrN_4O_2^+$, I^-, H_2O
J.J.Daly, F.Sanz *J. C. S. Dalton,* 2584, 1972
Residue 1 also classified in 83

71.84 **(Cyclo - octa - 1,5 - diene)(1,2 - bis(diphenylphosphino)ethane) methyl iridium(i)**
$C_{35}H_{39}IrP_2$
M.R.Churchill, S.A.Bezman *Inorg. Chem.*, **12**, 260, 1973
Also classified in 75, 86

71.85 **(Cyclo - octa - 1,5 - diene)(1,3 - bis(diphenylphosphino)propane)methyl iridium(i)**
$C_{36}H_{41}IrP_2$
M.R.Churchill, S.A.Bezman *Inorg. Chem.*, **12**, 531, 1973
Also classified in 75, 86

71.86 **(Di - (o - tolyl) - t - butylphosphine) - dibenzyl - platinum - o,o' - (t - butylphosphine)**
$C_{36}H_{44}P_2Pt$
A.J.Cheney, W.S.McDonald, K.O'Flynn, B.L.Shaw, B.L.Turtle
J. C. S. Chem. Comm., 128, 1973
Also classified in 86

71.87   (±) - cis - Di(o - ((o - tolyl)(t - butyl)phosphino)benzyl) platinum
$C_{36}H_{44}P_2Pt$
A.J.Cheney, W.S.McDonald, K.O'Flynn, B.L.Shaw, B.L.Turtle
J. C. S. Chem. Comm., 128, 1973
Also classified in 86

71.88   (±) - trans - Di(o - ((o - tolyl)(t - butyl)phosphino)benzyl) platinum
$C_{36}H_{44}P_2Pt$
A.J.Cheney, W.S.McDonald, K.O'Flynn, B.L.Shaw, B.L.Turtle
J. C. S. Chem. Comm., 128, 1973
Also classified in 86

71.89   Iodo - methyl - bis(triphenylphosphine) platinum - sulfur dioxide
$C_{37}H_{33}IP_2Pt$, O_2S
M.R.Snow, J.A.Ibers   *Inorg. Chem.*, **12,** 224, 1973
Residue 1 also classified in 86

71.90   2,5 - Di(trifluoromethyl) - 3 - (tricarbonyl - diphenylphosphino - iron) - 4 - diphenylphosphino - 1 - tricarbonyl - 1 - ferra - cyclopenta - 2,4 - diene dicarbonyl iron benzene solvate
$C_{38}H_{20}F_6Fe_3O_8P_2$, $0.5C_6H_6$
T.O'Connor, A.J.Carty, M.Mathew, G.J.Palenik
J. Organometal. Chem., **38,** C15, 1972
Residue 1 also classified in 86

71.91   Dichloro - carbonyl - bis(triphenylphosphine) - difluoromethyl - iridium
$C_{38}H_{31}Cl_2F_2IrP_2$
A.J.Schultz, G.P.Khare, J.V.McArdle, R.Eisenberg
J. Amer. Chem. Soc., **95,** 3434, 1973
Also classified in 86

71.92   bis(Triphenylphosphine)hexafluoroacetone nickel(0)
3,3 - Di(trifluoromethyl) - 1 - nickela - 2 - oxa - cyclopropane
bis(triphenylphosphine)
$C_{39}H_{30}F_6NiOP_2$
R.Countryman, B.R.Penfold   *J. Cryst. Mol. Struct.*, **2,** 281, 1972
Also classified in 86, 84

71.93   Chloro - carbonyl - bis(triphenylphosphine) - difluoromethyl - chlorodifluoroacetato - iridium benzene solvate
$C_{40}H_{31}Cl_2F_4IrO_3P_2$, C_6H_6
A.J.Schultz, G.P.Khare, J.V.McArdle, R.Eisenberg
J. Amer. Chem. Soc., **95,** 3434, 1973
Residue 1 also classified in 81, 86

71.94 **(1,4 - bis(o - Diphenylphosphino - phenyl) - buta - diyl) ruthenium(ii) dicarbonyl**
$C_{42}H_{34}O_2P_2Ru$
M.A.Bennett, G.B.Robertson, I.B.Tomkins, P.O.Whimp
Acta Cryst. (A), **28**, S82, 1972
Also classified in 86

71.95 **Osmium compound X**
$C_{43}H_{28}O_7Os_3P_2$
G.J.Gainsford, J.M.Guss, P.R.Ireland, R.Mason, C.W.Bradford,
R.S.Nyholm *J. Organometal. Chem.*, **40**, C70, 1972
Also classified in 86

71.96 **Osmium compound VI**
$C_{43}H_{30}O_7Os_3P_2$
G.J.Gainsford, J.M.Guss, P.R.Ireland, R.Mason, C.W.Bradford,
R.S.Nyholm *J. Organometal. Chem.*, **40**, C70, 1972
Also classified in 86

71.97 **Carbonyl - chloro - (4 - fluorophenyl - di - imide - 2C,N′)bis(triphenylphosphine) - iridium(iii) tetrafluoroborate acetone solvate**
$C_{43}H_{34}ClFIrN_2OP_2^+$, BF_4^- , C_3H_6O
F.W.B.Einstein, D.Sutton *J. C. S. Dalton*, 434, 1973
Residue 1 also classified in 83, 86

71.98 **Manganese compound 4**
$C_{44}H_{28}Mn_2O_8P_2$
R.J.McKinney, B.T.Huie, C.B.Knobler, H.D.Kaesz
J. Amer. Chem. Soc., **95,** 633, 1973
Also classified in 86

71.99 **Osmium compound IV**
$C_{44}H_{30}O_8Os_3P_2$
G.J.Gainsford, J.M.Guss, P.R.Ireland, R.Mason, C.W.Bradford,
R.S.Nyholm *J. Organometal. Chem.*, **40,** C70, 1972
Also classified in 86

71.100 **1,1 - bis(Triphenylphosphine) - 1 - platina - benzocyclopentenedione (monoclinic form)**
$C_{44}H_{34}O_2P_2Pt$
J.A.Evans, G.F.Everitt, R.D.W.Kemmitt, D.R.Russell
J. C. S. Chem. Comm., 158, 1973
Also classified in 86

71.101 **1,1 - bis(Triphenylphosphine) - 1 - platina - benzocyclopentenedione (triclinic form)**
$C_{44}H_{34}O_2P_2Pt$
J.A.Evans, G.F.Everitt, R.D.W.Kemmitt, D.R.Russell
J. C. S. Chem. Comm., 158, 1973
Also classified in 86

71.102 **(σ - Dicyanovinyl)(carbonyl)(π - dicyanoacetylene) - bis(triphenylphosphine) iridium(i)**
$C_{45}H_{31}IrN_4OP_2$
R.M.Kirchner, J.A.Ibers *J. Amer. Chem. Soc.*, **95**, 1095, 1973
Also classified in 72, 86

71.103 **trans - bis(Triphenylphosphine) - chloro - carbonyl - (2 - dicarbadodecaborane(12) - acetylide) - (trans - 2 - dicarbadodecaborane(12) - vinyl) - iridium dichloromethane solvate**
$C_{45}H_{54}B_{20}ClIrOP_2$, $2.5CH_2Cl_2$
K.P.Callahan, C.E.Strouse, S.W.Layten, M.F.Hawthorne
J. C. S. Chem. Comm., 465, 1973
Residue 1 also classified in 86

71.104 **Compound I**
$C_{46}H_{39}Au_2FeP_2^+$, BF_4^-
V.G.Andrianov, Yu.T.Struchkov, E.R.Rossinskaja
J. C. S. Chem. Comm., 338, 1973
Residue 1 also classified in 73, 86

71.105 **Bromo - bis(triphenylphosphine) - (1,2,3,4 - tetrakis(methoxycarbonyl) - buta - 1,3 - dienyl) palladium**
$C_{48}H_{43}BrO_8P_2Pd$
D.M.Roe, P.M.Bailey, K.Moseley, P.M.Maitlis
J. C. S. Chem. Comm., 1273, 1972
Also classified in 86

71.106 **Heptacarbonyl - μ - (1,2,3,4 - tetraphenylbutadiene - 1,4 - diyl) - μ - (diphenylacetylene) - triangulo - triosmium**
$C_{49}H_{30}O_7Os_3$
G.Gervasio, G.Ferraris *Abstr. Ital.-Yug. Congr*, 175, 1973
Also classified in 72

71.107 **bis(Triphenylphosphine) - cyclopentadienyl - ruthenium - phenylacetylide chloro - copper(i) acetone solvate**
$C_{49}H_{40}ClCuP_2Ru$, C_3H_6O
R.E.Davis, A.Garza, N.V.Raghavan
Amer. Cryst. Assoc., Abstr. Papers (Winter Meeting), 91, 1973
Residue 1 also classified in 72, 73, 86

71.C **Ruthenium complex with triphenyl phosphite**
$C_{51}H_{39}O_{12}P_3Ru_2$
For complete entry see 74.12

71.C **cis - Triphenylphosphine - phenyl - platinum - bis(triphenyl - lead)**
$C_{60}H_{50}PPb_2Pt$
For complete entry see 69.28

71.108 **Neodymium tri(methylcyclopentadienide) tetramer**
$C_{72}H_{84}Nd_4$
J.H.Burns, W.H.Baldwin, F.H.Fink
Amer. Cryst. Assoc., Abstr. Papers (Winter Meeting), 41, 1973
Also classified in 73

METAL PI-COMPLEXES (OPEN-CHAIN)

72.1 **Trichloro(pent - 4 - enylammonium) platinum(ii) (yellow form)**
$C_5H_{12}Cl_3NPt$
R.Spagna, L.Zambonelli *Acta Cryst. (B),* **28,** 2760, 1972

72.2 **Trichloro(pent - 4 - enylammonium) platinum(ii) (orange form)**
$C_5H_{12}Cl_3NPt$
R.Spagna, L.Zambonelli *Acta Cryst. (B),* **28,** 2760, 1972

72.3 **Trichloro - (π - trans - pent - 2 - enylammonium) platinum(ii)**
$C_5H_{12}Cl_3NPt$
R.Spagna, G.Ughetto, L.Zambonelli *Acta Cryst. (B),* **29,** 1151, 1973

72.4 **Acetylacetonato - chloro(1,2 - dihapto(vinyl alcohol)) platinum(ii)**
$C_7H_{11}ClO_3Pt$
F.A.Cotton, J.N.Francis, B.A.Frenz, M.Tsutsui
J. Amer. Chem. Soc., **95,** 2483, 1973
Also classified in 77

72.5 **trans - 2 - Bromovinyl - μ - bromo - bis(tricarbonyl iron)**
$C_8H_2Br_2Fe_2O_6$
C.Kruger, Y.H.Tsay, F.W.Grevels, E.Koerner von Gustorf
Israel J. Chem., **10,** 201, 1972

72.6 **bis(Butadiene) iron - trifluorophosphine**
$C_8H_{12}F_3FeP$
C.Kruger, Y.-H.Tsay *Eur. Cryst. Meeting,* 1973

72.7 **cis - Dichloro - (2,2' - oxydi - 3 - butene) platinum(ii)**
$C_8H_{14}Cl_2OPt$
J.Hubert, A.L.Beauchamp, T.Theophanides
Canad. J. Chem., **51,** 604, 1973

72.8 **π - 2,3 - Dichlorobutadiene - π - cyclopentadienyl - rhodium(i)**
$C_9H_9Cl_2Rh$
M.G.B.Drew, S.M.Nelson, M.Sloan *J. Organometal. Chem.,* **39,** C9, 1972
Also classified in 73

72.9 **Tricarbonyl(1 - 3 - η - hexen - 5 - one) iron hexafluorophosphate**
$C_9H_9FeO_4^+$, F_6P^-
A.D.U.Hardy, G.A.Sim *J. C. S. Dalton,* 2305, 1972
Residue 1 also classified in 84

72.10   **bis(Butadiene)monocarbonyl iron**
$C_9H_{12}FeO$
D.A.Whiting   *Cryst. Struct. Comm.*, **1,** 379, 1972

72.11   **Ethylene - platinum - dichloride di - t - butyl - sulfur di - imine**
$C_{10}H_{22}Cl_2N_2PtS$
R.T.Kops, E.van Aken, H.Schenk   *Acta Cryst. (B)*, **29,** 913, 1973
Also classified in 83

72.12   **π - 2,3 - Dimethylbutadiene - π - cyclopentadienyl - rhodium(i)**
$C_{11}H_{15}Rh$
M.G.B.Drew, S.M.Nelson, M.Sloan   *J. Organometal. Chem.*, **39,** C9, 1972
Also classified in 73

72.C   **trans - Chloromethyl - bis(trimethylarsine) - platinum(ii) hexafluorobut - 2 - yne**
$C_{11}H_{21}As_2ClF_6Pt$
For complete entry see 71.16

72.13   **μ - Butadiene - dimanganese - octacarbonyl**
$C_{12}H_6Mn_2O_8$
H.E.Sasse, M.L.Ziegler   *Z. Anorg. Allg. Chem.*, **392,** 167, 1972

72.14   **Tricarbonyl(cinnamaldehyde) iron**
$C_{12}H_8FeO_4$
A.de Cian, R.Weiss   *Acta Cryst. (B)*, **28,** 3273, 1972

72.C   **tris(Methyl - vinyl - ketone) tungsten**
$C_{12}H_{18}O_3W$
For complete entry see 71.20

72.15   **Potassium trichloroplatinum - (di(diethyl - hydroxy - methyl)acetylene)**
$C_{12}H_{22}Cl_3O_2Pt^-$, K^+
A.L.Beauchamp, F.D.Rochon, T.Theophanides
Canad. J. Chem., **51,** 126, 1973

72.16   **Butadiene(cyclo - octatetraene) iron monocarbonyl**
$C_{13}H_{14}FeO$
I.W.Bassi, R.Scordamaglia   *J. Organometal. Chem.*, **37,** 353, 1972
Also classified in 75

72.C   **trihapto - 1,1′,2′ - ((5′ - Oxocyclopenten - 1 - yl) - 1 - ethyl)(π - cyclopentadienyl) - dicarbonylmolybdenum**
$C_{14}H_{13}MoO_2$
For complete entry see 71.29

72.17   **bis - (π - 2 - Methylallyl) - bis(trimethylphosphite) ruthenium (monoclinic form)**
$C_{14}H_{32}O_6P_2Ru$
R.A.Marsh, J.Howard, P.Woodward   *J. C. S. Dalton*, 778, 1973
Also classified in 86

72.C   Hydrido - enneacarbonyl - μ - (3,3 - dimethyl - 1 - butynyl) - triangulo - triruthenium
$C_{15}H_{10}O_9Ru_3$
For complete entry see 71.32

72.18   bis(Methyl sorbate) iron carbonyl
$C_{15}H_{20}FeO_5$
C.Kruger, Y.-H.Tsay   *Eur. Cryst. Meeting,* 1973

72.19   Cyclo - octatetraenyl - (allyl) - t - butoxy - zirconium
$C_{15}H_{22}OZr$
D.J.Brauer, C.Kruger   *Eur. Cryst. Meeting,* 1973
Also classified in 75, 84

72.C   5 - (Tricarbonyl - iron - methyl) - 6 - allyl - cyclohexa - 1,3 - diene iron tricarbonyl
$C_{16}H_{12}Fe_2O_6$
For complete entry see 71.37

72.20   3,α - Dimethylstyrene - bis(tricarbonyl - iron)
$C_{16}H_{12}Fe_2O_6$
F.H.Herbstein, M.G.Reisner   *J. C. S. Chem. Comm.,* 1077, 1972
Also classified in 74

72.21 π - Methallyl - (2 - (R,S) - α - phenylethylimino - 3 - penten - 4 - olato) palladium(ii)
$C_{17}H_{23}NOPd$
R.Claverini, P.Ganis, C.Pedone   *J. Organometal. Chem.,* **50,** 327, 1973
Also classified in 83, 84

72.22   Tricarbonyl(N - cinnamylideneaniline) iron (orthorhombic form)
$C_{18}H_{13}FeNO_3$
A.de Cian, R.Weiss   *Acta Cryst. (B),* **28,** 3264, 1972

72.23   Tricarbonyl(N - cinnamylideneaniline) iron (monoclinic form)
$C_{18}H_{13}FeNO_3$
A.de Cian, R.Weiss   *Acta Cryst. (B),* **28,** 3264, 1972

72.24 μ - (2,6 - Dimethoxyphenyl(ethoxy)carbene)heptacarbonyl - di - iron
$C_{18}H_{14}Fe_2O_{10}$
G.Huttner, D.Regler   *Chem. Ber.,* **105,** 2726, 1972

72.C   (Di - isopropyl - phenyl - phosphine) - π - pentenyl - σ - methyl - nickel
$C_{18}H_{31}NiP$
For complete entry see 71.43

72.C   Complex III
$C_{19}H_9CoF_6N_2O_3$
For complete entry see 71.44

72.25 **1,4 - Diphenylbutadiene iron tricarbonyl - 1,4 - diphenylbutadiene**
$C_{19}H_{12}FeO_3$, $0.5C_{16}H_{12}$
A.de Cian, P.M.L'Huillier, R.Weiss *Bull. Soc. Chim. Fr.*, 451, 1973
Residue 2 classified in 19

72.26 **Tetracarbonyl(trans - α,β - dibenzoylethylene) iron**
$C_{20}H_{12}FeO_6$
V.G.Andrianov, Yu.T.Struchkov, M.I.Rybinskaya, L.V.Rybin,
N.T.Gubenko *Zh. Strukt. Khim.*, **13**, 86, 1972

72.27 **Dicarbonyl - (2,2' - bipyridyl) - (pyridine) - (π - allyl) - molybdenum tetrafluoroborate**
$C_{20}H_{18}MoN_3O_2^+$, BF_4^-
R.H.Fenn, A.J.Graham *J. Organometal. Chem.*, **37**, 137, 1972
Residue 1 also classified in 83

72.28 **bis(Dimethyl - phenyl - phosphine) - butadiene - carbonyl - iridium tetrafluoroborate**
$C_{21}H_{26}IrOP_2^+$, BF_4^-
I.A.G.Mustafa, J.H.Robertson *Acta Cryst. (A)*, **28**, S83, 1972
Residue 1 also classified in 86

72.29 **bis(Dimethyl - phenyl - phosphine) - butadiene - carbonyl - iridium perchlorate**
$C_{21}H_{26}IrOP_2^+$, ClO_4^-
I.A.G.Mustafa, J.H.Robertson *Acta Cryst. (A)*, **28**, S83, 1972
Residue 1 also classified in 86

72.30 **Ethylene - (dicyclohexyl - (3 - (ethyl - vinyl - boro)propyl) - phosphine) nickel**
$C_{21}H_{40}BNiP$
B.L.Barnett, D.Brauer, C.Krueger, Y.-H.Tsay
Amer. Cryst. Assoc., Abstr. Papers (Winter Meeting), 42, 1973
Also classified in 86, 62

72.31 **bis(Ethylene)(tricyclohexylphosphine) nickel**
$C_{22}H_{41}NiP$
C.Kruger, Y.-H.Tsay *J. Organometal. Chem.*, **34**, 387, 1972
Also classified in 86

72.32 **Monothiodibenzoylmethanato - (π - syn - 1 - t - butyl - 2 - methallyl) palladium(ii)**
$C_{23}H_{26}OPdS$
S.J.Lippard, S.M.Morehouse *J. Amer. Chem. Soc.*, **94**, 6956, 1972
Also classified in 85

72.33 **bis(t - Butyl - isocyanide)(diphenylacetylene) nickel(0)**
$C_{24}H_{28}N_2Ni$
R.S.Dickson, J.A.Ibers *J. Organometal. Chem.*, **36**, 191, 1972
Also classified in 71

72.C π - **Diphenylacetylene** - π - **cyclopentadienyl** - **phenyl** - **oxo** - **tungsten**
$C_{25}H_{20}OW$
For complete entry see 71.65

72.C **Di** - μ - **acetato** - **bis((2** - **methallyl** - **3** - **norbornyl) palladium(ii))**
$C_{26}H_{40}O_4Pd_2$
For complete entry see 71.71

72.34 μ - $(\alpha,\omega$ - **Octadi** - π - **enyl)bis(bromo** - **tri** - **isopropylphosphine** - **nickel(ii))**
$C_{26}H_{54}Br_2Ni_2P_2$
T.S.Cameron, C.K.Prout *Acta Cryst. (B),* **28,** 2021, 1972
Also classified in 86

72.C **(Dimenthenyl** - **methyl** - **phosphine)** - π - **pentenyl** - σ - **methyl** - **nickel**
$C_{27}H_{49}NiP$
For complete entry see 71.73

72.35 **Carbonyl** - **chloro** - **bis(trimethylphosphine)** - **(1,2,3** - **triphenylpropenylium** -
1,3 - **diyl) iridium tetrafluoroborate dichloromethane solvate**
$C_{28}H_{33}ClIrOP_2{}^+ , BF_4{}^- , CH_2Cl_2$
R.M.Tuggle, D.L.Weaver *Inorg. Chem.,* **11,** 2237, 1972
Residue 1 also classified in 83

72.36 **bis** - μ - **Dicyclohexylphosphido** - **bis(π** - **ethylene nickel)**
$C_{28}H_{52}Ni_2P_2$
B.L.Barnett, C.Kruger *Cryst. Struct. Comm.,* **2,** 85, 1973
Also classified in 86

72.C **Di** - μ - **chloro** - **bis(dicarbonyl** - **cyclopentadienyl** - **iron** - **phenylacetylide)**
$C_{30}H_{20}Cl_2Cu_2Fe_2O_4$
For complete entry see 71.74

72.C **1,2,3,4** - **Tetra(trifluoromethyl)butadienyl** - **(cyclopentadienyl)** - **triphenyl** -
phosphine - **ruthenium**
$C_{31}H_{21}F_{12}PRu$
For complete entry see 71.76

72.37 **1,2** - **bis(Dicyclohexylphosphino)ethane** - **(tetramethylethylene)** - **nickel**
$C_{32}H_{60}NiP_2$
B.L.Barnett, D.Brauer, C.Krueger, Y.-H.Tsay
Amer. Cryst. Assoc., Abstr. Papers (Winter Meeting), 42, 1973
Also classified in 86

72.C **(bis(Biphenylylidene)butatriene)hexacarbonyl** - **di** - **iron**
$C_{34}H_{16}Fe_2O_6$
For complete entry see 71.81

72.38 **Enneacarbonyl** - μ - **(1,2,3,4** - **tetraphenylbutadiene** - **1,4** - **diyl)** - **triangulo** -
triosmium (monoclinic form)
$C_{37}H_{20}O_9Os_3$
G.Gervasio, G.Ferraris *Abstr. Ital.-Yug. Congr,* 175, 1973

72.39   **Enneacarbonyl - μ - (1,2,3,4 - tetraphenylbutadiene - 1,4 - diyl) - triangulo - triosmium (orthorhombic form)**
$C_{37}H_{20}O_9Os_3$
G.Gervasio, G.Ferraris   *Abstr. Ital.-Yug. Congr,* 175, 1973

72.40   **Chloro - (2,2' - bis(diphenylphosphino)stilbene) - rhodium**
$C_{38}H_{30}ClP_2Rh$
M.A.Bennett, P.W.Clark, G.B.Robertson, P.O.Whimp
J. C. S. Chem. Comm., 1011, 1972
Also classified in 86

72.41 π - **Cyclopentadienyl -** π - **diphenylacetylene - monocarbonyl - triphenyl - stannyl - iron**
$C_{38}H_{30}FeOSn$
V.E.Shklober, V.V.Skripkin, A.I.Gusev, Yu.T.Struchkov
Zh. Strukt. Khim., **13,** 744, 1972
Also classified in 73, 69

72.42   **bis(Triphenylphosphine) - (ethylene) platinum**
$C_{38}H_{34}P_2Pt$
P.-T.Cheng, S.C.Nyburg   *Canad. J. Chem.,* **50,** 912, 1972
Also classified in 86

72.43   **bis(Triphenylphosphine) - hexafluoroacetone - nickel(0)**
$C_{39}H_{30}F_{12}NiOP_2$
R.Countryman, B.R.Penfold   *Acta Cryst. (A),* **28,** S83, 1972
Also classified in 86

72.44   **bis(Triphenylphosphine) - (N - methyl - hexafluoroacetonimide) platinum(0)**
$C_{40}H_{36}F_{12}NO_2Pt$
R.Countryman, B.R.Penfold   *Acta Cryst. (A),* **28,** S83, 1972
Also classified in 86

72.45   **(1,4 - bis(o - Diphenylphosphino - phenyl)buta - 1,3 - diene) ruthenium(0) carbonyl**
$C_{41}H_{32}OP_2Ru$
M.A.Bennett, G.B.Robertson, I.B.Tomkins, P.O.Whimp
Acta Cryst. (A), **28,** S82, 1972
Also classified in 86

72.46   **bis(Triphenylphosphine) - bis(π - allyl) ruthenium toluene solvate**
$C_{42}H_{40}P_2Ru$, C_7H_8
A.E.Smith   *Inorg. Chem.,* **11,** 2306, 1972
Residue 1 also classified in 86

72.47   **Carbonyl - chloro - (tetracyanoethylene)bis(triphenylarsine) iridium**
$C_{43}H_{30}As_2ClIrN_4O$
J.B.R.Dunn, R.Jacobs, C.J.Fritchie Junior   *J. C. S. Dalton,* 2007, 1972
Also classified in 86

72.48   **Tetrafluoroethylene - (1,1,1 - tris(diphenylphosphinomethyl)ethane) nickel**
$C_{43}H_{39}F_4NiP_3$
J.Browning, B.R.Penfold. *J. C. S. Chem. Comm.*, 198, 1973
Also classified in 86

72.49   **Ethylene - bis(tri - o - tolyl - phosphite) nickel**
$C_{44}H_{46}NiO_6P_2$
L.J.Guggenberger *Inorg. Chem.*, **12,** 499, 1973
Also classified in 86

72.C   **(σ - Dicyanovinyl)(carbonyl)(π - dicyanoacetylene) -**
bis(triphenylphosphine) iridium(i)
$C_{45}H_{31}IrN_4OP_2$
For complete entry see 71.102

72.50   **Acrylonitrile - bis(tri - o - phosphite) nickel**
$C_{45}H_{45}NNiO_6P_2$
L.J.Guggenberger *Inorg. Chem.*, **12,** 499, 1973
Also classified in 86

72.C   **Heptacarbonyl - μ - (1,2,3,4 - tetraphenylbutadiene - 1,4 - diyl) - μ -**
(diphenylacetylene) - triangulo - triosmium
$C_{49}H_{30}O_7Os_3$
For complete entry see 71.106

72.C   **bis(Triphenylphosphine) - cyclopentadienyl - ruthenium - phenylacetylide**
chloro - copper(i) acetone solvate
$C_{49}H_{40}ClCuP_2Ru$, C_3H_6O
For complete entry see 71.107

72.51   **μ - Diphenylacetylene - bis(triphenylphosphine - bis(trifluorophosphine)**
rhodium) diethyl ether solvate
$C_{50}H_{40}F_{12}P_6Rh_2$, $C_4H_{10}O$
M.A.Bennett, R.N.Johnson, G.B.Robertson, T.W.Turney, P.O.Whimp
J. Amer. Chem. Soc., **94,** 6540, 1972
Residue 1 also classified in 86

72.52   **tris(Dibenzylideneacetone) dipalladium(0) methylene chloride solvate**
$C_{51}H_{42}O_3Pd_2$, CH_2Cl_2
M.C.Mazza, C.G.Pierpont *J. C. S. Chem. Comm.*, 207, 1973

72.53   **μ - (α,ω - Octadi - π - enyl)bis(bromo - bis(diphenylphosphino)ethane -**
nickel(ii)) chloroform solvate
$C_{60}H_{62}Br_2Ni_2P_4$, $2CHCl_3$
T.S.Cameron, C.K.Prout *Acta Cryst. (B)*, **28,** 2021, 1972
Residue 1 also classified in 86

72.54 μ - (α,ω - **Octadi** - π - **enyl**) - **bis(bromo** - **(bis(diphenylphosphino)ethane)**
nickel) chloroform solvate
$C_{60}H_{62}Br_2Ni_2P_4$, $8CHCl_3$
T.S.Cameron, M.L.H.Green, H.Munakata, C.K.Prout, M.J.Smith
J. Coord. Chem., **2**, 43, 1972
Residue 1 also classified in 86

72.55 μ - **Diphenylacetylene** - **bis(π** - **pentaphenylcyclopentadienyl) dipalladium(i)**
$C_{84}H_{60}Pd_2$
E.Ban, P.-T.Cheng, T.Jack, S.C.Nyburg, J.Powell
J. C. S. Chem. Comm., 368, 1973
Also classified in 73

METAL PI-COMPLEXES (CYCLOPENTADIENE)

73.C μ - **1,4** - **Dioxane** - **bis(di(tetrahydrofuran) lithium) di(trichloro** - **cyclopentadienyl** - **chromium)**
$2C_5H_5Cl_3Cr^-$, $C_{20}H_{40}Li_2O_6{}^{2+}$
For complete entry see 67.16

73.1   **Dicarbonyl** - π - **cyclopentadienyl cobalt tri(mercury(ii) chloride)**
$C_7H_5ClCoHgO_2{}^+$, Cl^- , $2Cl_2Hg$
I.W.Nowell, D.R.Russell   *J. C. S. Dalton,* 2396, 1972

73.2   **Dicarbonyl** - π - **cyclopentadienyl** - **cobalt mercury(ii) chloride**
$C_7H_5Cl_2CoHgO_2$
I.W.Nowell, D.R.Russell   *J. C. S. Dalton,* 2393, 1972

73.3   **1** - (π - **Cyclopentadienyl)dodecahydro** - **1** - **cobalta** - **2,4** - **dicarba** - **closo** - **tridecaborane**
$C_7H_{17}B_{10}Co$
M.R.Churchill, B.G.DeBoer   *J. C. S. Chem. Comm.*, 1326, 1972
Also classified in 62

73.C π - **2,3** - **Dichlorobutadiene** - π - **cyclopentadienyl** - **rhodium(i)**
$C_9H_9Cl_2Rh$
For complete entry see 72.8

73.4   **1,2** - **Di** - **iodoferrocene**
$C_{10}H_8FeI_2$
Z.Kaluski, A.I.Gusiev, Yu.T.Struchkov
Bull. Acad. Polon. Sci., Ser. Sci. Chim., **20,** 875, 1972

73.5   **1,2,3** - **Trithia** - **(3)** - **ferrocenophane**
$C_{10}H_8FeS_3$
B.R.Davis, I.Bernal   *J. Cryst. Mol. Struct.*, **2,** 107, 1972

73.6   **Bromo** - **bis(π** - **cyclopentadienyl)(tribromostannyl) molybdenum(iv)**
$C_{10}H_{10}Br_4MoSn$
T.S.Cameron, C.K.Prout   *J. C. S. Dalton,* 1447, 1972

73.7   **bis(Ruthenocene)** - **tri(mercury dichloride)**
$C_{10}H_{10}Cl_2HgRu$, $C_{10}H_{10}ClHgRu^+$, Cl^- , Cl_2Hg
A.I.Gusev, Yu.T.Struchkov   *Zh. Strukt. Khim.*, **12,** 1121, 1971

73.8   **Dichloro - dicyclopentadienyl - titanium**
$C_{10}H_{10}Cl_2Ti$
V.V.Tkachev, L.O.Atovmyan   *Zh. Strukt. Khim.*, **13**, 287, 1972

73.9   **bis(Cyclopentadienyl) tungsten tetrasulfide**
$C_{10}H_{10}S_4W$
B.R.Davis, I.Bernal   *J. Cryst. Mol. Struct.*, **2**, 135, 1972

73.10   **Tetrahydroborato - bis(cyclopentadienyl) titanium(iii)**
$C_{10}H_{10}Ti^+$, H_4B^-
K.M.Melmed, D.Coucouvanis, S.J.Lippard   *Inorg. Chem.*, **12**, 232, 1973

73.11   **π - (Cyclopentadienyl) - trans - iododicarbonyl(trimethyl phosphite) molybdenum**
$C_{10}H_{14}IMoO_5P$
A.D.U.Hardy, G.A.Sim   *J. C. S. Dalton,* 1900, 1972
Also classified in 86

73.C   **Cyclopentadienyl - carbonyl - iron - bis(methylisocyanide) borohydride adduct**
$C_{10}H_{15}BFeN_2O$
For complete entry see 71.11

73.12   **Di(cyclopentadienyl) - niobium carbonyl hydride**
$C_{11}H_{11}NbO$
N.I.Kirillova, A.I.Gusev, Yu.T.Struchkov   *Zh. Strukt. Khim.*, **13**, 473, 1972

73.C   **π - 2,3 - Dimethylbutadiene - π - cyclopentadienyl - rhodium(i)**
$C_{11}H_{15}Rh$
For complete entry see 72.12

73.13   **π - (Methylcyclopentadienyl) - trans - iododicarbonyl(trimethyl phosphite) molybdenum**
$C_{11}H_{16}IMoO_5P$
A.D.U.Hardy, G.A.Sim   *J. C. S. Dalton,* 1900, 1972
Also classified in 86

73.14   **Iodo - π - indenyl - tricarbonyl - molybdenum(ii)**
$C_{12}H_7IMoO_3$
A.Mawby, G.E.Pringle   *J. Inorg. Nucl. Chem.*, **34**, 525, 1972
Also classified in 75

73.15   **1,1' - Dimethyl - ferricenium tri - iodide**
$C_{12}H_{14}Fe^+$, I_3^-
J.W.Bats, J.J.de Boer, D.Bright   *Inorg. Chim. Acta,* **5**, 605, 1971

73.16   **bis(π - Cyclopentadienyl) - glycinato - molybdenum(iv) chloride monohydrate**
$C_{12}H_{14}MoNO_2{}^+$, Cl^- , H_2O
C.K.Prout, G.B.Allison, L.T.J.Delbaere, E.Gore
Acta Cryst. (B), **28**, 3043, 1972
Residue 1 also classified in 81

73.C   **Hydrogen - di(bis(π - cyclopentadienyl) - L - cysteinato - molybdenum(iv)) chloride**
$C_{13}H_{15}MoNO_2S$, $C_{13}H_{16}MoNO_2S^+$, Cl^-
For complete entry see 82.21

73.17   **Hydrogen - di(bis(π - cyclopentadienyl) - L - cysteinato - molybdenum(iv)) hexafluorophosphate**
$C_{13}H_{15}MoNO_2S$, $C_{13}H_{16}MoNO_2S^+$, F_6P^-
C.K.Prout, G.B.Allison, L.T.J.Delbaere, E.Gore
Acta Cryst. (B), **28**, 3043, 1972
Residue 1 also classified in 82

73.18   **bis(π - Cyclopentadienyl) - sarcosinato - molybdenum(iv) chloride methanol solvate**
$C_{13}H_{16}MoNO_2{}^+$, Cl^- , CH_4O
C.K.Prout, G.B.Allison, L.T.J.Delbaere, E.Gore
Acta Cryst. (B), **28**, 3043, 1972
Residue 1 also classified in 81

73.C   **Compound III**
$C_{13}H_{17}B_3Co_2$
For complete entry see 62.9

73.C   **Cyclopentadienyl - (2 - hydroxycyclo - octa - 5 - enyl) - palladium(ii)**
$C_{13}H_{18}OPd$
For complete entry see 71.27

73.19   **1,1' - Diethoxy - octachloroferrocene**
$C_{14}H_{10}Cl_8FeO_2$
G.M.Brown, F.L.Hedberg, H.Rosenberg
Amer. Cryst. Assoc., Abstr. Papers (Winter Meeting), 90, 1973

73.20   **μ - (Sulfur dioxide) - bis(π - cyclopentadienyl - dicarbonyl - iron)**
$C_{14}H_{10}Fe_2O_6S$
M.R.Churchill, B.G.DeBoer, K.L.Kalra, P.Reich-Rohrwig, A.Wojcicki
J. C. S. Chem. Comm., 981, 1972

73.C   **trihapto - 1,1',2' - ((5' - Oxocyclopenten - 1 - yl) - 1 - ethyl)(π - cyclopentadienyl) - dicarbonylmolybdenum**
$C_{14}H_{13}MoO_2$
For complete entry see 71.29

73.C   **Di - (μ - methylisonitrile) di - (h^5 - cyclopentadienyl) dinickel**
$C_{14}H_{16}N_2Ni_2$
For complete entry see 71.30

73.21   **Di - μ - thioethyl - μ - dithio - bis(cyclopentadienyl - iron)**
$C_{14}H_{20}Fe_2S_4$
G.T.Kubas, T.G.Spiro, A.Terzis   *J. Amer. Chem. Soc.*, **95,** 273, 1973
Also classified in 85

73.22   **Tri(cyclopentadienyl) - tricobalt disulfide (room temp. form)**
$C_{15}H_{15}Co_3S_2$
P.D.Frisch, L.F.Dahl   *J. Amer. Chem. Soc.*, **94,** 5082, 1972

73.23   **Tri(cyclopentadienyl) - tricobalt disulfide (room temp.form)**
$C_{15}H_{15}Co_3S_2$
N.Kamijyo, T.Watanabe   *Acta Cryst. (A),* **28,** S81, 1972

73.24   **Tri(cyclopentadienyl) - tricobalt disulfide (low temp.form)**
$C_{15}H_{15}Co_3S_2$
N.Kamijyo, T.Watanabe   *Acta Cryst. (A),* **28,** S81, 1972

73.25   **Tri(cyclopentadienyl) - tricobalt disulfide iodide**
$C_{15}H_{15}Co_3S_2^+$, I^-
P.D.Frisch, L.F.Dahl   *J. Amer. Chem. Soc.*, **94,** 5082, 1972

73.C   **(−) - Ferrocene - 2,3 (5 - exo - methyl - cyclohexen - 2 - one)**
$C_{15}H_{15}FeO$
For complete entry see 75.21

73.C   **Tricyclopentadienyl - scandium**
$(C_{15}H_{15}Sc)_n$
For complete entry see 71.33

73.26   **Tricyclopentadienyl - titanium**
$C_{15}H_{15}Ti$
C.R.Lucas, M.Green, R.A.Forder, K.Prout
J. C. S. Chem. Comm., 97, 1973

73.C   **Cyclopentadienyl(duroquinone) cobalt dihydrate**
$C_{15}H_{17}CoO_2$, $2H_2O$
For complete entry see 74.6

73.C   **π - Cyclopentadienyl - iridium - π - duroquinone**
$C_{15}H_{17}IrO_2$
For complete entry see 74.7

73.27   **Benzene - 1,2 - dithiolato - di(π - cyclopentadienyl) - titanium(iv)**
$C_{16}H_{14}S_2Ti$
A.Kutoglu   *Z. Anorg. Allg. Chem.*, **390,** 195, 1972
Also classified in 85

73.28 **Tri(cyclopentadienyl) - tricobalt carbonyl sulfide**
$C_{16}H_{15}Co_3OS$
P.D.Frisch, L.F.Dahl   *J. Amer. Chem. Soc.*, **94,** 5082, 1972

73.29 **Di - μ - dimethylstannylene - bis(carbonyl - π - cyclopentadienyl - cobalt)**
$C_{16}H_{22}Co_2O_2Sn_2$
J.Weaver, P.Woodward   *J. C. S. Dalton,* 1060, 1973
Also classified in 69

73.30 **(6,6 - Dimethyl - tricyclo(5.3.1^{2,5}.0)undeca - 3 - ene - 1(7),9 - dienyl) - cyclopentadienyl - iron**
$C_{18}H_{20}Fe$
T.S.Cameron, J.F.Maguire, T.D.Turbitt, W.E.Watts
J. Organometal. Chem., **49,** C79, 1973

73.31 **Ethylmercapto - cyclopentadienyl - iron - acetonitrile dimer hexafluorophosphate**
$C_{18}H_{26}Fe_2N_2S_2^{2+}$, $2F_6P^-$
P.J.Vergamini, R.R.Ryan, G.J.Kubas
Amer. Cryst. Assoc., Abstr. Papers (Winter Meeting), 45, 1973
Residue 1 also classified in 85

73.C **π - Cyclopentadienyl - azotoluene - 2 - yl - nickel**
$C_{19}H_{18}N_2Ni$
For complete entry see 71.45

73.32 **1,1'(α - Keto - γ - phenyltrimethylene) - 2' - methyl - ferrocene (mp. 228°C)**
$C_{20}H_8FeO$
C.Lecomte, Y.Dusausoy, J.Protas, C.Moise
Acta Cryst. (B), **29,** 1127, 1973

73.33 **Di - (π - cyclopentadienyl) ruthenium - mercury dibromide dimer**
Ruthenocene - mercury dibromide dimer
$C_{20}H_{20}Br_4Hg_2Ru_2$
A.I.Gusev, Yu.T.Struchkov   *Zh. Strukt. Khim.*, **12,** 1121, 1971

73.34 **Cyclopentadienyl - cobalt phosphide tetramer**
$C_{20}H_{20}Co_4P_4$
G.L.Simon, L.F.Dahl   *J. Amer. Chem. Soc.*, **95,** 2175, 1973

73.35 **Cyclopentadienyl - cobalt sulfide tetramer**
$C_{20}H_{20}Co_4S_4$
G.L.Simon, L.F.Dahl   *J. Amer. Chem. Soc.*, **95,** 2164, 1973

73.36 **Cyclopentadienyl - cobalt sulfide tetramer hexafluorophosphate**
$C_{20}H_{20}Co_4S_4^+$, F_6P^-
G.L.Simon, L.F.Dahl   *J. Amer. Chem. Soc.*, **95,** 2164, 1973

73.37 **Dicyclopentadienyl - di - iron trisulfide dimer**
$C_{20}H_{20}Fe_4S_6$
P.J.Vergamini, R.R.Ryan, G.J.Kubas
Amer. Cryst. Assoc., Abstr. Papers (Winter Meeting), 45, 1973
Also classified in 85

73.C **Tetracyclopentadienyl - hafnium (at 0°C)**
$C_{20}H_{20}Hf$
For complete entry see 71.47

73.C **Niobocene dimer**
$C_{20}H_{20}Nb_2$
For complete entry see 71.48

73.38 **Tetra - cyclopentadienyl - uranium(iv)**
$C_{20}H_{20}U$
J.H.Burns *J. Amer. Chem. Soc.*, **95**, 3815, 1973

73.39 **μ - Chloro - μ - hydrido - bis(chloro(pentamethylcyclopentadienyl) rhodium(iii))**
$C_{20}H_{31}Cl_3Rh_2$
M.R.Churchill, S.W.Ni *J. Amer. Chem. Soc.*, **95**, 2150, 1973

73.40 **2,1' - Trimethylene - 1 - (α - phenyl - α - hydroxypropyl) ferrocene (m.p. 122°C)**
$C_{22}H_{24}FeO$
C.Lecomte, Y.Dusausoy, J.Protas, C.Moise, J.Tirouflet
Acta Cryst. (B), **29**, 488, 1973

73.41 **(S,R,S) - 2 - (p - Methoxyphenyl)hydroxymethyl - N,N - dimethyl - 1 - ferrocenyl - ethylamine (absolute configuration)**
$C_{22}H_{27}FeNO_2$
L.F.Battelle, R.Bau, G.W.Gokel, R.T.Oyakawa, I.K.Ugi
J. Amer. Chem. Soc., **95**, 482, 1973

73.C **Tricyclopentadienyl - phenylethynyl - uranium(iv)**
$C_{23}H_{20}U$
For complete entry see 71.56

73.C **tris(Cyclopentadienyl - tricarbonyl - molybdenum) thallium(iii)**
$C_{24}H_{15}Mo_3O_9Tl$
For complete entry see 68.27

73.42 **bis(Dicarbonyl - (tricyclo(6.1.0.0$^{2.6}$)deca - 2,5 - dienyl) iron)**
$C_{24}H_{22}Fe_2O_4$
F.A.Cotton, J.M.Troup *J. Amer. Chem. Soc.*, **95**, 3798, 1973

73.43   **1,12 - Dimethyl(1.1)ferrocenophane**
1,1″.1′,1‴ - Diethylidene - diferrocene
$C_{24}H_{24}Fe_2$
J.S.McKechnie, C.A.Maier, B.Bersted, I.C.Paul
J. C. S. Perkin II, 138, 1973

73.44   **Dicarbonyl - pentamethylcyclopentadienyl - chromium dimer**
$C_{24}H_{30}Cr_2O_4$
J.Potenza, P.Giordano, D.Mastropaolo, A.Efraty, R.B.King
J. C. S. Chem. Comm., 1333, 1972

73.C π **- Diphenylacetylene -** π **- cyclopentadienyl - phenyl - oxo - tungsten**
$C_{25}H_{20}OW$
For complete entry see 71.65

73.C   **Dimethylaluminium - dihydride - bis -** μ **- (**π **- cyclopentadienyl molybdenum) - (methylaluminium - di -** μ **- (1,1 - cyclopentadienyl) - dimethylaluminium)**
$C_{25}H_{35}Al_3Mo_2$
For complete entry see 68.29

73.C   **Tri - indenyl - samarium**
$C_{27}H_{21}Sm$
For complete entry see 75.39

73.C π **- Cyclopentadienyl - (**π **- 1,3 - diphenyl - bis(trimethylsilyl) cyclobutadiene) - cobalt**
$C_{27}H_{33}CoSi_2$
For complete entry see 75.40

73.C π **- Cyclopentadienyl - (**π **- 1,3 - diphenyl - 2,4 - bis(trimethylsilyl) cyclobutadiene) cobalt**
$C_{27}H_{33}CoSi_2$
For complete entry see 75.41

73.C π **- Cyclopentadienyl - (**π **- 1,2 - diphenyl - 3,4 - bis(trimethylsilyl) cyclobutadiene) cobalt**
$C_{27}H_{33}CoSi_2$
For complete entry see 75.42

73.C   **Di -** μ **- chloro - bis(dicarbonyl - cyclopentadienyl - iron - phenylacetylide)**
$C_{30}H_{20}Cl_2Cu_2Fe_2O_4$
For complete entry see 71.74

73.C   **1,2,3,4 - Tetra(trifluoromethyl)butadienyl - (cyclopentadienyl) - triphenyl - phosphine - ruthenium**
$C_{31}H_{21}F_{12}PRu$
For complete entry see 71.76

73.C **Triphenylphosphine - (1 - (di(trifluoromethyl) - hydroxymethyl) - cyclopentadienyl) - (1,2 - di(carboxymethyl)ethylene - 1 - yl) - ruthenium**
$C_{32}H_{27}F_6O_5PRu$
For complete entry see 71.78

73.C **π - (1,2 - Di - iodocyclopentadienyl)tetraphenylcyclobutadiene - cobalt**
$C_{33}H_{23}CoI_2$
For complete entry see 75.44

73.C **π - (Iodocyclopentadienyl) - tetraphenylcyclobutadiene - cobalt**
$C_{33}H_{24}CoI$
For complete entry see 75.45

73.C **π - Cyclopentadienyl - π - tetraphenylcyclobutadiene - rhodium(i)**
$C_{33}H_{25}Rh$
For complete entry see 75.46

73.C **π - (Cyanocyclopentadienyl)tetraphenylcyclobutadiene - cobalt**
$C_{34}H_{24}CoN$
For complete entry see 75.47

73.C **Structure III**
$C_{34}H_{25}FeNiO_3P$
For complete entry see 71.82

73.45 **tris(Benzylcyclopentadienide)chloro - uranium(iv)**
$C_{36}H_{33}ClU$
J.Leong, K.O.Hodgson, K.N.Raymond   *Inorg. Chem.*, **12,** 1329, 1973

73.C **π - Cyclopentadienyl - π - diphenylacetylene - monocarbonyl - triphenyl - stannyl - iron**
$C_{38}H_{30}FeOSn$
For complete entry see 72.41

73.46 **bis(μ - Diphenylphosphido - μ' - carbonyl - π - methylcyclopentadienyl - carbonyl - iron) rhodium hexafluorophosphate**
$C_{40}H_{34}F_6Fe_2O_4P_3Rh$
R.J.Haines, R.Mason, J.A.Zubieta, C.R.Nolte
J. C. S. Chem. Comm., 990, 1972
Also classified in 86

73.C **Compound I**
$C_{46}H_{39}Au_2FeP_2^+$, BF_4^-
For complete entry see 71.104

73.C **bis(Triphenylphosphine) - cyclopentadienyl - ruthenium - phenylacetylide chloro - copper(i) acetone solvate**
$C_{49}H_{40}ClCuP_2Ru$, C_3H_6O
For complete entry see 71.107

73.C   **Neodymium tri(methylcyclopentadienide) tetramer**
$C_{72}H_{84}Nd_4$
For complete entry see 71.108

73.C μ - **Diphenylacetylene - bis(π - pentaphenylcyclopentadienyl) dipalladium(i)**
$C_{84}H_{60}Pd_2$
For complete entry see 72.55

METAL PI-COMPLEXES (ARENE)

74.1   Benzene - chromium tricarbonyl (neutron study, at −195°C)
$C_9H_6CrO_3$
B.Rees, P.Coppens *J. Organometal. Chem.*, **42**, C102, 1972

74.2   Benzene - chromium tricarbonyl (at −195°C)
$C_9H_6CrO_3$
B.Rees, P.Coppens *J. Organometal. Chem.*, **42**, C102, 1972

74.3   o - Hydroxy - acetylbenchrotrene
$C_{11}H_8CrO_5$
Y.Dusausoy, J.Protas, J.Besancon, J.Tirouflet
Acta Cryst. (B), **29**, 469, 1973

74.4   DL - o - Methoxyacetylbenchrotrene
$C_{12}H_{10}CrO_5$
Y.Dusausoy, J.Protas, J.Besancon, J.Tirouflet
Acta Cryst. (B), **29**, 469, 1973

74.5   (+) - o - Methoxy - (1' - hydroxyethyl)benzene chromium tricarbonyl
(m.p.70°C,absolute configuration)
$C_{12}H_{12}CrO_5$
Y.Dusausoy, J.Protas, J.Besancon, J.Tirouflet
Acta Cryst. (B), **28**, 3183, 1972

74.6   Cyclopentadienyl(duroquinone) cobalt dihydrate
$C_{15}H_{17}CoO_2$, $2H_2O$
V.A.Uchtman, L.F.Dahl *J. Organometal. Chem.*, **40**, 403, 1972
Residue 1 also classified in 73

74.7 π - Cyclopentadienyl - iridium - π - duroquinone
$C_{15}H_{17}IrO_2$
G.G.Aleksandrov, Yu.T.Struchkov *Zh. Strukt. Khim.*, **12**, 1037, 1971
Also classified in 73

74.C   3,α - Dimethylstyrene - bis(tricarbonyl - iron)
$C_{16}H_{12}Fe_2O_6$
For complete entry see 72.20

74.8   **Dichloro - (diphenyl - methyl - phosphine) ruthenium benzene**
$C_{19}H_{19}Cl_2PRu$
M.A.Bennett, G.B.Robertson, A.K.Smith
J. Organometal. Chem., **43,** C41, 1972
Also classified in 86

74.C   **Phenylmethinyl - tricobalt hexacarbonyl - mesitylene**
$C_{22}H_{17}Co_3O_6$
For complete entry see 71.52

74.9   **Dichloro - (diphenyl - methyl - phosphine) ruthenium p - cymene**
$C_{23}H_{27}Cl_2PRu$
M.A.Bennett, G.B.Robertson, A.K.Smith
J. Organometal. Chem., **43,** C41, 1972
Also classified in 86

74.10   **bis(Triethylphosphine) - (hexakis(trifluoromethyl)benzene) platinum**
$C_{24}H_{30}F_{18}P_2Pt$
J.Browning, M.Green, B.R.Penfold, J.L.Spencer, F.G.A.Stone
J. C. S. Chem. Comm., 31, 1973
Also classified in 86

74.11   **bis(Hexamethylbenzene) ruthenium(0)**
$C_{24}H_{36}Ru$
G.Huttner, S.Lange   *Acta Cryst. (B)*, **28,** 2049, 1972

74.12   **Ruthenium complex with triphenyl phosphite**
$C_{51}H_{39}O_{12}P_3Ru_2$
M.I.Bruce, J.Howard, I.W.Nowell, G.Shaw, P.Woodward
J. C. S. Chem. Comm., 1041, 1972
Also classified in 71, 86

METAL PI-COMPLEXES
(MISCELLANEOUS RING SYSTEMS)

75.1   **Azepine - iron - tricarbonyl**
$C_9H_7FeNO_3$
A.Gieren, W.Hoppe   *Acta Cryst. (B)*, **28**, 2766, 1972

75.2   **π - Cycloheptatrienylium - tricarbonyl - molybdenum(0) tetrafluoroborate**
$C_{10}H_7MoO_3^+$, BF_4^-
G.R.Clark, G.J.Palenik   *J. Organometal. Chem.*, **50**, 185, 1973

75.3   **Tricarbonyl - (1,4 - dimethyl - 1,2 - dihydropyridine) chromium(0)**
$C_{10}H_{11}CrNO_3$
G.Huttner, O.S.Mills   *Chem. Ber.*, **105**, 3924, 1972

75.4   **π - endo - Dicyclopentadiene - platinum dichloride (absolute configuration)**
π - endo - Tricyclo(4.3.3)deca - 3,7 - diene - platinum dichloride
$C_{10}H_{12}ClPt$
P.Ganis, U.Lepore, G.Avitabile, A.Panunzi   *Acta Cryst. (A)*, **28**, S81, 1972

75.5   **1,6,7,8 - tetrahapto - Heptafulvene - iron tricarbonyl**
$C_{11}H_8FeO_3$
M.R.Churchill, B.G.DeBoer   *Inorg. Chem.*, **12**, 525, 1973

75.6   **1H - 1 - Carboethoxy - 1,2 - diazepine iron tricarbonyl**
$C_{11}H_{10}FeN_2O_5$
A.de Cian, P.M.L'Huillier, R.Weiss   *Bull. Soc. Chim. Fr.*, 457, 1973

75.7   **N - Methyl - 3 - ethyl - 1,6 - dihydropyridine - tricarbonyl - chromium**
$C_{11}H_{13}CrNO_3$
C.A.Bear, W.R.Cullen, J.P.Kutney, V.E.Ridaura, J.Trotter, A.Zanarotti
J. Amer. Chem. Soc., **95**, 3058, 1973

75.8   **N - Methyl - 3 - ethyl - 1,2 - dihydropyridine - tricarbonyl - chromium**
$C_{11}H_{13}CrNO_3$
C.A.Bear, W.R.Cullen, J.P.Kutney, V.E.Ridaura, J.Trotter, A.Zanarotti
J. Amer. Chem. Soc., **95**, 3058, 1973

75.C   **Iodo - π - indenyl - tricarbonyl - molybdenum(ii)**
$C_{12}H_7IMoO_3$
For complete entry see 73.14

75.C   **Iron carbonyl derivative of barbalone (triclinic form)**
$C_{12}H_8FeO_4$
For complete entry see 71.17

75.9   **1,2,3,6 - tetrahapto - (5 - Cyanocyclo - octadienyl)tricarbonyl - ruthenium**
$C_{12}H_{11}NO_3Ru$
F.A.Cotton, M.D.LaPrade   *J. Organometal. Chem.*, **39,** 345, 1972

75.10   **bis(1 - Methylborinato) cobalt**
$C_{12}H_{16}B_2Co$
G.Huttner, B.Krieg, W.Gartzke   *Chem. Ber.*, **105,** 3424, 1972

75.11   **bis(1 - Methoxyborinato) cobalt**
$C_{12}H_{16}B_2CoO_2$
G.Huttner, B.Krieg, W.Gartzke   *Chem. Ber.*, **105,** 3424, 1972

75.12   **Cyclo - octatetraenyl(dichloro)(tetrahydrofuran) zirconium**
$C_{12}H_{16}Cl_2OZr$
D.J.Brauer, C.Kruger   *Eur. Cryst. Meeting,* 1973
Also classified in 84

75.13   **trans,trans,trans - 1,5,9 - Cyclododecatriene - nickel**
$C_{12}H_{18}Ni$
D.J.Brauer, C.Kruger   *J. Organometal. Chem.*, **44,** 397, 1972

75.14   **Di - μ - carbonyl - tetracarbonyl - (π - norbornadiene) dicobalt**
$C_{13}H_8Co_2O_6$
F.S.Stephens   *J. C. S. Dalton,* 1754, 1972

75.15   **Tricarbonyl - iron cyclodeca - 1,3,5,7 - tetraene**
$C_{13}H_{12}FeO_3$
F.A.Cotton, J.M.Troup   *J. Amer. Chem. Soc.*, **95,** 3798, 1973

75.16   **Tricyclo(6.2.0.0^{2,7})decadiene iron tricarbonyl**
$C_{13}H_{12}FeO_3$
B.A.Frenz, J.M.Troup, F.A.Cotton
Amer. Cryst. Assoc., Abstr. Papers (Winter Meeting), 90, 1973

75.C   **Butadiene(cyclo - octatetraene) iron monocarbonyl**
$C_{13}H_{14}FeO$
For complete entry see 72.16

75.C   **Compound III**
$C_{13}H_{17}B_3Co_2$
For complete entry see 62.9

75.C   **Cyclopentadienyl - (2 - hydroxycyclo - octa - 5 - enyl) - palladium(ii)**
$C_{13}H_{18}OPd$
For complete entry see 71.27

75.17 **(5,7 - Dimethyl - 4H - cyclohepta(c)thiophene) chromium tricarbonyl**
$C_{14}H_{12}CrO_3S$
Y.Dusausoy, J.Protas, R.Guilard *Acta Cryst. (B)*, **29,** 726, 1973

75.18 **Tricyclo(6.3.0^{2,7})undecadiene iron tricarbonyl**
$C_{14}H_{14}FeO_3$
B.A.Frenz, J.M.Troup, F.A.Cotton
Amer. Cryst. Assoc., Abstr. Papers (Winter Meeting), 90, 1973

75.19 **(5,7,8 - Trimethyl - 8H - cyclohepta(b)thiophene) chromium tricarbonyl**
$C_{15}H_{14}CrO_3S$
Y.Dusausoy, J.Protas, R.Guilard *Acta Cryst. (B)*, **29,** 477, 1973

75.20 **(4,5,7 - Trimethyl - 4H - cyclohepta(b)thiophene) chromium tricarbonyl**
$C_{15}H_{14}CrO_3S$
Y.Dusausoy, J.Protas, R.Guilard *Acta Cryst. (B)*, **29,** 477, 1973

75.21 **(−) - Ferrocene - 2,3 (5 - exo - methyl - cyclohexen - 2 - one)**
$C_{15}H_{15}FeO$
C.Lecomte, Y.Dusausoy, R.Broussier, B.Gautheron, J.Protas
C. R. Acad. Sci., Fr., C, **275,** 1263, 1972
Also classified in 73

75.22 **Tricyclo(6.4.0.0^{2,7})dodecadiene iron tricarbonyl**
$C_{15}H_{16}FeO_3$
B.A.Frenz, J.M.Troup, F.A.Cotton
Amer. Cryst. Assoc., Abstr. Papers (Winter Meeting), 90, 1973

75.C **Cyclo - octatetraenyl - (allyl) - t - butoxy - zirconium**
$C_{15}H_{22}OZr$
For complete entry see 72.19

75.23 **Tricarbonyl(hexaethylborazine) chromium(0)**
$C_{15}H_{30}B_3CrN_3O_3$
G.Huttner, B.Krieg *Chem. Ber.,* **105,** 3437, 1972

75.24 **Azulene dimolybdenum hexacarbonyl (triclinic form)**
$C_{16}H_8Mo_2O_6$
A.W.Schlueter, R.A.Jacobson *Inorg. Chim. Acta,* **2,** 241, 1968

75.C **(Bullvalene - 2,8 - diyl - iron tricarbonyl) iron tricarbonyl**
$C_{16}H_{10}Fe_2O_6$
For complete entry see 71.36

75.C **5 - (Tricarbonyl - iron - methyl) - 6 - allyl - cyclohexa - 1,3 - diene iron tricarbonyl**
$C_{16}H_{12}Fe_2O_6$
For complete entry see 71.37

75.25   cis(1,2,6 - Trihapto - 3,4,5 - trihapto - bicyclo(6.2.0)deca - 1,3,5 - triene) -
hexacarbonyl - di - iron
$C_{16}H_{12}Fe_2O_6$
F.A.Cotton, B.A.Frenz, G.Deganello, A.Shaver
J. Organometal. Chem., **50,** 227, 1973

75.26   Potassium 2,2' - di(methoxy)diethylether bis(cyclo - octatetraenyl)
cerium(iii)
$C_{16}H_{16}Ce^-$, $C_6H_{14}KO_3^+$
K.O.Hodgson, K.N.Raymond   *Inorg. Chem.*, **11,** 3030, 1972
Residue 2 classified in 67

75.27   cis - Di - μ - carbonyl - bis(carbonyl - (π - cyclohexa - 1,3 - diene) cobalt)
$C_{16}H_{16}Co_2O_4$
F.S.Stephens   *J. C. S. Dalton*, 1752, 1972

75.28   bis(Cyclo - octatetraenyl - nickel)
$C_{16}H_{16}Ni_2$
D.J.Brauer, C.Kruger   *Eur. Cryst. Meeting,* 1973

75.29   Azulene - triruthenium heptacarbonyl
$C_{17}H_8O_7Ru_3$
M.R.Churchill, J.Wormald   *Inorg. Chem.*, **12,** 191, 1973

75.30   Cyclo - oct - 1 - ene - 5 - yne tetraruthenium undecacarbonyl
$C_{19}H_{10}O_{11}Ru_4$
R.Mason, K.M.Thomas   *J. Organometal. Chem.*, **43,** C39, 1972

75.31   (Dihydro - bis(3,5 - dimethyl - 1 - pyrazolyl)borate) - (trihapto -
cycloheptatrienyl)dicarbonyl - molybdenum
$C_{19}H_{23}BMoN_4O_2$
F.A.Cotton, M.Jeremic, A.Shaver   *Inorg. Chim. Acta*, **6,** 543, 1972
Also classified in 83, 62

75.32   bis(Cyclo - octatetraene)(tetrahydrofuran) zirconium (absolute
configuration)
$C_{20}H_{24}OZr$
D.J.Brauer, C.Kruger   *J. Organometal. Chem.*, **42,** 129, 1972
Also classified in 84

75.C   Decacarbonyl(cyclododecatrienyl) - tetrahedro - tetraruthenium
$C_{22}H_{16}O_{10}Ru_4$
For complete entry see 71.51

75.33   3,3' - bis(Tricarbonyl - iron - bicyclo(4.2.0)octa - 2,4 - diene)
$C_{22}H_{18}Fe_2O_6$
F.A.Cotton, J.M.Troup   *J. Amer. Chem. Soc.*, **95,** 3798, 1973

75.C   1 - Tricarbonyl - 2,3,4,5 - tetra(trimethylsiloxy) - 1 - ferracyclopentadiene
tricarbonyl - iron
$C_{22}H_{36}Fe_2O_{10}Si_4$
For complete entry see 71.54

75.34   bis(1,3,5,7 - Tetramethyl - cyclotetraenyl) uranium(iv)
$C_{24}H_{32}U$
K.O.Hodgson, K.N.Raymond   *Inorg. Chem.*, **12,** 458, 1973

75.C   Tricarbonyl - π - (1,1,1 - tricarbonyl - 2 - methyl - 3 - diphenylmethylene - 6 -
methoxy - ferra - 2 - oxacyclohexenyl) iron
$C_{25}H_{18}Fe_2O_8$
For complete entry see 71.63

75.C   Tricarbonyl - π - (1,1,1 - tricarbonyl - 2,3 - dimethoxy - 5 - (diphenylmethyl)
ferracyclopentadiene) iron
$C_{25}H_{18}Fe_2O_8$
For complete entry see 71.64

75.35   (2,2' - bis(Dimethylarsino) - octafluoro - 1,1' - bi(cyclobut - 1 - enyl)) -
bis(dimethyl - arsino) - nonacarbonyl - tetracobalt - dihydride
$C_{25}H_{26}As_4Co_4F_8O_9$
F.W.B.Einstein, R.D.G.Jones   *J. C. S. Dalton,* 2568, 1972
Also classified in 86

75.36   (Cyclo - octa - 1,5 - diene) - bis(dimethylphenylphosphine) methyl iridium(i)
$C_{25}H_{37}IrP_2$
M.R.Churchill, S.A.Bezman   *Inorg. Chem.*, **11,** 2243, 1972
Also classified in 86

75.37   2,4,6 - Triphenyl - phosphorin chromium(0) tricarbonyl
$C_{26}H_{17}CrO_3P$
H.Vahrenkamp, H.Noth   *Chem. Ber.*, **105,** 1148, 1972

75.C   Norbornadiene - (binorbornenyl - diyl) iridium acetylacetonate
$C_{26}H_{31}IrO_2$
For complete entry see 71.70

75.38   bis(6 - t - Butyl - 1,3,5 - trimethylcyclohexadienyl) iron(ii)
$C_{26}H_{38}Fe$
M.Mathew, G.J.Palenik   *Inorg. Chem.*, **11,** 2809, 1972

75.C   Nonacarbonyl - μ - dimethylarsino - μ - (2 -
(diphenylphosphino)tetrafluorocyclobut - 1 - enyl) - tri - iron
$C_{27}H_{16}AsF_4Fe_3O_9P$
For complete entry see 71.72

75.39   **Tri - indenyl - samarium**
$C_{27}H_{21}Sm$
J.L.Atwood, J.H.Burns, P.G.Laubereau
J. Amer. Chem. Soc., **95,** 1830, 1973
Also classified in 73

75.40 π - **Cyclopentadienyl - (π - 1,3 - diphenyl - bis(trimethylsilyl)**
cyclobutadiene) - cobalt
$C_{27}H_{33}CoSi_2$
I.Bernal, B.R.Davis, M.Rausch, A.Siegel
J. C. S. Chem. Comm., 1169, 1972
Also classified in 73, 63

75.41 π - **Cyclopentadienyl - (π - 1,3 - diphenyl - 2,4 - bis(trimethylsilyl)**
cyclobutadiene) cobalt
$C_{27}H_{33}CoSi_2$
C.Kabuto, J.Hayashi, H.Sakurai, Y.Kitahara
J. Organometal. Chem., **43,** C23, 1972
Also classified in 73, 63

75.42 π - **Cyclopentadienyl - (π - 1,2 - diphenyl - 3,4 - bis(trimethylsilyl)**
cyclobutadiene) cobalt
$C_{27}H_{33}CoSi_2$
C.Kabuto, J.Hayashi, H.Sakurai, Y.Kitahara
J. Organometal. Chem., **43,** C23, 1972
Also classified in 73, 63

75.43   **Tetraphenylarsole - manganese - tricarbonyl**
$C_{31}H_{20}AsMnO_3$
E.W.Abel, I.W.Nowell, A.G.J.Modinos, C.Towers
J. C. S. Chem. Comm., 258, 1973

75.44 π - **(1,2 - Di - iodocyclopentadienyl)tetraphenylcyclobutadiene - cobalt**
$C_{33}H_{23}CoI_2$
A.C.Villa, A.G.Manfredotti, C.Guastini   *Abstr. Ital.-Yug. Congr,* 179, 1973
Also classified in 73

75.45 π - **(Iodocyclopentadienyl) - tetraphenylcyclobutadiene - cobalt**
$C_{33}H_{24}CoI$
A.C.Villa, A.G.Manfredotti, C.Guastini   *Abstr. Ital.-Yug. Congr,* 179, 1973
Also classified in 73

75.46 π - **Cyclopentadienyl - π - tetraphenylcyclobutadiene - rhodium(i)**
$C_{33}H_{25}Rh$
G.G.Cash, J.F.Helling, M.Mathew, G.J.Palenik
J. Organometal. Chem., **50,** 277, 1973
Also classified in 73

75.47 π - **(Cyanocyclopentadienyl)tetraphenylcyclobutadiene - cobalt**
$C_{34}H_{24}CoN$
A.C.Villa, A.G.Manfredotti, C.Guastini   *Abstr. Ital.-Yug. Congr,* 179, 1973
Also classified in 73

75.C   **(Cyclo - octa - 1,5 - diene)(1,2 - bis(diphenylphosphino)ethane) methyl iridium(i)**
$C_{35}H_{39}IrP_2$
For complete entry see 71.84

75.C   **(Cyclo - octa - 1,5 - diene)(1,3 - bis(diphenylphosphino)propane)methyl iridium(i)**
$C_{36}H_{41}IrP_2$
For complete entry see 71.85

75.48   **1 - (Diphenylphosphino) - 2 - (dimethylarsino)tetrafluorocyclobut - 1 - ene - μ - (1 - (diphenylphosphino) - 2 - (dimethylarsino)tetrafluorocyclobut - 1 - ene) - tetracarbonyl - di - iron**
$C_{40}H_{32}As_2F_8Fe_2O_4P_2$
F.W.B.Einstein, R.D.G.Jones   *Inorg. Chem.,* **12,** 255, 1973
Also classified in 86

METAL COMPLEXES (ETHYLENEDIAMINE)

76.C catena - μ - Ethylenediamine - cadmium(ii) tetracyanonickelate(ii) dibenzene
$(C_4H_8CdN_4Ni)_n$, $2nC_6H_6$
For complete entry see 61.2

76.1 Nitro(ethylenetriamine) platinum(ii) iodide
$C_4H_{13}N_4O_2Pt^+$, I^-
S.H.Simonsen, T.G.A.Stewart
Amer. Cryst. Assoc., Abstr. Papers (Winter Meeting), 91, 1973

76.2 trans - bis(Ethylenediamine) - dichlorocobalt(iii) tetrachlorothallate(iii)
$C_4H_{16}Cl_2CoN_4^+$, Cl_4Tl^-
K.Brodersen, J.Rath, G.Thiele *Z. Anorg. Allg. Chem.*, **394**, 13, 1972

76.3 μ - Oxo - bis(dichloro - oxo - rhenium(v) - ethylenediamine)
$C_4H_{16}Cl_4N_4O_3Re_2$
T.Glowiak, T.Lis, B.Jezowska-Trzebiatowska
Bull. Acad. Polon. Sci., Ser. Sci. Chim., **20**, 199, 1972

76.4 $(-)_{589}$ - Dinitro - bis(ethylenediamine) cobalt(iii) $(+)_{589}$ - bis(malonato) -
ethylenediamine cobaltate(iii) (absolute configuration)
$C_4H_{16}CoN_6O_4^+$, $C_8H_{12}CoN_2O_8^-$
K.Matsumoto, H.Kuroya *Bull. Chem. Soc. Jap.*, **45**, 1755, 1972
Residue 2 classified in 81, 76

76.5 bis(Ethylenediamine) copper(ii) thiocyanate perchlorate
$C_4H_{16}CuN_4^{2+}$, CNS^- , ClO_4^-
M.Cannas, G.Carta, G.Marongiu *J. C. S. Dalton*, 251, 1973

76.6 (Di(ethylenediamine) copper(ii)) silicate hydrate
$4C_4H_{16}CuN_4^{2+}$, $O_{20}Si_8^{8-}$, $38H_2O$
Yu.I.Smolin, Yu.F.Shepelev, I.K.Butikova *Kristallografija*, **17**, 15, 1972

76.7 trans - Dioxo - bis(ethylenediamine) rhenium(v) chloride
$C_4H_{16}N_4O_2Re^+$, Cl^-
T.Glowiak, T.Lis, B.Jezowska-Trzebiatowska
Bull. Acad. Polon. Sci., Ser. Sci. Chim., **20**, 957, 1972

76.8 Iodo - bis(ethylenediamine) - aquo - nickel(ii) iodide
$C_4H_{18}IN_4NiO^+$, I^-
L.Kh.Minacheva, A.S.Antsyshkina, M.A.Porai-Koshits
Zh. Strukt. Khim., **13**, 81, 1972

76.9   Ethylene - bis(biguanide) - silver(iii) perchlorate

$C_6H_{16}AgN_{10}^{3+}$, $3ClO_4^-$

M.L.Simms, J.L.Atwood, D.A.Zatko   *J. C. S. Chem. Comm.*, 46, 1973

Residue 1 also classified in 79

76.10   Dichloro - (N,N,N',N' - tetramethylethylenediamine) zinc(ii)

$C_6H_{16}Cl_2N_2Zn$

S.Htoon, M.F.C.Ladd   *J. Cryst. Mol. Struct.*, **3**, 95, 1973

76.11   Iodo - (triethylenetetramine) zinc(ii) iodide

$C_6H_{18}IN_4Zn^+$, I^-

G.Marongiu, M.Cannas, G.Carta   *J. Coord. Chem.*, **2**, 167, 1973

76.12   Triethylenetetramine - nickel(ii) perchlorate

$C_6H_{18}N_4Ni^{2+}$, $2ClO_4^-$

A.McPherson Junior, M.G.Rossmann, D.W.Margerum, M.R.James

J. Coord. Chem., **1**, 39, 1971

76.13   bis(1,2 - Propanediamine) copper(ii) perchlorate

$C_6H_{20}CuN_4^{2+}$, $2ClO_4^-$

A.Pajunen, J.Lehtonen   *Suomen Kemistil. (B)*, **45**, 43, 1972

76.14 π - Chloro(ethylenediamine)(diethylenetriamine) cobalt(iii)
tetrachlorozincate(ii)

$C_6H_{21}ClCoN_5^{2+}$, Cl_4Zn^{2-}

A.R.Gainsford, D.A.House, W.T.Robinson

Inorg. Chim. Acta, **5**, 595, 1971

76.15 χ - Chloro(ethylenediamine)(diethylenetriamine) cobalt(iii)
tetrachlorozincate(ii)

$C_6H_{21}ClCoN_5^{2+}$, Cl_4Zn^{2-}

A.R.Gainsford, D.A.House, W.T.Robinson

Inorg. Chim. Acta, **5**, 595, 1971

76.16   tris(Ethylenediamine) cadmium(ii) selenosulfate

$C_6H_{24}CdN_6^{2+}$, O_3SSe^{2-}

N.V.Podberezskaya, S.V.Borisov   *Zh. Strukt. Khim.*, **12**, 1114, 1971

76.17   (+)D - tris(Ethylenediamine) cobalt(iii) nitrate (absolute configuration)

$C_6H_{24}CoN_6^{3+}$, $3NO_3^-$

D.Witiak, J.C.Clardy, D.S.Martin Junior   *Acta Cryst. (B)*, **28**, 2694, 1972

76.18   tris(Ethylenediamine) cobalt(iii) hexacyanoferrate(iii) dihydrate

$C_6H_{24}CoN_6^{3+}$, $C_6FeN_6^{3-}$, $2H_2O$

L.D.C.Bok, J.G.Leipoldt, S.S.Basson

Z. Anorg. Allg. Chem., **389**, 307, 1972

76.19   tris(Ethylenediamine) zinc(ii) thiosulfate

$C_6H_{24}N_6Zn^{2+}$, $O_3S_2^{2-}$

N.V.Podberezskaya, S.V.Borisov   *Zh. Strukt. Khim.*, **12**, 1114, 1971

76.20 $(+)_{589}$ - cis - β - **Dinitro** - **(R - 5 - methyltriethyltetramine) cobalt(iii) chloride (absolute configuration)**
$C_7H_{20}CoN_6O_4{}^+$, Cl^-
K.Tanaka, F.Marumo, Y.Saito *Acta Cryst. (B)*, **29**, 733, 1973

76.21   **bis(Ethylenediamine) - trimethylenediamine - cobalt(iii) bromide (absolute configuration)**
$C_7H_{26}CoN_6{}^{3+}$, $3Br^-$
H.V.F.Schousboe-Jensen *Acta Chem. Scand.*, **26**, 3413, 1972
Residue 1 also classified in 83

76.22   **Potassium nitro - (ethylenediaminetriacetato) - cobaltate(iii) sesquihydrate**
$C_8H_{11}CoN_3O_8{}^-$, K^+ , $1.5H_2O$
J.D.Bell, G.L.Blackmer *Inorg. Chem.*, **12**, 836, 1973
Residue 1 also classified in 81, 82

76.C $(-)_{589}$ - **Dinitro** - **bis(ethylenediamine) cobalt(iii)** $(+)_{589}$ - **bis(malonato) - ethylenediamine cobaltate(iii) (absolute configuration)**
$C_8H_{12}CoN_2O_8{}^-$, $C_4H_{16}CoN_6O_4{}^+$
For complete entry see 76.4

76.23   **Aquo - bis(ethylenediamine)(tetrafluoroborato) nickel(ii) tetrafluoroborate**
$C_8H_{18}BF_4N_4NiO^+$, $BF_4{}^-$
A.A.G.Tomlinson, M.Bonamico, G.Dessy, V.Fares, L.Scaramuzza
J, C. S. Dalton, 1671, 1972

76.24   **bis(N - Methylethylenediamine) - di(thiocyanato) copper(ii)**
$C_8H_{20}CuN_6S_2$
A.Pajunen, R.Hamalainen *Suomen Kemistil. (B)*, **45**, 122, 1972

76.25 $(-)_{589}$ - **trans** - **Dichloro** - **(1,10 - diamino - 4,7 - diazadecane) cobalt(iii) nitrate (absolute configuration)**
$C_8H_{22}Cl_2CoN_4{}^+$, $NO_3{}^-$
N.C.Payne *Inorg. Chem.*, **12**, 1151, 1973

76.26 $\alpha\beta S$ - **Chloro(tetraethylenepentamine) cobalt(iii) perchlorate (absolute configuration)**
$C_8H_{23}ClCoN_5{}^{2+}$, $2ClO_4{}^-$
M.R.Snow *J. C. S. Dalton*, 1627, 1972

76.27 $\alpha\beta R$ - **Chloro(tetraethylenepentamine) cobalt(iii) perchlorate (absolute configuration)**
$C_8H_{23}ClCoN_5{}^{2+}$, $2ClO_4{}^-$
M.R.Snow *J. C. S. Dalton*, 1627, 1972

76.28   **bis(N - (2 - Hydroxyethyl) - ethylenediamine) copper(ii) chloride**
$C_8H_{24}CuN_4O_2{}^{2+}$, $2Cl^-$
A.Pajunen, M.Nasakkala *Suomen Kemistil. (B)*, **45**, 47, 1972

76.29 $(-)_{589}$ - u - facial - bis(Diethylenetriamine) cobalt(iii) hexacyanocobaltate(iii) dihydrate (absolute configuration)
$C_8H_{26}CoN_6^{3+}$, $C_6CoN_6^{3-}$, $2H_2O$
M.Konno, F.Marumo, Y.Saito *Acta Cryst. (B),* **29,** 739, 1973

76.30 μ - Amido - μ - peroxo - bis(bis(ethylenediamine) cobalt(iii)) tri - thiocyanate monohydrate
$C_8H_{34}Co_2N_9O_2^{3+}$, $3CNS^-$, H_2O
U.Thewalt *Z. Anorg. Allg. Chem.,* **393,** 1, 1972

76.31 $(+)_{440}$ - (Chloro - 4 - (2 - aminoethyl) - 7 - methyl - 1,4,7,10 - tetra - azadecane) - cobalt(iii) perchlorate
$C_9H_{25}ClCoN_5^{2+}$, $2ClO_4^-$
D.A.Buckingham, M.Dwyer, A.M.Sargeson, K.J.Watson
Acta Chem. Scand., **26,** 2813, 1972

76.32 Magnesium ethylenediaminetetra - acetato - zinc hexahydrate
$C_{10}H_{12}N_2O_8Zn^{2-}$, $H_8MgO_4^{2+}$, $2H_2O$
A.I.Pozhidaev, N.N.Neronova, T.N.Polynova, M.A.Porai-Koshits,
V.A.Logvinenko *Zh. Strukt. Khim.,* **13,** 738, 1972
Residue 1 also classified in 81, 82

76.33 Hexa - aquo - magnesium ethylenediaminetetra - acetato - aquo - cadmium trihydrate
$C_{10}H_{14}CdN_2O_9^{2-}$, $H_{12}MgO_6^{2+}$, $3H_2O$
A.I.Pozhidaev, T.N.Polynova, M.A.Porai-Koshits
Acta Cryst. (A), **28,** S76, 1972
Residue 1 also classified in 81, 82

76.34 Guanidinium iron(iii) ethylenediaminetetra - acetate dihydrate
$C_{10}H_{14}FeN_2O_9^-$, $CH_6N_3^+$, $2H_2O$
Ya.M.Nesterova, T.N.Polynova, L.I.Martynenko, N.I.Pechurova
Zh. Strukt. Khim., **12,** 1110, 1971
Residue 1 also classified in 81, 82; residue 2 classified in 8

76.35 Aquo(hydrogen ethenelenediaminetetra - acetato) rhodium(iii)
$C_{10}H_{15}N_2O_9Rh$
G.H.Y.Lin, J.D.Leggett, R.M.Wing *Acta Cryst. (B),* **29,** 1023, 1973
Also classified in 81, 83

76.36 Dicaesium μ - ethylenediaminetetra - acetato - di - μ - sulfido - bis(oxomolybdate(v)) dihydrate
$C_{10}H_{16}Mo_2N_2O_{12}S_2^{2-}$, $2Cs^+$, $2H_2O$
B.Spivack, Z.Dori *J. C. S. Dalton,* 1173, 1973
Residue 1 also classified in 81, 82

76.37 **Guanidinium ethylenediaminetetra - acetato - neodymium trihydrate**
$C_{10}H_{18}N_2NdO_{11}^-$, $CH_6N_3^+$
Ya.M.Nesterova, S.G.Zbryskaya, T.N.Polynova, M.A.Porai-Koshits
Zh. Strukt. Khim., **13**, 739, 1972
Residue 1 also classified in 82, 81; residue 2 classified in 8

76.38 **Di - μ - thiocyanato - bis(di(2 - aminoethyl)amine copper(ii)) perchlorate**
$C_{10}H_{26}Cu_2N_8S_2^{2+}$, $2ClO_4^-$
M.Cannas, G.Carta, G.Marongiu *Acta Cryst. (A)*, **28**, S92, 1972
Residue 1 also classified in 83

76.39 **N,N' - Ethylene - bis(acetylacetoniminato) cobalt(ii) benzene solvate**
$C_{12}H_{18}CoN_2O_2$, C_6H_6
S.Bruckner, M.Calligaris, G.Nardin, L.Randaccio
Inorg. Chim. Acta, **2**, 386, 1968
Residue 1 also classified in 77

76.40 **N,N' - Ethylene - bis(acetylacetoniminato) - nitrosyl - cobalt(ii)**
$C_{12}H_{18}CoN_3O_3$
R.Wiest, R.Weiss *Rev. Chim. Miner., Fr.*, **9**, 655, 1972
Also classified in 77

76.41 **Bromoazido - 1,1,7,7 - tetraethyl - diethylenetriamine - copper(ii)**
$C_{12}H_{29}BrCuN_6$
R.F.Ziolo, M.Allen, D.D.Titus, H.B.Gray, Z.Dori
Inorg. Chem., **11**, 3044, 1972

76.42 **bis(Iodo - (tris(2 - dimethylaminoethyl)amine) cadmium) hexaiodo -
dicadmium**
$2C_{12}H_{30}CdIN_4^+$, $Cd_2I_6^{2-}$
P.L.Orioli, M.Ciampolini *J. C. S. Chem. Comm.*, 1280, 1972

76.43 **N,N' - Ethylene - bis(acetylacetoniminato) methyl - cobalt(iii)**
$C_{13}H_{23}CoN_2O_2$
S.Bruckner, M.Calligaris, G.Nardin, L.Randaccio
Inorg. Chim. Acta, **4**, 308, 1970
Also classified in 77

76.44 **Isothiocyanato - (tris(2 - dimethylaminoethyl)amine) nickel(ii) thiocyanate
monohydrate**
$C_{13}H_{30}N_5NiS^+$, CNS^- , H_2O
P.Dapporto, D.Gatteschi *Cryst. Struct. Comm.*, **2**, 137, 1973

76.45 **Chloro - (1,6 - bis(2' - pyridyl) - 2,5 - diazahexane) copper(ii) chloride
hydrate**
$C_{14}H_{18}ClCuN_4^+$, Cl^- , xH_2O
N.A.Bailey, E.D.McKenzie, J.M.Worthington *J. C. S. Dalton*, 1227, 1973
Residue 1 also classified in 83

76.C N,N' - Ethylene - bis(acetylacetoniminato) vinyl - aquo - cobalt(iii)
$C_{14}H_{23}CoN_2O_3$
For complete entry see 71.31

76.46 Di - μ - cyanato - bis(2,2',2'' - triaminotriethylamine) dinickel(ii)
tetraphenylborate (form I)
$C_{14}H_{36}N_{10}Ni_2O_2^{2+}$, $2C_{24}H_{20}B^-$
D.M.Duggan, D.N.Hendrickson *J. C. S. Chem. Comm.*, 411, 1973
Residue 2 classified in 62

76.47 N,N' - Dimethyl - N,N' - bis(β - mercaptoethyl) ethylenediamine zinc(ii)
chloride dihydrate
$C_{16}H_{36}Cl_4N_4S_4Zn_4$, $2H_2O$
W.J.Hu, D.Barton, S.J.Lippard *J. Amer. Chem. Soc.*, **95,** 1170, 1973
Residue 1 also classified in 85

76.C N,N' - bis(Salicylidene) - 1,5 - diamino - 3 - azapentane - dioxouranium(vi)
$C_{18}H_{19}N_3O_4U$
For complete entry see 78.7

76.C Salicylaldehydato - (N - (2' - dimethylaminoethyl) - salicylaldiminato)
copper(ii)
$C_{18}H_{20}CuN_3O_3$
For complete entry see 78.8

76.C N,N' - Ethylene - bis(acetylacetoniminato) - methyl - pyridino - cobalt(iii)
$C_{18}H_{26}CoN_3O_2$
For complete entry see 71.42

76.C N,N' - Ethylene - bis(3 - methoxysalicylideneiminato) cobalt(ii)
monohydrate (yellow form)
$C_{18}H_{28}CoN_2O_4$, H_2O
For complete entry see 78.9

76.48 $(+)_{589}$ - tris((−) - trans - 1,2 - Diaminocyclohexane) cobalt(iii) chloride
monohydrate (absolute configuration)
$C_{18}H_{42}CoN_6^{3+}$, $3Cl^-$, H_2O
A.Kobayashi, F.Marumo, Y.Saito *Acta Cryst. (B),* **28,** 2709, 1972

76.49 N,N' - Ethylene - bis(benzoylacetoniminato) - nitrosyl - cobalt(ii)
$C_{22}H_{22}CoN_3O_3$
R.Wiest, R.Weiss *Rev. Chim. Miner., Fr.,* **9,** 655, 1972
Also classified in 83, 84

76.C bis(Isothiocyanato) - (N - (diphenylphosphinoethyl) - N' -
diethylethylenediamine) cobalt(ii)
$C_{22}H_{28}CoN_4PS_2$
For complete entry see 86.28

76.50 **((2 - ((2 - (Diethylamino)ethyl)amino)ethyl)diphenylphosphine oxide) di - isothiocyanato - cobalt(ii)**
$C_{22}H_{29}CoN_4OPS_2$
C.A.Ghilardi, A.B.Orlandini *J. C. S. Dalton,* 1698, 1972
Also classified in 83, 84

76.C **N,N' - (1,2 - Dimethyl - ethylenediamine) - bis(3 - methoxy - salicylideneiminato) - cobalt(ii) pyridine**
$C_{23}H_{23}CoN_3O_2$
For complete entry see 78.16

76.C **bis(Isothiocyanato) - (N - methyl - N - (diphenylphosphinoethyl) - N' - diethyl - ethylenediamine) cobalt(ii)**
$C_{23}H_{32}CoN_4PS_2$
For complete entry see 86.32

76.51 **Isothiocyanato - (bis(2 - (diethylamino)ethyl) - 2 - (diphenylarsino)ethylamine) nickel(ii) tetraphenylborate**
$C_{27}H_{42}AsN_4NiS^+$, $C_{24}H_{20}B^-$
M.di Vaira, A.B.Orlandini *J. C. S. Dalton,* 1704, 1972
Residue 1 also classified in 83; residue 2 classified in 62

76.52 **Iodo - (N,N - bis(2 - diphenylphosphinoethyl) - N - (2 - diethylaminoethyl)amine) cobalt(ii) iodide**
$C_{34}H_{42}CoIN_2P_2^+$, I^-
A.Bianchi, P.Dapporto, G.Fallani, C.A.Ghilardi, L.Sacconi
J. C. S. Dalton, 641, 1973
Residue 1 also classified in 86

76.53 **Iodo - (N,N - bis(2 - diphenylphosphinoethyl) - N - (2 - diethylaminoethyl)amine) nickel(ii) iodide**
$C_{34}H_{42}IN_2NiP_2^+$, I^-
A.Bianchi, P.Dapporto, G.Fallani, C.A.Ghilardi, L.Sacconi
J. C. S. Dalton, 641, 1973
Residue 1 also classified in 86

76.C **Sodium - bis(tetrahydrofuran) - bis(N,N' - ethylenebis(salicylideneiminato) cobalt(ii)) tetraphenylborate**
$C_{40}H_{44}Co_2N_4NaO_6^+$, $C_{24}H_{20}B^-$
For complete entry see 67.22

METAL COMPLEXES (ACETYLACETONE)

77.C   Acetylacetonato - chloro(1,2 - dihapto(vinyl alcohol)) platinum(ii)
$C_7H_{11}ClO_3Pt$
For complete entry see 72.4

77.C   bis(1,1,1,5,5,5 - Hexafluoropentaen - 2,4 - dionato) copper(ii) - 1,4 - diazabicyclo(2.2.2)octane complex
$C_{10}H_2CuF_{12}O_4$, $C_6H_{12}N_2$
For complete entry see 60.12

77.1   trans - bis(Acetylacetonato) - di - iodo - platinum(iv)
$C_{10}H_{14}I_2O_4Pt$
P.M.Cook, L.F.Dahl, D.Hopgood, R.A.Jenkins *J. C. S. Dalton,* 294, 1973

77.2   Dioxo - bis(acetylacetonato) molybdenum(vi)
$C_{10}H_{14}MoO_6$
B.Kamenar, M.Penavic, C.K.Prout *Cryst. Struct. Comm.*, **2,** 41, 1973

77.3   (−)$_{589}$ - Acetylacetonato - bis(trimethylenediamine) cobalt(iii) diarsenic - ditartrate monohydrate (absolute configuration)
$C_{11}H_{27}CoN_4O_2^{2+}$, $C_8H_8As_2O_{12}^{2-}$, H_2O
H.Kawaguchi, K.Matsumoto, H.Kuroya, S.Kawaguchi
Chem. Letters, 125, 1972
Residue 1 also classified in 83; residue 2 classified in 65

77.C   N,N' - Ethylene - bis(acetylacetoniminato) cobalt(ii) benzene solvate
$C_{12}H_{18}CoN_2O_2$, C_6H_6
For complete entry see 76.39

77.C   N,N' - Ethylene - bis(acetylacetoniminato) - nitrosyl - cobalt(ii)
$C_{12}H_{18}CoN_3O_3$
For complete entry see 76.40

77.C   N,N' - Ethylene - bis(acetylacetoniminato) methyl - cobalt(iii)
$C_{13}H_{23}CoN_2O_2$
For complete entry see 76.43

77.C   N,N' - Ethylene - bis(acetylacetoniminato) vinyl - aquo - cobalt(iii)
$C_{14}H_{23}CoN_2O_3$
For complete entry see 71.31

77.C   **Iron(iii) acetylacetonate silver perchlorate**
$C_{15}H_{21}AgFeO_6^+$, ClO_4^-
For complete entry see 71.34

77.4   **tris(Acetylacetonato) rhodium(iii)**
$C_{15}H_{21}O_6Rh$
J.C.Morrow, E.B.Parker Junior   *Acta Cryst. (B)*, **29**, 1145, 1973

77.5   **tris(Acetylacetonato) scandium(iii)**
$C_{15}H_{21}O_6Sc$
T.J.Anderson, M.A.Neuman, G.A.Melson   *Inorg. Chem.*, **12**, 927, 1973

77.6   **cis - bis(Hexafluoroacetylacetonato) - bis(pyridine) copper(ii)**
$C_{20}H_{12}CuF_{12}N_2O_4$
J.Pradilla-Sorzano, J.P.Fackler Junior   *Inorg. Chem.*, **12**, 1174, 1973
Also classified in 83

77.7   **cis - bis(Hexafluoroacetylacetonato) - bis(pyridine) zinc(ii)**
$C_{20}H_{12}F_{12}N_2O_4Zn$
J.Pradilla-Sorzano, J.P.Fackler Junior   *Inorg. Chem.*, **12**, 1174, 1973
Also classified in 83

77.8   **Thorium(iv) tetrakis(trifluoroacetylacetonate)**
$C_{20}H_{16}F_{12}O_8Th$
G.F.S.Wessels, J.G.Leipoldt, L.D.C.Bok
Z. Anorg. Allg. Chem., **393**, 284, 1972

77.9   **tetrakis(Acetylacetonato) neptunium(iv) (β form)**
$C_{20}H_{28}NpO_8$
B.Allard   *Acta Chem. Scand.*, **26**, 3492, 1972

77.10   **Di - μ - oxo - bis(diacetylacetonato - titanium(iv))**
$C_{20}H_{28}O_{10}Ti_2$
G.D.Smith, C.N.Caughlan, J.A.Campbell   *Inorg. Chem.*, **11**, 2989, 1972

77.11   **Di - μ - oxo - bis(diacetylacetonato - titanium(iv)) dioxane solvate**
$C_{20}H_{28}O_{10}Ti_2$, $2C_4H_8O_2$
G.D.Smith, C.N.Caughlan, J.A.Campbell   *Inorg. Chem.*, **11**, 2989, 1972

77.12   **Acetylacetonato - C - meso - (5,5,7,12,12,14 - hexamethyl - 1,4,8,11 - tetra - aza - cyclotetradecane) nickel(ii) perchlorate**
$C_{21}H_{43}N_4NiO_2^+$, ClO_4^-
N.F.Curtis, D.A.Swann, T.N.Waters   *J. C. S. Dalton*, 1408, 1973
Residue 1 also classified in 83

77.C   **Compound II**
$C_{23}H_{25}ClO_{14}Pd$, $0.66CHCl_3$
For complete entry see 71.58

77.13   **Di - μ - phenylmethoxo - bis(pentane - 2,4 - dionato) dicopper(ii)**
$C_{24}H_{28}Cu_2O_6$
J.E.Andrew, A.B.Blake   *J. C. S. Dalton,* 1102, 1973
Also classified in 84

77.C   **(Tricyclohexylphosphine)methyl - nickel(ii) 2,4 - pentanedionate**
$C_{24}H_{43}NiO_2P$
For complete entry see 71.62

77.C   **Norbornadiene - (binorbornenyl - diyl) iridium acetylacetonate**
$C_{26}H_{31}IrO_2$
For complete entry see 71.70

77.14   **tris(Acetylacetonato)(1,10 - phenanthroline) europium(iii)**
$C_{27}H_{29}EuN_2O_6$
W.H.Watson, R.J.Williams, N.R.Stemple
J. Inorg. Nucl. Chem., **34,** 501, 1972
Also classified in 83

77.15   **bis(Acetylacetonato) - triphenylphosphine - palladium benzene solvate**
$C_{28}H_{29}O_4PPd$, $0.5C_6H_6$
N.Kasai, N.Yasuoka, Y.Kai, M.Horike, H.Tokunan, S.Kawaguchi
Acta Cryst. (A), **28,** S83, 1972
Residue 1 also classified in 86

77.16   **Di - μ - acetato - tetrakis(μ_3 - methoxo - 2,4 - pentanedionato) cobalt(ii,iii)**
$C_{28}H_{46}Co_4O_{16}$
J.A.Bertrand, T.C.Hightower   *Inorg. Chem.,* **12,** 206, 1973
Also classified in 81

77.17   **bis(Acetylacetonato) - bis(2,6 - di - isopropylphenoxo) titanium(iv)**
$C_{34}H_{48}O_6Ti$
P.H.Bird, A.R.Fraser, C.F.Lau   *Inorg. Chem.,* **12,** 1322, 1973
Also classified in 84

77.18   **bis(μ - O - 1,3 - Diphenylpropane - 1,3 - dionato - tricarbonyl rhenium(i))**
$C_{36}H_{22}O_{10}Re_2$
J.C.Barrick, M.Fredette, C.J.L.Lock   *Canad. J. Chem.,* **51,** 317, 1973

77.19   **tris(Dipivalomethanato) europium(iii) 3,3 - dimethylthietane - 1 - oxide**
$C_{38}H_{67}EuO_7S$
J.J.Uebel, R.M.Wing   *J. Amer. Chem. Soc.,* **94,** 8910, 1972
Also classified in 84

77.20   **3 - Methylpyridine - tris(2,2,6,6 - tetramethyl - 3,5 - heptanedionato) lutetium(iii)**
$C_{39}H_{64}LuNO_6$
S.J.S.Wasson, D.E.Sands, W.F.Wagner   *Inorg. Chem.,* **12,** 187, 1973
Also classified in 83

77.21   **Piperidinium tetrakis(benzoylacetonato) europate**
$C_{40}H_{36}EuO_8{}^-$, $C_5H_{12}N^+$
A.L.Il'inskii, M.A.Porai-Koshits, L.A.Aslanov, P.I.Lazarev
Zh. Strukt. Khim., **13,** 277, 1972
Residue 2 classified in 33

77.C   **Hydrogen tetrakis(benzoylacetonato) europate diethylamine complex**
$C_{40}H_{37}EuO_8$, $xC_4H_{11}N$
For complete entry see 61.10

77.22   **tris - (2,2,6,6 - Tetramethylheptane - 3,5 - dionato) europium(iii) di - pyridine**
$C_{43}H_{67}EuN_2O_6$
R.E.Cramer, K.Seff   *Acta Cryst. (B),* **28,** 3281, 1972
Also classified in 83

77.23   **tetrakis(Dipivaloylmethanato) niobium(iv)**
$C_{44}H_{76}NbO_8$
T.J.Pinnavaia, G.Podolsky, P.W.Codding
J. C. S. Chem. Comm., 242, 1973

77.24   **bis(1,5 - Diphenyl - 1,3,5 - pentanetrionato)tetrapyridine - dicobalt(ii) pyridine solvate**
$C_{54}H_{44}Co_2N_4O_6$, $4C_5H_5N$
J.M.Kuszaj, B.Tomlonovic, D.P.Murtha, R.L.Lintvedt, M.D.Glick
Inorg. Chem., **12,** 1297, 1973
Residue 1 also classified in 83

77.25   **Basic erbium 2,2,6,6 - tetramethylhepta - 3,5 - dionate**
$C_{110}H_{202}Er_8O_{33}$
J.C.A.Boeyens, J.P.R.de Villiers   *J. Cryst. Mol. Struct.,* **2,** 197, 1972

METAL COMPLEXES
(SALICYLIC DERIVATIVES)

78.1   N - (Carbamoylmethyl) - salicylideneiminato - aquo - copper(ii) sulfate
hydrate
$2C_9H_{17}CuN_2O_3^+ , O_4S^{2-} , 2H_2O$
H.Tamura, K.Ogawa   *Cryst. Struct. Comm.*, **2**, 103, 1973

78.2   (N - Picolinylidene - N' - salicylhydrazinato) - μ - isothiocyanato - copper(ii)
$(C_{14}H_{10}CuN_4O_2S)_n$
P.Domiano, A.Musatti, M.Nardelli, C.Pelizzi
Abstr. Ital.-Yug. Congr, 208, 1973

78.3   bis(N - Methylsalicylaldiminato) copper(ii) (α form,new crystal data)
$C_{16}H_{16}CuN_2O_2$
R.R.Bartkowski, B.Morosin   *Phys. Rev.*, **6**, 4209, 1972

78.4   Uranyl N,N' - ethylene - bis(salicylideneiminato) methanol complex
$C_{17}H_{18}N_2O_5U$
G.Bandoli, D.A.Clemente, U.Croatto, M.Vidali, P.A.Vigato
Inorg. Nucl. Chem. Letters, **8**, 961, 1972
Also classified in 84

78.5   (+) - N,N' - Butylene - bis(salicylideneiminato) cobalt(ii)
$C_{18}H_{18}CoN_2O_2$
M.Calligaris, G.Nardin, L.Randaccio   *J. C. S. Dalton,* 419, 1973

78.6   meso - N,N' - Butylene - bis(salicylideneiminato) cobalt(ii)
$C_{18}H_{18}CoN_2O_2$
M.Calligaris, G.Nardin, L.Randaccio   *J. C. S. Dalton,* 419, 1973

78.7   N,N' - bis(Salicylidene) - 1,5 - diamino - 3 - azapentane - dioxouranium(vi)
$C_{18}H_{19}N_3O_4U$
M.N.Akhtar, A.J.Smith   *Acta Cryst. (B),* **29,** 275, 1973
Also classified in 76

78.8   Salicylaldehydato - (N - (2' - dimethylaminoethyl) - salicylaldiminato)
copper(ii)
$C_{18}H_{20}CuN_3O_3$
R.Tewari, R.C.Srivastava, R.H.Balundgi, A.Chakravorty
Inorg. Nucl. Chem. Letters, **9,** 583, 1973
Also classified in 76

78.9 N,N' - Ethylene - bis(3 - methoxysalicylideneiminato) cobalt(ii) monohydrate (yellow form)
$C_{18}H_{28}CoN_2O_4$, H_2O
M.Calligaris, G.Nardin, L.Randaccio *Abstr. Ital.-Yug. Congr*, 202, 1973
Residue 1 also classified in 76

78.10 Chloro - (N - (3 - n - propanolato) - salicylaldiminato) iron(iii) dimer
$C_{20}H_{22}Cl_2Fe_2N_2O_4$
P.G.Eller, J.L.Breece, J.A.Bertrand
Amer. Cryst. Assoc., Abstr. Papers (Winter Meeting), 70, 1973

78.11 bis(N - 3 - Hydroxypropyl - salicylideneiminato) dicopper(ii)
$C_{20}H_{22}Cu_2N_2O_4$
K.Ogawa, H.Tamura *Acta Cryst. (A)*, **28**, S87, 1972
Also classified in 84

78.12 Chloro - bis(N - n - propylsalicylaldiminato) iron(iii)
$C_{20}H_{24}ClFeN_2O_2$
J.E.Davies, B.M.Gatehouse *Acta Cryst. (B)*, **28**, 3641, 1972

78.13 α,α' - (2 - (2' - Pyridyl)ethyl)ethylenebis(salicylideneiminato) cobalt(ii) ethanol solvate
$C_{23}H_{21}CoN_3O_2$, C_2H_6O
J.P.Collman, H.Takaya, B.Winkler, L.Libit, S.S.Koon, G.A.Rodley,
W.T.Robinson *J. Amer. Chem. Soc.*, **95**, 1656, 1973

78.14 Dioxygen - $(\alpha,\alpha'$ - (2 - (2 - pyridyl)ethyl)ethylene - bis(salicylideneiminato)) cobalt(ii) solvate
$C_{23}H_{21}CoN_3O_4$
J.P.Collman, H.Takaya, B.Winkler, L.Libit, S.S.Koon, G.A.Rodley,
W.T.Robinson *J. Amer. Chem. Soc.*, **95**, 1656, 1973

78.15 α,α' - (2 - (2 - Pyridyl)ethyl)ethylene - bis(salicylideneiminato) iron(ii)
$C_{23}H_{21}FeN_3O_2$
W.T.Robinson
Amer. Cryst. Assoc., Abstr. Papers (Winter Meeting), 70, 1973

78.C N,N' - Ethylene - bis(salicylideneiminato) - pyridine - vinyl - cobalt(iii)
$C_{23}H_{22}CoN_3O_2$
For complete entry see 71.57

78.16 N,N' - (1,2 - Dimethyl - ethylenediamine) - bis(3 - methoxy - salicylideneiminato) - cobalt(ii) pyridine
$C_{23}H_{23}CoN_3O_2$
M.Calligaris, G.Nardin, L.Randaccio *Abstr. Ital.-Yug. Congr*, 202, 1973
Also classified in 76, 83

78.17 bis(N - Picolinylidene - N' - salicylhydrazinato) nickel(ii)
$C_{26}H_{20}N_6NiO_4$
P.Domiano, A.Musatti, M.Nardelli, C.Pelizzi
Abstr. Ital.-Yug. Congr, 208, 1973

78.18 **(N - Picolinylidene - N' - salicylhydrazino) - (N - picolinylidene - N' - salicylhydrazine) copper(ii) perchlorate ethanol solvate**
$C_{26}H_{21}CuN_6O_4^+ , ClO_4^- , C_2H_6O$
P.Domiano, A.Musatti, M.Nardelli, C.Pelizzi
Abstr. Ital.-Yug. Congr, 208, 1973

78.19 **tetrakis(Salicylato) tetra - aquo - dicadmium(ii)**
$C_{28}H_{28}Cd_2O_{16}$
K.Venkatasubramanian, A.C.Villa, A.G.Manfredotti, C.Guastini
Cryst. Struct. Comm., **1,** 427, 1972

78.20 **N,N' - (Tetramethylethylene) - bis(3 - t - butyl - salicylaldiminato) cobalt**
$C_{28}H_{38}CoN_2O_2$
G.G.Christoph, J.F.Rogers, W.P.Schaefer
Amer. Cryst. Assoc., Abstr. Papers (Winter Meeting), 39, 1973

78.21 **Chloro - bis(N - (2 - phenylethyl)salicylaldiminato) iron(iii)**
$C_{30}H_{28}ClFeN_2O_2$
P.G.Eller, J.L.Breece, J.A.Bertrand
Amer. Cryst. Assoc., Abstr. Papers (Winter Meeting), 70, 1973

78.22 **Di - μ - methoxo - bis(salicylaldehyde anthraniloylhydrazonato) - dimanganese(iii) methanol solvate**
$C_{30}H_{28}Mn_2N_6O_6 , 2CH_4O$
A.Mangia, M.Nardelli, C.Pelizzi, G.Pelizzi   *J. C. S. Dalton,* 1141, 1973
Residue 1 also classified in 83, 84

78.C **Sodium - bis(tetrahydrofuran) - bis(N,N' - ethylenebis(salicylideneiminato) cobalt(ii)) tetraphenylborate**
$C_{40}H_{44}Co_2N_4NaO_6^+ , C_{24}H_{20}B^-$
For complete entry see 67.22

METAL COMPLEXES (THIOUREA)

79.1   **Chloro - bis(thiourea) mercury(ii) chloride**
$C_2H_8ClHgN_4S_2^+$, Cl^-
P.D.Brotherton, P.C.Healy, C.L.Raston, A.H.White
J. C. S. Dalton, 334, 1973

79.2   **cis - Nickel(ii) - dithiosemicarbazide dinitrate**
$C_2H_{10}N_6Ni^{2+}$, $2NO_3^-$
R.G.Hazell   *Acta Chem. Scand.,* **26,** 1365, 1972

79.3   **trans - Nickel(ii) - dithiosemicarbazide dinitrate**
$C_2H_{10}N_6Ni^{2+}$, $2NO_3^-$
R.G.Hazell   *Acta Chem. Scand.,* **26,** 1365, 1972

79.4   **bis(Thiosemicarbazide) platinum(ii) sulfate trihydrate**
$C_2H_{10}N_6PtS_2^{2+}$, O_4S^{2-} , $3H_2O$
L.Gastaldi, P.Porta   *Cryst. Struct. Comm.,* **1,** 353, 1972

79.5   **Tetra - aquo - bis(urea) cobalt(ii) nitrate**
$C_2H_{16}CoN_4O_6^{2+}$, $2NO_3^-$
T.F.Rau, E.N.Kurkutova   *Dokl. Akad. Nauk S. S. S. R.,* **204,** 600, 1972

79.6   **bis(2,4 - Dithiobiureto) nickel(ii) glycol solvate**
$C_4H_8N_6NiS_4$, $C_2H_6O_2$
A.Pignedoli, G.Peyronel, L.Antolini   *Gazz. Chim. Ital.,* **102,** 679, 1972
Residue 1 also classified in 85

79.7   **Dichloro - (acetone - thiosemicarbazone) zinc(ii)**
$C_4H_9Cl_2N_3SZn$
M.Mathew, G.J.Palenik   *Inorg. Chim. Acta,* **5,** 349, 1971
Also classified in 83

79.8   **bis(Dithiobiuret) nickel(ii) perchlorate ethanol solvate**
$C_4H_{10}N_6NiS_4^{2+}$, $2ClO_4^-$, C_2H_6O
A.Pignedoli, G.Peyronel, L.Antolini   *Abstr. Ital.-Yug. Congr,* 212, 1973

79.9   **bis(Dithiobiureto) nickel(ii) perchlorate ethanol solvate**
$C_4H_{10}N_6NiS_4^{2+}$, $2ClO_4^-$, C_2H_6O
A.Pignedoli, G.Peyronel, L.Antolini   *Eur. Cryst. Meeting,* 1973

79.10   **tetrakis(Thiourea) mercury(ii) chloride**
$C_4H_{16}HgN_8S_4{}^{2+}$, $2Cl^-$
A.Korczynski, M.Nardelli, M.A.Pellinghelli
Cryst. Struct. Comm., **1**, 327, 1972

79.11   **Monoaquo - tetra(urea) - dioxouranium(vi) nitrate (neutron and X - ray study)**
$C_4H_{18}N_8O_7U^{2+}$, $2NO_3{}^-$
N.K.Dalley, M.H.Mueller, S.H.Simonsen   *Inorg. Chem.*, **11**, 1840, 1972

79.12   **Diaquo - cobalt(ii) tetra - urea nitrate**
$C_4H_{20}CoN_8O_6{}^{2+}$, $2NO_3{}^-$
T.F.Rau, E.N.Kurkutova   *Dokl. Akad. Nauk S. S. S. R.*, **200**, 134, 1971

79.13   **Chloro - (diacetylmonoxime - thiosemicarbazonato) copper(ii) monohydrate**
$C_5H_9ClCuN_4OS$, $0.5H_2O$
T.J.Malinowski, W.K.Rotaru, E.W.Suntzov, G.A.Kiosse, A.W.Ablov
Eur. Cryst. Meeting, 1973

79.C   **Ethylene - bis(biguanide) - silver(iii) perchlorate**
$C_6H_{16}AgN_{10}{}^{3+}$, $3ClO_4{}^-$
For complete entry see 76.9

79.14   **Chloro - tris(monomethyl - thiourea) silver(i)**
$C_6H_{18}AgClN_6S_3$
T.C.Lee, E.L.Amma   *J. Cryst. Mol. Struct.*, **2**, 125, 1972

79.15   **hexakis(Thiourea)tetra - copper(i) nitrate tetrahydrate**
$C_6H_{24}Cu_4N_{12}S_6{}^{4+}$, $4NO_3{}^-$, $4H_2O$
G.W.Hunt, E.L.Amma
Amer. Cryst. Assoc., Abstr. Papers (Winter Meeting), 81, 1973

79.16   **Titanium(iii) iodide - hexa - urea complex (at 90°K)**
$C_6H_{24}N_{12}O_6Ti^{3+}$, $3ClO_4{}^-$
B.N.Figgis, L.G.B.Wadley   *Aust. J. Chem.*, **25**, 223, 1972
Residue 1 also classified in 83

79.17   **hexakis(Urea) vanadium tri - iodide (at 300°K)**
$C_6H_{24}N_{12}O_6V^{3+}$, $3I^-$
B.N.Figgis, L.G.B.Wadley   *J. C. S. Dalton*, 2182, 1972

79.18   **hexakis(Urea) vanadium tri - iodide (at 90°K)**
$C_6H_{24}N_{12}O_6V^{3+}$, $3I^-$
B.N.Figgis, L.G.B.Wadley   *J. C. S. Dalton*, 2182, 1972

79.19   **Nitrato - aquo - (o - aminobenzaldehyde - thiosemicarbazone) copper(ii) nitrate monohydrate**
$C_7H_{12}CuN_5O_4S^+$, $NO_3{}^-$, H_2O
T.J.Malinowski, W.K.Rotaru, E.W.Suntzov, G.A.Kiosse, A.W.Ablov
Eur. Cryst. Meeting, 1973

79.20 **Chloro - bis(acetone thiosemicarbazone) nickel(ii) chloride monohydrate**
$C_8H_{18}ClN_6NiS_2{}^+$, Cl^- , H_2O
M.Mathew, G.J.Palenik, G.R.Clark   *Inorg. Chem.*, **12**, 446, 1973
Residue 1 also classified in 83, 85

79.21 **Nitrato - bis(acetone thiosemicarbazone) nickel(ii) nitrate monohydrate**
$C_8H_{18}N_7NiO_3S_2{}^+$, $NO_3{}^-$, H_2O
M.Mathew, G.J.Palenik, G.R.Clark   *Inorg. Chem.*, **12**, 446, 1973
Residue 1 also classified in 83, 85

79.22 **tris(Ethylenethiourea) copper(i) sulfate**
$2C_9H_{18}CuN_6S_3{}^+$, O_4S^{2-}
M.S.Weininger, G.W.Huny, E.L.Amma   *J. C. S. Chem. Comm.*, 1140, 1972

79.23 **Dichloro - bis(N,N' - diethylthiourea) cobalt(ii) (triclinic form)**
$C_{10}H_{24}Cl_2CoN_4S_2$
M.Bonamico, G.Dessy, V.Fares, L.Scaramuzza   *J. C. S. Dalton,* 876, 1973

79.24 **decakis(Thiourea) tetracopper(i) hexafluorosilicate monohydrate**
$(C_{10}H_{40}Cu_4N_{20}S_{10}{}^{4+})_n$, $2nF_6Si^{2-}$, nH_2O
A.G.Gash, E.H.Griffith, W.A.Spofford III, E.L.Amma
J. C. S. Chem. Comm., 256, 1973

79.C **bis(Picolinato)bis(thiourea) copper(ii)**
$C_{14}H_{16}CuN_6O_4S_2$
For complete entry see 81.47

79.25 **tris(Tetramethylthiourea) copper(i) tetrafluoroborate**
$C_{15}H_{36}CuN_6S_3{}^+$, $BF_4{}^-$
M.S.Weininger, G.W.Hunt, E.L.Amma   *J. C. S. Chem. Comm.*, 1140, 1972

79.C **bis(Picolinato)bis(allylthiourea) copper(ii)**
$C_{20}H_{24}CuN_6O_4S_2$
For complete entry see 81.54

79.26 **Dibromo - bis(N,N' - diphenylthiourea) cobalt(ii)**
$C_{26}H_{24}Br_2CoN_4S_2$
A.C.Bonamartini, A.Mangia, G.Pelizzi   *Cryst. Struct. Comm.*, **2**, 73, 1973

79.27 **Dibromo - bis(N,N' - diphenylthiourea) nickel(ii)**
$C_{26}H_{24}Br_2N_4NiS_2$
A.Mangia, G.Pelizzi   *Cryst. Struct. Comm.*, **2**, 77, 1973

METAL COMPLEXES
(THIOCARBAMATE OR XANTHATE)

80.1 bis(N - Methyldithiocarbamato) nickel(ii)
$C_4H_8N_2NiS_4$
P.W.G.Newman, A.H.White *J. C. S. Dalton,* 1460, 1972

80.2 Cadmium ethylxanthate
$(C_6H_{10}CdO_2S_4)_n$
Y.Iimura, T.Ito, H.Hagihara *Acta Cryst. (B),* **28,** 2271, 1972

80.3 Mercury ethylxanthate
$(C_6H_{10}HgO_2S_4)_n$
Y.Watanabe, H.Hagihara *Acta Cryst. (A),* **28,** S89, 1972

80.4 bis(Dithiocarbazato) nickel(ii)
$C_6H_{14}N_2NiS_4$
P.Porta, L.Gastaldi *Acta Cryst. (A),* **28,** S78, 1972
Also classified in 83

80.5 tris(Ethylthioxanthato) cobalt(iii)
$C_9H_{15}CoS_9$
A.C.Villa, A.G.Manfredotti, C.Guastini, M.Nardelli
Acta Cryst. (B), **28,** 2231, 1972

80.6 bis(Pyrrolidine - carbodithioato) copper(ii)
$C_{10}H_{16}CuN_2S_4$
P.W.G.Newman, C.L.Raston, A.H.White *J. C. S. Dalton,* 1332, 1973
Also classified in 85

80.7 bis(Pyrrolidine - carbodithioato) nickel(ii)
$C_{10}H_{16}N_2NiS_4$
P.W.G.Newman, C.L.Raston, A.H.White *J. C. S. Dalton,* 1332, 1973
Also classified in 85

80.8 bis(N,N - Diethyldithiocarbamato) mercury(ii) (form II)
$C_{10}H_{20}HgN_2S_4$
H.Iwasaki *Acta Cryst. (A),* **28,** S85, 1972

80.9 bis(N,N - Diethyldithiocarbamato) mercury(ii)
$C_{10}H_{20}HgN_2S_4$
P.C.Healy, A.H.White *J. C. S. Dalton,* 284, 1973

80.10 **Dioxomolybdenum(vi) diethyldithiocarbamate**
$C_{10}H_{20}MoN_2O_2S_4$
A.Kopwillem *Acta Chem. Scand.*, **26**, 2941, 1972

80.11 **N,N - Di - n - butyldithiocarbamato - (1,2 - dicyanoethene - 1,2 - dithiolato) gold(iii)**
$C_{13}H_{18}AuN_3S_4$
J.H.Noordik, T.W.Hummelink, J.G.M.van der Linden
J. Coord. Chem., **2**, 185, 1973
Also classified in 85

80.12 **bis(N,N - Diethyldithiocarbamato) - dioxo - trimethylamineoxide - uranium(vi)**
$C_{13}H_{29}N_3O_3S_4U$
E.Forsellini, G.Bombieri, R.Graziani, B.Zarli
Inorg. Nucl. Chem. Letters, **8**, 461, 1972
Also classified in 84

80.13 **bis(Cyclopentamethylene - dithiocarbamato) iron(ii) dicarbonyl**
$C_{14}H_{20}FeN_2O_2S_4$
J.S.Ricci Junior, C.A.Eggers, I.Bernal *Inorg. Chim. Acta*, **6**, 97, 1972

80.14 **Nickel hexamethylenedithiocarbamate**
$C_{14}H_{24}N_2NiS_4$
Z.A.Starikova, E.A.Shugam, V.M.Agre, Yu.V.Oboznenko
Kristallografija, **17**, 111, 1972

80.15 **Gold(i) di - n - propyldithiocarbamate dimer**
$C_{14}H_{28}Au_2N_2S_4$
R.Hesse, P.Jennische *Acta Chem. Scand.*, **26**, 3855, 1972

80.16 **Copper(ii) bis(N,N - di - n - propyldithiocarbamate)**
$C_{14}H_{28}CuN_2S_4$
G.Peyronel, A.Pignedoli, L.Antolini *Acta Cryst. (B)*, **28**, 3596, 1972

80.17 **bis(N,N - Di - isopropyldithiocarbamato) nickel(ii)**
$C_{14}H_{28}N_2NiS_4$
P.W.G.Newman, A.H.White *J. C. S. Dalton*, 2239, 1972

80.18 **tris(N,N - Diethyldithiocarbamato) gold(iii)**
$C_{15}H_{30}AuN_3S_6$
J.H.Noordik *Cryst. Struct. Comm.*, **2**, 81, 1973

80.19 **tris(N,N - Diethyldithiocarbamato) manganese(iii)**
$C_{15}H_{30}MnN_3S_6$
P.C.Healy, A.H.White *J. C. S. Dalton*, 1883, 1972

80.20 **tris(N,N - Di - n - butyldithiocarbamato) nickel(iv) bromide**
$C_{15}H_{54}N_3NiS_6^+$, Br^-
J.P.Fackler Junior, A.Avdeef, R.G.Fischer Junior
J. Amer. Chem. Soc., **95**, 774, 1973

80.21 bis(N - Methyl - N - phenyl - dithiocarbamato) copper(ii)
$C_{16}H_{16}CuN_2S_4$
J.M.Martin, P.W.G.Newman, B.W.Robinson, A.H.White
J. C. S. Perkin II, 2233, 1972

80.22 bis(N - Methyl - N - phenyl - dithiocarbamato) nickel(ii)
$C_{16}H_{16}N_2NiS_4$
J.M.Martin, P.W.G.Newman, B.W.Robinson, A.H.White
J. C. S. Perkin II, 2233, 1972

80.23 bis(Diethyldithiocarbamato) - bis(bis(cis - 1,2 - perfluoromethyl)ethene - 1,2 - dithiolato) - dinickel
$C_{18}H_{20}F_{12}N_2Ni_2S_8$
A.Hermann, R.M.Wing *Inorg. Chem.*, **11**, 1415, 1972
Also classified in 85

80.24 bis(N,N - Di - n - butyldithiocarbamato) copper(iii) tri - iodide
$C_{18}H_{36}CuN_2S_4{}^+$, $I_3{}^-$
J.G.Wijnhoven, T.E.M.van den Hark, P.T.Beurskens
J. Cryst. Mol. Struct., **2**, 189, 1972

80.25 Dichloro - tetrakis(diethyldithiocarbamato) trimercury
$(C_{20}H_{40}Cl_2Hg_3N_4S_8)_n$
H.Iwasaki *Chem. Letters*, 1105, 1972

80.26 Di - μ - N,N - diethyldithiocarbamato - bis(N,N - diethyldithiocarbamato - mercury) (form I)
$C_{20}H_{40}Hg_2N_4S_8$
H.Iwasaki *Acta Cryst. (A)*, **28**, S85, 1972

80.27 bis(Pentamethylenethiocarbamato) zinc(ii) bis(piperidine)
$C_{22}H_{42}N_4O_2S_2Zn$
C.G.Pierpont, D.L.Greene, B.J.McCormick
J. C. S. Chem. Comm., 960, 1972

80.28 Cadmium hexamethylene - dithiocarbamate
$C_{28}H_{48}Cd_2N_4S_8$
V.M.Agre, E.A.Shugam *Kristallografija*, **17**, 303, 1972

80.29 Zinc hexamethylene - dithiocarbamate
$C_{28}H_{48}N_4S_8Zn_2$
V.M.Agre, E.A.Shugam *Zh. Strukt. Khim.*, **13**, 660, 1972

METAL COMPLEXES (CARBOXYLIC ACID)

81.1   Copper formate tetrahydrate (antiferroelectric form)
$C_2H_2CuO_4$, $4H_2O$
M.I.Kay, R.Kleinberg   *Acta Cryst. (A)*, **28**, S186, 1972

81.2   Iron(ii) thioglycolate monohydrate
$(C_2H_2FeO_2S)_n$, nH_2O
S.Jeannin, Y.Jeannin, G.Lavigne   *J. Organometal. Chem.*, **40**, 187, 1972
Residue 1 also classified in 85

81.3   Aquo - uranium oxalate dihydrate
$(C_2H_2O_7U)_n$, $2nH_2O$
N.C.Jayadevan, D.M.Chackrabartty   *Acta Cryst. (B)*, **28**, 3178, 1972

81.4   Oxalato - diammine - copper(ii) dihydrate
$(C_2H_6CuN_2O_4)_n$, $2nH_2O$
J.Garaj, H.Langfelderova, G.Lundgren, J.Gazo
Collect. Czechosl. Chem. Communic., **37**, 3181, 1972

81.5   catena - μ - Acetato - bromo - diammine - copper(ii)
$C_2H_9BrCuN_2O_2$
M.B.Ferrari, L.C.Capacchi, G.G.Fava, M.Nardelli
J. Cryst. Mol. Struct., **2**, 291, 1972

81.6   (−) - Malato - diaquo - cobalt(ii) monohydrate
$(C_3H_6CoO_7)_n$, nH_2O
A.Karipides, A.T.Reed
Amer. Cryst. Assoc., Abstr. Papers (Winter Meeting), 69, 1973

81.7   Silver trifluoroacetate
$C_4Ag_2F_6O_4$
R.G.Griffin, J.D.Ellett Junior, M.Mehring, J.G.Bullitt, J.S.Waugh
J. Chem. Phys., **57**, 2147, 1972

81.8   Mercury(i) trifluoroacetate
$C_4F_6Hg_2O_4$
D.Grdenic, B.Kamenar, M.Sikirica, B.Kaitner
Abstr. Ital.-Yug. Congr, 76, 1973

81.9   **Ammonium aquo - cis - bis(oxalato) - oxovanadium(iv) monohydrate**
$C_4H_2O_{10}V^{2-}$, $2H_4N^+$, H_2O
G.E.Form, E.S.Raper, R.E.Oughtred, H.M.M.Shearer
J. C. S. Chem. Comm., 945, 1972

81.10   **Cesium oxo - bis(oxalato) - bis(aquo) niobate(v) dihydrate**
$C_4H_4NbO_{11}^-$, Cs^+ , $2H_2O$
B.Kojic-Prodic, R.Liminga, S.Scavnicar *Acta Cryst. (B)*, **29,** 864, 1973

81.11   **Uranyl oxydiacetate**
$(C_4H_4O_7U)_n$
G.Bombieri, E.Forsellini, R.Graziani, G.Tomat, L.Magon
Inorg. Nucl. Chem. Letters, **8,** 1003, 1972

81.12   **Copper(i) acetate**
$C_4H_6Cu_2O_4$
M.G.B.Drew, D.A.Edwards, R.Richards *J. C. S. Chem. Comm.*, 124, 1973

81.13   **Mercury(ii) acetate**
$C_4H_6HgO_4$
R.Allmann *Eur. Cryst. Meeting,* 1973

81.14   **Diaquo - cobalt(ii) (−) - maleate monohydrate (absolute configuration)**
$(C_4H_8CoO_7)_n$, nH_2O
L.Kryger, S.E.Rasmussen *Acta Chem. Scand.*, **26,** 2349, 1972

81.15   **Cesium aquo - dimanganese - trioxalate dihydrate**
$(C_6H_2Mn_2O_{13}^{2-})_n$, $2nCS^+$, $2nH_2O$
H.Siems, J.Lohn *Z. Anorg. Allg. Chem.*, **393,** 97, 1972

81.16   **Diaquo cobalt(ii) 2,3 - pyrazinedicarboxylate**
$(C_6H_6CoN_2O_6)_n$
P.Richard, D.T.Qui, E.F.Bertaut *Acta Cryst. (B)*, **29,** 1111, 1973
Also classified in 83

81.C   **Sodium copper(ii) nitrilotriacetate monohydrate**
$C_6H_6CuNO_6^-$, Na^+ , H_2O
For complete entry see 82.4

81.C   **Lithium nitrilotriacetato - copper(ii) trihydrate**
$(C_6H_6CuNO_6^-)_n$, nLi^+ , $3nH_2O$
For complete entry see 82.5

81.17   **Manganese(ii) citrate decahydrate**
$2C_6H_7MnO_8^-$, $H_{12}MnO_6^{2+}$, $2H_2O$
H.L.Carrell, J.P.Glusker *Acta Cryst. (B)*, **29,** 638, 1973

81.18   **Manganese(ii) citrate decahydrate (further discussion)**
$2C_6H_7MnO_8^-$, $H_{12}MnO_6^{2+}$, $2H_2O$
J.P.Glusker, H.L.Carrell *J. Molec. Struct.*, **15,** 151, 1973

81.19   **Tetra - aquo - tris(oxalato) discandium(iii) dihydrate**
$(C_6H_8O_{16}Sc_2)_n$, $2nH_2O$
E.Hansson *Acta Chem. Scand.*, **26,** 1337, 1972

81.20   **Tetra - aquo tris(oxalato) - diytterbium(iii) dihydrate**
$(C_6H_8O_{16}Yb_2)_n$, $2nH_2O$
E.Hansson *Acta Chem. Scand.*, **27,** 823, 1973

81.21   **tris(Hydroxyacetato) gadolinium(iii) (orthorhombic form)**
tris(Glycolato) gadolinium(iii)
$(C_6H_9GdO_9)_n$
I.Grenthe *Acta Chem. Scand.*, **26,** 1479, 1972

81.22   **tris(Hydroxyacetato) lanthanum(iii) (orhorhombic form)**
tris(Glycolato) lanthanum(iii)
$(C_6H_9LaO_9)_n$
I.Grenthe *Acta Chem. Scand.*, **26,** 1479, 1972

81.C   **Nitrilotriacetato - diaquo - dysprosium(iii) dihydrate**
$C_6H_{10}DyNO_8$, $2H_2O$
For complete entry see 82.6

81.C   **Nitrilotriacetato - diaquo - praseodymium(iii) monohydrate**
$C_6H_{10}NO_8Pr$, H_2O
For complete entry see 82.8

81.23   **Potassium difluoro - peroxo - titanium dipicolinate dihydrate**
$C_7H_3F_2NO_6Ti^{2-}$, $2K^+$, $2H_2O$
D.Schwarzenbach *Helv. Chim. Acta,* **55,** 2990, 1972
Residue 1 also classified in 83

81.24   **Ammonium oxo - peroxo - (pyridine - 2,6 - dicarboxylato) vanadate(v) hydrate (at −100°C)**
$C_7H_5NO_8V^-$, H_4N^+ , $1.3H_2O$
R.E.Drew, F.W.B.Einstein *Inorg. Chem.*, **12,** 829, 1973
Residue 1 also classified in 83

81.25   **(Pyridine - 2,6 - dicarboxylato)diaquo - copper(ii)**
$C_7H_7CuNO_6$
M.B.Cingi, A.C.Villa, C.Guastini, M.Nardelli
Gazz. Chim. Ital., **101,** 825, 1971
Also classified in 83

81.26   **Diaquo - peroxo - titanium dipicolinate dihydrate**
$C_7H_7NO_8Ti$, $2H_2O$
D.Schwarzenbach *Helv. Chim. Acta,* **55,** 2990, 1972
Residue 1 also classified in 83

81.27 **Diaquo - (peroxo) - dipicolinato - titanium(iv) dihydrate (orthorhombic form)**
$C_7H_7NO_8Ti$, $2H_2O$
D.Schwarzenbach, H.Manchar   *Eur. Cryst. Meeting,* 1973
Residue 1 also classified in 83

81.28 **Molybdenum(ii) trifluoroacetate dimer**
$C_8F_{12}Mo_2O_8$
F.A.Cotton, J.G.Norman Junior   *J. Coord. Chem.*, **1**, 161, 1971

81.29 **Tetrasodium dicopper(ii) - DL - tartrate decahydrate**
$C_8H_4Cu_2O_{12}{}^{4-}$, $4Na^+$, $10H_2O$
R.J.Missavage, R.L.Belford, I.C.Paul   *J. Coord. Chem.*, **2**, 145, 1972

81.C **Phenylmercury(ii) acetate**
$C_8H_8HgO_2$
For complete entry see 71.6

81.30 **Sodium bis(oxydiacetato)dioxo - uranium(vi) dihydrate**
$C_8H_8O_{12}U^{2-}$, $2Na^+$, $2H_2O$
G.Bombieri, R.Graziani, E.Forsellini
Inorg. Nucl. Chem. Letters, **9**, 551, 1973

81.C **Potassium nitro - (ethylenediaminetriacetato) - cobaltate(iii) sesquihydrate**
$C_8H_{11}CoN_3O_8{}^-$, K^+ , $1.5H_2O$
For complete entry see 76.22

81.C **(−)$_{589}$ - Dinitro - bis(ethylenediamine) cobalt(iii) (+)$_{589}$ - bis(malonato) - ethylenediamine cobaltate(iii) (absolute configuration)**
$C_8H_{12}CoN_2O_8{}^-$, $C_4H_{16}CoN_6O_4{}^+$
For complete entry see 76.4

81.31 **bis(Iminodiacetato)dioxouranium(vi)**
$(C_8H_{12}N_2O_{10}U)_n$
G.Bombieri, R.Graziani, E.Forsellini   *Eur. Cryst. Meeting,* 1973
Also classified in 82

81.32 **Copper(ii) di(γ - aminobutyrate)**
$(C_8H_{16}CuN_2O_3)_n$
A.Takenaka, E.Oshima, S.Yamada, T.Watanabe
Acta Cryst. (B), **29**, 503, 1973

81.33 **Copper(ii) di(γ - aminobutyrate) dihydrate**
$(C_8H_{16}CuN_2O_4)_n$, $2nH_2O$
A.Takenaka, E.Oshima, S.Yamada, T.Watanabe
Acta Cryst. (B), **29**, 503, 1973

81.34 **Dichloro - bis(N,N - dimethylacetamide) zinc**
$C_8H_{18}Cl_2N_2O_2Zn$
M.Herzeg, J.Fischer   *Abstr. Ital.-Yug. Congr.*, 163, 1973
Also classified in 84

81.35 **Acetato - (di - (3 - aminopropyl)amine) copper(ii) perchlorate**
$C_8H_{20}CuN_3O_2{}^+$, $ClO_4{}^-$
B.W.Skelton, T.N.Waters, N.F.Curtis *J. C. S. Dalton,* 2133, 1972
Residue 1 also classified in 83

81.C **Magnesium ethylenediaminetetra - acetato - zinc hexahydrate**
$C_{10}H_{12}N_2O_8Zn^{2-}$, $H_8MgO_4{}^{2+}$, $2H_2O$
For complete entry see 76.32

81.C **Hexa - aquo - magnesium ethylenediaminetetra - acetato - aquo - cadmium trihydrate**
$C_{10}H_{14}CdN_2O_9{}^{2-}$, $H_{12}MgO_6{}^{2+}$, $3H_2O$
For complete entry see 76.33

81.C **Guanidinium iron(iii) ethylenediaminetetra - acetate dihydrate**
$C_{10}H_{14}FeN_2O_9{}^-$, $CH_6N_3{}^+$, $2H_2O$
For complete entry see 76.34

81.C **Aquo(hydrogen ethenelenediaminetetra - acetato) rhodium(iii)**
$C_{10}H_{15}N_2O_9Rh$
For complete entry see 76.35

81.C **Dicaesium μ - ethylenediaminetetra - acetato - di - μ - sulfido - bis(oxomolybdate(v)) dihydrate**
$C_{10}H_{16}Mo_2N_2O_{12}S_2{}^{2-}$, $2Cs^+$, $2H_2O$
For complete entry see 76.36

81.C **Guanidinium ethylenediaminetetra - acetato - neodymium trihydrate**
$C_{10}H_{18}N_2NdO_{11}{}^-$, $CH_6N_3{}^+$
For complete entry see 76.37

81.36 **Diaquo - bis(picolinato) cobalt(ii) dihydrate**
$C_{12}H_{12}CoN_2O_6$, $2H_2O$
S.C.Chang, J.K.H.Ma, J.T.Wang, N.C.Li *J. Coord. Chem.,* **2,** 31, 1972
Residue 1 also classified in 83

81.C **Potassium bis(triacetatonitrido) nickelate(ii) octahydrate**
$C_{12}H_{12}N_2NiO_{12}{}^{4-}$, $4K^+$, $8H_2O$
For complete entry see 82.19

81.37 **Sodium μ - oxo - bis(peroxo - (nitrilotriacetato) titanium(iv)) undecahydrate**
$C_{12}H_{12}N_2O_{15}Ti_2{}^{4-}$, $4Na^+$, $11H_2O$
D.Schwarzenbach, H.Manohar *Eur. Cryst. Meeting,* 1973
Residue 1 also classified in 83

81.C **bis(π - Cyclopentadienyl) - glycinato - molybdenum(iv) chloride monohydrate**
$C_{12}H_{14}MoNO_2{}^+$, Cl^- , H_2O
For complete entry see 73.16

81.38   **bis(Dioxazolecarboxylato) copper(ii)**
$C_{12}H_{16}CuN_2O_8$
J.R.Brush, R.J.Magee, M.J.O'Connor, S.B.Teo, R.J.Geue, M.R.Snow
J. Amer. Chem. Soc., **95,** 2034, 1973
Also classified in 83

81.39   **Cyclo - (bis(μ - acetato - μ - nitrosyl) - bis(di - μ - acetato - diplatinum))
acetic acid solvate**
$C_{12}H_{18}N_2O_{14}Pt_4$, $2C_2H_4O_2$
P.de Meester, A.C.Skapski, J.P.Heffer   *J. C. S. Chem. Comm.*, 1039, 1972

81.40   **bis(6 - Aminohexanoato) copper(ii) dihydrate**
$(C_{12}H_{24}CuN_2O_4)_n$, $2nH_2O$
B.Sjoberg, R.Osterberg, R.Soderquist   *Acta Cryst. (B),* **29,** 1136, 1973
Residue 1 also classified in 82

81.41   **bis(Acetato) - bis(morpholine) copper(ii) dihydrate**
$C_{12}H_{24}CuN_2O_6$
L.P.Battaglia, A.B.Corradi, M.Nardelli, C.Palmieri
Abstr. Ital.-Yug. Congr, 200, 1973
Also classified in 83

81.C   **bis(π - Cyclopentadienyl) - sarcosinato - molybdenum(iv) chloride methanol
solvate**
$C_{13}H_{16}MoNO_2{}^+$, Cl^- , CH_4O
For complete entry see 73.18

81.C   **Dipivaloylmethyl - mercury(ii) acetate**
$C_{13}H_{22}HgO_4$
For complete entry see 71.28

81.42   **(Pyridine - 2,6 - dicarboxylic acid)(pyridine - 2,6 - dicarboxylato) copper(ii)
hydrate**
$C_{14}H_8CuN_2O_8$, xH_2O
C.Sarchet, H.Loiseleur   *Acta Cryst. (B),* **29,** 1345, 1973
Residue 1 also classified in 83

81.43   **bis(Hydrogen pyridine - 2,6 - dicarboxylato) copper(ii) trihydrate**
$C_{14}H_8CuN_2O_8$, $3H_2O$
M.B.Cingi, A.C.Villa, C.Guastini, M.Nardelli
Gazz. Chim. Ital., **102,** 1026, 1972
Residue 1 also classified in 83

81.44   **bis(Hydrogen pyridine - 2,6 - dicarboxylato) nickel(ii) trihydrate**
$C_{14}H_8N_2NiO_8$, $3H_2O$
A.C.Villa, C.Guastini, A.Musatti, M.Nardelli
Gazz. Chim. Ital., **102,** 226, 1972
Residue 1 also classified in 83

81.45   **bis(Hydrogen pyridine - 2,6 - dicarboxylato) nickel(ii) trihydrate**
$C_{14}H_8N_2NiO_8$, $3H_2O$
P.Quaglieri, H.Loiseleur, G.Thomas   *Acta Cryst. (B)*, **28**, 2583, 1972
Residue 1 also classified in 83

81.46   **Copper(ii) 2 - pyridylacetate dihydrate**
$C_{14}H_{12}CuN_2O_4$, $2H_2O$
R.Faure, H.Loiseleur   *Acta Cryst. (B)*, **28**, 2733, 1972
Residue 1 also classified in 83

81.47   **bis(Picolinato)bis(thiourea) copper(ii)**
$C_{14}H_{16}CuN_6O_4S_2$
M.B.Ferrari, L.C.Capacchi, G.G.Fava, A.Montenero, M.Nardelli
Kristallografija, **17**, 22, 1972
Also classified in 83, 79

81.48   **bis((Acetato)aquo(pyridine)) nickel(ii)**
$C_{14}H_{20}N_2NiO_6$
J.Drew, M.B.Hursthouse, P.Thornton   *J. C. S. Dalton*, 1658, 1972
Also classified in 83

81.49   **Cyclo - bis(μ - acetato - μ - nitrosyl) - bis(di - μ - acetato - diplatinum(ii))**
$C_{16}H_{26}N_2O_{18}Pt_4$
P.de Meester, A.C.Skapski   *J. C. S. Dalton*, 1194, 1973

81.50   **Tetra - μ - trifluoroacetato - di(pyridine - molybdenum(ii))**
$C_{18}H_{10}F_{12}Mo_2N_2O_8$
F.A.Cotton, J.G.Norman Junior   *J. Amer. Chem. Soc.*, **94**, 5697, 1972

81.51   **Diaquo - tris(isonicotinato) lanthanum(iii)**
$(C_{18}H_{16}LaN_3O_8)_n$
J.Kay, J.W.Moore, M.D.Glick   *Inorg. Chem.*, **11**, 2818, 1972

81.52   **Diaquo - tri(nicotinic acid) holmium(iii) hexa(isothiocyanato) chromate(iii) dihydrate**
$(C_{18}H_{19}HoN_3O_8^{3+})_n$, $nC_6CrN_6S_6^{3-}$, $2nH_2O$
J.Kay, J.W.Moore, M.D.Glick   *Inorg. Chem.*, **11**, 2818, 1972

81.53   **Aquo - 2 - benzothiazolin - 2 - ylideneaminomethylpyridine(picolinato) copper(ii) perchlorate**
$C_{19}H_{17}CuN_4O_3S^+$, ClO_4^-
A.Mangia, M.Nardelli, C.Pelizzi, G.Pelizzi   *J. C. S. Dalton*, 2483, 1972
Residue 1 also classified in 83

81.54   **bis(Picolinato)bis(allylthiourea) copper(ii)**
$C_{20}H_{24}CuN_6O_4S_2$
M.B.Ferrari, L.C.Capacchi, G.G.Fava, A.Montenero, M.Nardelli
Kristallografija, **17**, 22, 1972
Also classified in 83, 79

81.55 **Tetra - μ - betaine - bis(perchlorato - copper(ii)) perchlorate**
$C_{20}H_{44}Cl_2Cu_2N_4O_{12}{}^{2+}$, $2ClO_4{}^-$
R.S.McEwen   *J. C. S. Chem. Comm.*, 68, 1973
Residue 1 also classified in 82

81.56 **Copper(ii) tetra - betaine perchlorate**
$C_{20}H_{44}CuN_4O_8{}^{2+}$, $2ClO_4{}^-$
R.S.McEwen   *J. C. S. Chem. Comm.*, 68, 1973
Residue 1 also classified in 82

81.57 **Di - μ - propionato - bis(propionato - p - toluidine copper(ii))**
$C_{23}H_{38}Cu_2N_2O_8$
D.B.W.Yawney, J.A.Moreland, R.J.Doedens
J. Amer. Chem. Soc., **95,** 1164, 1973
Also classified in 83

81.C **Acetato - bis(phenylazophenyl - 2C,N') rhodium(ii)**
$C_{26}H_{21}N_4O_2Rh$
For complete entry see 71.69

81.C **Di - μ - acetato - bis((2 - methallyl - 3 - norbornyl) palladium(ii))**
$C_{26}H_{40}O_4Pd_2$
For complete entry see 71.71

81.58 **Tetra - μ - o - bromobenzoate - bis(aquo - copper(ii))**
$C_{28}H_{20}Br_4Cu_2O_{10}$
W.Harrison, S.Rettig, J.Trotter   *J. C. S. Dalton,* 1852, 1972

81.C **Di - μ - acetato - tetrakis(μ_3 - methoxo - 2,4 - pentanedionato) cobalt(ii,iii)**
$C_{28}H_{46}Co_4O_{16}$
For complete entry see 77.16

81.59 **Diaquo - praseodymium(iii) nicotinate dimer**
$C_{36}H_{32}N_6O_{16}Pr_2$
L.A.Aslanov, I.K.Abdul'minev, M.A.Porai-Koshits
Zh. Strukt. Khim., **13,** 468, 1972

81.C **Chloro - carbonyl - bis(triphenylphosphine) - difluoromethyl - chlorodifluoroacetato - iridium benzene solvate**
$C_{40}H_{31}Cl_2F_4IrO_3P_2$, C_6H_6
For complete entry see 71.93

81.60 **Acetato - bis(triphenylphosphine)dicarbonyl - manganese(i)**
$C_{40}H_{33}MnO_4P_2$
W.K.Dean, G.L.Simon, P.M.Treichel, L.F.Dahl
J. Organometal. Chem., **50,** 193, 1973
Also classified in 86

81.61   Tetra - μ - benzoato - bis(quinoline - cobalt(ii))
$C_{46}H_{34}Co_2N_2O_8$
J.Drew, M.B.Hursthouse, P.Thornton, A.J.Welch
J. C. S. Chem. Comm., 52, 1973
Also classified in 83

81.62   bis(μ - Hydroxy - μ - acetato - (4 - hydroxy - bis(3,5 - (N - (2' - hydroxyphenyl) - formimidoyl)toluene) - cobalt(ii) - cobalt(iii))) dihydrate ethanol solvate
$C_{46}H_{38}Co_4N_4O_{12}$, $2C_2H_6O$, $2H_2O$
B.F.Hoskins, R.Robson, D.Vince   *J. C. S. Chem. Comm.*, 392, 1973
Residue 1 also classified in 83, 84

81.63   Hexa - μ - acetato - μ_3 - oxo - tri(triphenylphosphine - ruthenium)
$C_{66}H_{63}O_{13}P_3Ru_3$
F.A.Cotton, J.G.Norman Junior   *Inorg. Chim. Acta,* **6,** 411, 1972
Also classified in 86

METAL COMPLEXES (AMINO-ACID)

82.1 **Dichloro - (S - methyl - L - cysteine) palladium(ii) monohydrate**
$C_4H_9Cl_2NO_2PdS$, H_2O
L.P.Battaglia, A.B.Corradi, C.G.Palmieri, M.Nardelli, M.E.V.Tani
Acta Cryst. (B), **29**, 762, 1973
Residue 1 also classified in 85

82.2 **Glycine silver(i) nitrate (ferroelectric form)**
$C_4H_{10}Ag_2N_2O_4^{2+}$, $2NO_3^-$
S.Guha *Indian J. Phys.*, **46**, 255, 1972

82.3 **(Glycyl - L - alaninato) copper(ii) hydrate**
$(C_5H_8CuN_2O_3)_n$, nH_2O
H.C.Freeman, M.J.Healy, G.H.W.Milburn, M.L.Scudder
Stockholm Symposium, 69, 1973

82.4 **Sodium copper(ii) nitrilotriacetate monohydrate**
$C_6H_6CuNO_6^-$, Na^+ , H_2O
S.H.Whitlow *Acta Cryst. (A)*, **28**, S89, 1972
Residue 1 also classified in 81

82.5 **Lithium nitrilotriacetato - copper(ii) trihydrate**
$(C_6H_6CuNO_6^-)_n$, nLi^+ , $3nH_2O$
V.V.Fomenko, L.I.Kopaneva, T.N.Polynova, M.A.Porai-Koshits,
N.D.Mitrofanova, L.I.Martynenko *Zh. Strukt. Khim.*, **13**, 166, 1972
Residue 1 also classified in 81, 83

82.6 **Nitrilotriacetato - diaquo - dysprosium(iii) dihydrate**
$C_6H_{10}DyNO_8$, $2H_2O$
L.L.Martin, R.A.Jacobson *Inorg. Chem.*, **11**, 2789, 1972
Residue 1 also classified in 81, 83

82.7 **Sodium di - μ - sulfido - bis((L - cysteinato)oxomolybdate(v)) dihydrate**
$C_6H_{10}Mo_2N_2O_6S_2^{2-}$, $2Na^+$, $2H_2O$
D.H.Brown, J.A.D.Jeffreys *J. C. S. Dalton*, 732, 1973

82.8 **Nitrilotriacetato - diaquo - praseodymium(iii) monohydrate**
$C_6H_{10}NO_8Pr$, H_2O
L.L.Martin, R.A.Jacobson *Inorg. Chem.*, **11**, 2785, 1972
Residue 1 also classified in 81, 83

82.9   **Aquo(glycyl - L - glutamato) copper(ii) hydrate**
$(C_7H_{12}CuN_2O_6)_n$, $2.5H_2O$
H.C.Freeman, M.J.Healy, G.H.W.Milburn, M.L.Scudder
Stockholm Symposium, 69, 1973

82.10   **Potassium bis - bis(iminodiacetato) cobaltate(iii) hydrate**
$C_8H_{10}CoN_2O_8^-$, K^+ , $2.5H_2O$
A.B.Corradi, C.G.Palmieri, M.Nardelli, M.A.Pellinghelli, M.E.V.Tani
J. C. S. Dalton, 655, 1973

82.C   **Potassium nitro - (ethylenediaminetriacetato) - cobaltate(iii) sesquihydrate**
$C_8H_{11}CoN_3O_8^-$, K^+ , $1.5H_2O$
For complete entry see 76.22

82.C   **bis(Iminodiacetato)dioxouranium(vi)**
$(C_8H_{12}N_2O_{10}U)_n$
For complete entry see 81.31

82.11   **Chloro - (glycylglycinato) - aquo - copper(ii) dimer**
$C_8H_{18}Cl_2Cu_2N_4O_8$
M.Shiro, Y.Nakao, O.Yamauchi, A.Nakahara   *Chem. Letters,* 123, 1972

82.12   **Di - μ - hydroxo - bis(bis(glycinato) chromium(iii))**
$C_8H_{18}Cr_2N_4O_{10}$
J.T.Veal, W.E.Hatfield, D.Y.Jeter, J.C.Hempel, D.J.Hodgson
Inorg. Chem., **12,** 342, 1973

82.C   **Magnesium ethylenediaminetetra - acetato - zinc hexahydrate**
$C_{10}H_{12}N_2O_8Zn^{2-}$, $H_8MgO_4^{2+}$, $2H_2O$
For complete entry see 76.32

82.13   **(Glycyl - L - histidylglycinato) copper(ii) sodium perchlorate monohydrate**
$C_{10}H_{13}CuN_5O_4$, Na^+ , ClO_4^- , H_2O
R.Osterberg, B.Sjoberg, R.Soderquist   *Acta Chem. Scand.,* **26,** 4184, 1972

82.14   **Glycyl - L - histidylglycinato - copper(ii) hydrate**
$C_{10}H_{13}CuN_5O_4$, $12H_2O$
R.Osterberg, B.Sjoberg, R.Soderquist   *J. C. S. Chem. Comm.,* 983, 1972

82.C   **Hexa - aquo - magnesium ethylenediaminetetra - acetato - aquo - cadmium trihydrate**
$C_{10}H_{14}CdN_2O_9^{2-}$, $H_{12}MgO_6^{2+}$, $3H_2O$
For complete entry see 76.33

82.C   **Guanidinium iron(iii) ethylenediaminetetra - acetate dihydrate**
$C_{10}H_{14}FeN_2O_9^-$, $CH_6N_3^+$, $2H_2O$
For complete entry see 76.34

82.C **Dicaesium** μ - **ethylenediaminetetra** - **acetato** - **di** - μ - **sulfido** -
bis(oxomolybdate(v)) dihydrate
$C_{10}H_{16}Mo_2N_2O_{12}S_2{}^{2-}$, $2Cs^+$, $2H_2O$
For complete entry see 76.36

82.15 **bis(Glycinato) - bis(imidazole) nickel(ii)**
$C_{10}H_{16}N_6NiO_4$
H.C.Freeman, J.M.Guss *Acta Cryst. (B),* **28,** 2090, 1972
Also classified in 83

82.C **Guanidinium ethylenediaminetetra - acetato - neodymium trihydrate**
$C_{10}H_{18}N_2NdO_{11}{}^-$, $CH_6N_3{}^+$
For complete entry see 76.37

82.16 **bis(L - Prolinamidato) nickel(ii) dihydrate**
$C_{10}H_{18}N_4NiO_2$, $2H_2O$
T.Tsukihara, Y.Katsube, K.Fujimori, Y.Ishimura
Bull. Chem. Soc. Jap., **45,** 1367, 1972

82.17 **Copper diaquo proline**
$C_{10}H_{20}CuN_2O_6$
N.Shamala, K.Venkatesan *Cryst. Struct. Comm.,* **2,** 5, 1973

82.18 **bis(L - Ornithinato) palladium(ii)**
$C_{10}H_{22}N_4O_4Pd$
Y.Nakayama, K.Matsumoto, S.Ooi, H.Kuroya
J. C. S. Chem. Comm., 170, 1973

82.19 **Potassium bis(triacetatonitrido) nickelate(ii) octahydrate**
$C_{12}H_{12}N_2NiO_{12}{}^{4-}$, $4K^+$, $8H_2O$
V.V.Fomenko, T.N.Polynova, M.A.Porai-Koshits, N.D.Mitrofanova
Zh. Strukt. Khim., **13,** 343, 1972
Residue 1 also classified in 81, 83

82.C **bis(6 - Aminohexanoato) copper(ii) dihydrate**
$(C_{12}H_{24}CuN_2O_4)_n$, $2nH_2O$
For complete entry see 81.40

82.20 **(+) - Dinitro - bis(L - arginine) cobalt(iii) nitrate dihydrate (absolute**
configuration)
$C_{12}H_{28}CoN_{10}O_{10}{}^+$, $NO_3{}^-$, $2H_2O$
W.H.Watson, D.R.Johnson, M.B.Celap, B.Kamberi
Inorg. Chim. Acta, **6,** 591, 1972

82.21 **Hydrogen - di(bis(π - cyclopentadienyl) - L - cysteinato - molybdenum(iv))**
chloride
$C_{13}H_{15}MoNO_2S$, $C_{13}H_{16}MoNO_2S^+$, Cl^-
C.K.Prout, G.B.Allison, L.T.J.Delbaere, E.Gore
Acta Cryst. (B), **28,** 3043, 1972
Residue 1 also classified in 73

82.C **Hydrogen - di(bis(π - cyclopentadienyl) - L - cysteinato - molybdenum(iv)) hexafluorophosphate**
$C_{13}H_{15}MoNO_2S$, $C_{13}H_{16}MoNO_2S^+$, F_6P^-
For complete entry see 73.17

82.22 **bis(o - (β - D - Xylopyranosyl) - L - serinato) copper(ii)**
$C_{16}H_{28}CuN_2O_{14}$
L.T.J.Delbaere, M.Higham, B.Kamenar, P.W.Kent, C.K.Prout
Biochim. Biophys. Acta, **286,** 441, 1972

82.23 **bis - (L - Tyrosinato) copper(ii)**
$C_{18}H_{20}CuN_2O_6$
D.van der Helm, C.E.Tatsch *Acta Cryst. (B),* **28,** 2307, 1972

82.C **Tetra - μ - betaine - bis(perchlorato - copper(ii)) perchlorate**
$C_{20}H_{44}Cl_2Cu_2N_4O_{12}^{2+}$, $2ClO_4^-$
For complete entry see 81.55

82.C **Copper(ii) tetra - betaine perchlorate**
$C_{20}H_{44}CuN_4O_8^{2+}$, $2ClO_4^-$
For complete entry see 81.56

82.24 **(N - Benzyl - D - prolinato) - (N - benzyl - L - prolinato) copper(ii)**
$C_{24}H_{28}CuN_2O_4$
G.G.Aleksandrov, Yu.T.Struchkov, A.A.Kurganov, S.V.Rogozhin,
V.A.Davankov *Zh. Strukt. Khim.,* **13,** 671, 1972

82.25 **bis(N - Benzyl - L - prolinato) copper(ii)**
$C_{24}H_{28}CuN_2O_4$
G.G.Aleksandrov, Yu.T.Struchkov, A.A.Kurganov, S.V.Rogozhin,
V.A.Davankov *J. C. S. Chem. Comm.,* 1328, 1972

82.26 **bis(DL - Valine - pyridoxylidene) nickel(ii) heptahydrate**
$C_{26}H_{34}N_4NiO_8$, $7H_2O$
S.Capasso, F.Giordano, C.Mattia, L.Mazzarella, A.Ripamonti
Acta Cryst. (A), **28,** S49, 1972

82.27 **bis(L - Valine - pyridoxylidene) zinc(ii) heptahydrate**
$C_{26}H_{34}N_4O_8Zn$, $7H_2O$
S.Capasso, F.Giordano, C.Mattia, L.Mazzarella, A.Ripamonti
Acta Cryst. (A), **28,** S49, 1972

82.28 **Ferrichrysin monohydrate**
$C_{29}H_{46}FeN_9O_{14}$, H_2O
R.Norrestam, B.Stensland *Acta Cryst. (A),* **28,** S39, 1972

METAL COMPLEXES (NITROGEN LIGAND)

83.1 **Penta - ammine - (acetamide) - cobalt(ii) perchlorate**
$C_2H_{19}CoN_6O^{2+}$, $2ClO_4^-$
M.L.Schneider, G.Ferguson, R.J.Balahura
Amer. Cryst. Assoc., Abstr. Papers (Winter Meeting), 74, 1973

83.2 **Trichloro - bis(acetonitrile) - oxo - niobium**
$C_4H_6Cl_3N_2NbO$
G.Constant, J.-C.Daran, Y.Jeannin *Eur. Cryst. Meeting,* 1973

83.3 **bis(Glyoximato) palladium(ii)**
$C_4H_6N_4O_4Pd$
M.Calleri, G.Ferris, D.Viterbo *Inorg. Chim. Acta,* **1,** 297, 1967

83.C **Dichloro - (acetone - thiosemicarbazone) zinc(ii)**
$C_4H_9Cl_2N_3SZn$
For complete entry see 79.7

83.C **Reaction product of bis(S - amino - dithionitrito) nickel(ii) with ammonia, formaldehyde and methanol**
$C_4H_9N_5NiOS_4$
For complete entry see 85.2

83.4 **Trichloro - adeninium - zinc**
$C_5H_6Cl_3N_5Zn$
M.R.Taylor *Acta Cryst. (B),* **29,** 884, 1973

83.C **Diaquo cobalt(ii) 2,3 - pyrazinedicarboxylate**
$(C_6H_6CoN_2O_6)_n$
For complete entry see 81.16

83.C **Lithium nitrilotriacetato - copper(ii) trihydrate**
$(C_6H_6CuNO_6^-)_n$, nLi^+ , $3nH_2O$
For complete entry see 82.5

83.5 **catena - Di - μ - chloro - bis(imidazole) cadmium(ii)**
$(C_6H_8CdCl_2N_4)_n$
R.J.Flook, H.C.Freeman, F.Huq, J.M.Rosalky
Acta Cryst. (B), **29,** 903, 1973

83.6 **Dichloro - bis(imidazole) cobalt(ii)**
$C_6H_8Cl_2CoN_4$
C.-J.Antti, B.K.S.Lundberg *Acta Chem. Scand.*, **26,** 3995, 1972

83.7 **Dichloro - bis(imidazole) copper(ii)**
$C_6H_8Cl_2CuN_4$
B.K.S.Lundberg *Acta Chem. Scand.*, **26,** 3977, 1972

83.8 **bis(Isocyanurato)diammine - copper(ii)**
$C_6H_{10}CuN_8O_6$
P.G.Slade, M.Raupach, E.W.Radoslovich *Acta Cryst. (B),* **29,** 279, 1973

83.C **Nitrilotriacetato - diaquo - dysprosium(iii) dihydrate**
$C_6H_{10}DyNO_8$, $2H_2O$
For complete entry see 82.6

83.C **Nitrilotriacetato - diaquo - praseodymium(iii) monohydrate**
$C_6H_{10}NO_8Pr$, H_2O
For complete entry see 82.8

83.C **bis(Dithiocarbazato) nickel(ii)**
$C_6H_{14}N_2NiS_4$
For complete entry see 80.4

83.9 **bis(1,3 - Propanediamine) copper(ii) benzoate**
$C_6H_{20}CuN_4{}^{2+}$, $2C_7H_5O_2{}^-$
R.Uggla, M.Klinga *Suomen Kemistil. (B),* **45,** 10, 1972

83.10 **Diaquo - bis(propylenediamine) nickel(ii) bromide**
$C_6H_{24}N_4NiO_2{}^{2+}$, $2Br^-$
J.M.Franco, J.Pardo, S.Garcia-Blanco *Eur. Cryst. Meeting,* 1973

83.C **Titanium(iii) iodide - hexa - urea complex (at 90°K)**
$C_6H_{24}N_{12}O_6Ti^{3+}$, $3ClO_4{}^-$
For complete entry see 79.16

83.C **Potassium difluoro - peroxo - titanium dipicolinate dihydrate**
$C_7H_3F_2NO_6Ti^{2-}$, $2K^+$, $2H_2O$
For complete entry see 81.23

83.C **Ammonium oxo - peroxo - (pyridine - 2,6 - dicarboxylato) vanadate(v) hydrate (at −100°C)**
$C_7H_5NO_8V^-$, H_4N^+ , $1.3H_2O$
For complete entry see 81.24

83.C **(Pyridine - 2,6 - dicarboxylato)diaquo - copper(ii)**
$C_7H_7CuNO_6$
For complete entry see 81.25

83.C **Diaquo - peroxo - titanium dipicolinate dihydrate**
$C_7H_7NO_8Ti$, $2H_2O$
For complete entry see 81.26

83.C **Diaquo - (peroxo) - dipicolinato - titanium(iv) dihydrate (orthorhombic form)**
$C_7H_7NO_8Ti$, $2H_2O$
For complete entry see 81.27

83.11 **Dibromo - (2 - (2 - aminoethyl)pyridine) copper(ii)**
$C_7H_8Br_2CuN_2$
V.C.Copeland, P.Singh, W.E.Hatfield, D.J.Hodgson
Inorg. Chem., **11,** 1826, 1972

83.C **cis - (N - Methylcarboxamido) - (methylamine) - tetracarbonyl - manganese(i) (monoclinic form)**
$C_7H_9MnN_2O_5$
For complete entry see 71.5

83.12 **Dichloro - (2 - (2' - aminoethyl) - pyridine) copper(ii)**
$(C_7H_{10}Cl_2CuN_2)_n$
V.C.Copeland, W.E.Hatfield, D.J.Hodgson   *Inorg. Chem.*, **12,** 1340, 1973

83.C **bis(Ethylenediamine) - trimethylenediamine - cobalt(iii) bromide (absolute configuration)**
$C_7H_{26}CoN_6^{3+}$, $3Br^-$
For complete entry see 76.21

83.13 **(Pyrrole - 2 - (N - 3 - propanolato) - carboxaldimino) copper(ii)**
$C_8H_{10}CuN_2O$
J.A.Bertrand, C.E.Kirkwood   *Inorg. Chim. Acta,* **6,** 248, 1972
Also classified in 84

83.14 **Mercury(ii) chloride - collidine complex**
Mercury(ii) chloride - 2,4,6 - trimethylpyridine complex
$C_8H_{11}Cl_2HgN$
S.Kulpe   *Z. Anorg. Allg. Chem.*, **349,** 314, 1967

83.15 **bis(4 - Aminoimidazole - 5 - carboxamidoxime) copper(ii) perchlorate**
$C_8H_{14}CuN_{10}O_2^{2+}$, $2ClO_4^-$
C.D.Stout, M.Sundaralingam, G.H.-Y.Lin   *Acta Cryst. (B),* **28,** 2136, 1972

83.16 **Triaquo - nitrato - caffeine - copper(ii) nitrate**
$C_8H_{16}CuN_5O_8^+$, NO_3^-
M.B.Cingi, A.C.Villa, A.G.Manfredotti, C.Guastini
Cryst. Struct. Comm., **1,** 363, 1972

83.C **Chloro - bis(acetone thiosemicarbazone) nickel(ii) chloride monohydrate**
$C_8H_{18}ClN_6NiS_2^+$, Cl^- , H_2O
For complete entry see 79.20

83.C **Nitrato - bis(acetone thiosemicarbazone) nickel(ii) nitrate monohydrate**
$C_8H_{18}N_7NiO_3S_2^+$, NO_3^- , H_2O
For complete entry see 79.21

83.C Acetato - (di - (3 - aminopropyl)amine) copper(ii) perchlorate
$C_8H_{20}CuN_3O_2^+$, ClO_4^-
For complete entry see 81.35

83.17 bis(1,3 - Diaminobutane) copper(ii) perchlorate (blue - violet form)
$C_8H_{24}CuN_4^{2+}$, $2ClO_4^-$
A.Pajunen, K.Smolander, I.Belinskij *Suomen Kemistil. (B)*, **45**, 317, 1972

83.18 bis(1,3 - Diaminobutane) copper(ii) perchlorate (red - violet form)
$C_8H_{24}CuN_4^{2+}$, $2ClO_4^-$
A.Pajunen, K.Smolander, I.Belinskij *Suomen Kemistil. (B)*, **45**, 317, 1972

83.19 Chloro - imidazolato - bis(imidazole) copper(ii)
$C_9H_{11}ClCuN_6$
B.K.S.Lundberg *Acta Chem. Scand.*, **26**, 3902, 1972

83.20 bis((3 - Chloropyridine) mercury(i)) diperchlorate
$C_{10}H_8Cl_2Hg_2N_2^{2+}$, $2ClO_4^-$
D.L.Kepert, D.Taylor, A.H.White *J. C. S. Dalton*, 893, 1973

83.21 Trinitrato - 2,2′ - bipyridyl manganese(iii)
$C_{10}H_8MnN_5O_9$
F.W.B.Einstein, D.W.Johnson, D.Sutton *Canad. J. Chem.*, **50**, 3332, 1972

83.22 Dinitrato - (2,2′ - bipyridyl) silver(ii)
$C_{10}H_{10}AgN_4O_6$
G.W.Bushnell, M.A.Khan *Canad. J. Chem.*, **50**, 315, 1972

83.23 bis(Pyrazine - 2 - carboxamide) copper(ii) perchlorate
$C_{10}H_{10}CuN_6O_2^{2+}$, $2ClO_4^-$
M.Sekizaki *Acta Cryst. (B)*, **29**, 327, 1973
Residue 1 also classified in 84

83.24 Dibromo - di(adeninium) copper(ii) dibromide
$C_{10}H_{12}Br_2CuN_{10}^{2+}$, $2Br^-$
P.de Meester, A.C.Skapski *J. C. S. Dalton*, 424, 1973

83.25 Octachloro - bis(adeninium) tricopper(ii) tetrahydrate
$C_{10}H_{12}Cl_8Cu_3N_{10}$, $4H_2O$
P.de Meester, A.C.Skapski *J. C. S. Dalton*, 2400, 1972

83.C Aquo(hydrogen ethenelenediaminetetra - acetato) rhodium(iii)
$C_{10}H_{15}N_2O_9Rh$
For complete entry see 76.35

83.C bis(Glycinato) - bis(imidazole) nickel(ii)
$C_{10}H_{16}N_6NiO_4$
For complete entry see 82.15

83.26   **bis(Adeninium) tetra - aquo - bis(adenine) cobalt(ii) sulfate hexahydrate**
$C_{10}H_{18}CoN_{10}O_4{}^{2+}$, $2C_5H_6N_5{}^+$, $2O_4S^{2-}$, $6H_2O$
P.De Meester, D.M.L.Goodgame, D.J.Richman, A.C.Skapski
Nature, **242**, 257, 1973
Residue 2 classified in 44

83.27   **Dicarbonyl - diacetylbis(dimethylhydrazone) nickel(0)**
$C_{10}H_{18}N_4NiO_2$
H.-D.Hausen, K.Krogmann   *Z. Anorg. Allg. Chem.*, **389**, 247, 1972

83.28   **Nickel(ii) methylethylglyoxime (α form)**
$C_{10}H_{18}N_4NiO_4$
R.H.Bowers, C.V.Banks, R.A.Jacobson   *Acta Cryst. (B)*, **28**, 2318, 1972

83.29   **Dibromo - diaquo - (N,N,N′,N′ - tetramethyl - o - phenylenediamine) nickel(ii)**
$C_{10}H_{20}Br_2N_2NiO_2$
G.Bombieri, E.Forsellini, G.Bandoli, L.Sindellari, R.Graziani, C.Panattoni
Inorg. Chim. Acta, **2**, 27, 1968

83.C   **Chloro - ethyl - (dimethylglyoximato)(dimethylglyoxime) cobalt(iii) monohydrate**
$C_{10}H_{20}ClCoN_4O_2$, H_2O
For complete entry see 71.13

83.30   **Dichloro - diaquo - (N,N,N′,N′ - tetramethyl - o - phenylenediamine) nickel(ii)**
$C_{10}H_{20}Cl_2N_2NiO_2$
G.Bombieri, E.Forsellini, G.Bandoli, L.Sindellari, R.Graziani, C.Panattoni
Inorg. Chim. Acta, **2**, 27, 1968

83.C   **Ethylene - platinum - dichloride di - t - butyl - sulfur di - imine**
$C_{10}H_{22}Cl_2N_2PtS$
For complete entry see 72.11

83.C   **Di - μ - thiocyanato - bis(di(2 - aminoethyl)amine copper(ii)) perchlorate**
$C_{10}H_{26}Cu_2N_8S_2{}^{2+}$, $2ClO_4{}^-$
For complete entry see 76.38

83.31   **(3,3 - Dimethyl - 1,5,8,11 - tetra - azacyclotridecane) nickel(ii) perchlorate**
$C_{11}H_{26}N_4Ni^{2+}$, $2ClO_4{}^-$
J.M.Waters, K.R.Whittle   *J. Inorg. Nucl. Chem.*, **34**, 155, 1972

83.C   **(−)$_{589}$ - Acetylacetonato - bis(trimethylenediamine) cobalt(iii) diarsenic - ditartrate monohydrate (absolute configuration)**
$C_{11}H_{27}CoN_4O_2{}^{2+}$, $C_8H_8As_2O_{12}{}^{2-}$, H_2O
For complete entry see 77.3

83.C   **o - Aminothiophenol - di - iron hexacarbonyl**
$C_{12}H_5Fe_2NO_6S$
For complete entry see 85.16

83.32   **4 - Cyanopyridine - mercury(i) perchlorate**
$C_{12}H_8Hg_2N_4^{2+}$, $2ClO_4^-$
D.L.Kepert, D.Taylor, A.H.White   *Inorg. Chem.*, **11,** 1639, 1972

83.33   **Dichloro - bis(nitrobenzene) palladium(ii)**
$C_{12}H_{10}Cl_2N_2O_2Pd$
R.G.Little, R.J.Doedens   *Inorg. Chem.*, **12,** 537, 1973

83.C   **Diaquo - bis(picolinato) cobalt(ii) dihydrate**
$C_{12}H_{12}CoN_2O_6$, $2H_2O$
For complete entry see 81.36

83.C   **Potassium bis(triacetatonitrido) nickelate(ii) octahydrate**
$C_{12}H_{12}N_2NiO_{12}^{4-}$, $4K^+$, $8H_2O$
For complete entry see 82.19

83.C   **Sodium μ - oxo - bis(peroxo - (nitrilotriacetato) titanium(iv)) undecahydrate**
$C_{12}H_{12}N_2O_{15}Ti_2^{4-}$, $4Na^+$, $11H_2O$
For complete entry see 81.37

83.34   **Dibromo - bis(2 - methylpyridine) copper(ii)**
$C_{12}H_{14}Br_2CuN_2$
P.Singh, D.Y.Jeter, W.E.Hatfield, D.J.Hodgson
Inorg. Chem., **11,** 1657, 1972

83.35   **bis(Dihydro - bis(1 - pyrazolyl)borato) cobalt(ii)**
$C_{12}H_{16}B_2CoN_8$
L.J.Guggenberger, C.T.Prewitt, P.Meakin, S.Trofimenko, J.P.Jesson
Inorg. Chem., **12,** 508, 1973

83.36   **bis(Isonicotinato) - tetra - aquo - cadmium(ii)**
$C_{12}H_{16}CdN_3O_8$
M.B.Cingi, A.G.Manfredotti, C.Guastini, A.Musatti, M.Nardelli
Gazz. Chim. Ital., **101,** 815, 1971

83.C   **bis(Dioxazolecarboxylato) copper(ii)**
$C_{12}H_{16}CuN_2O_8$
For complete entry see 81.38

83.37   **tetrakis(Imidazole) copper(ii) sulfate**
$C_{12}H_{16}CuN_8^{2+}$, O_4S^{2-}
G.Fransson, B.K.S.Lundberg   *Acta Chem. Scand.*, **26,** 3969, 1972

83.38   **bis(Isonicotinato) - tetra - aquo - zinc(ii)**
$C_{12}H_{16}N_2O_8Zn$
M.B.Cingi, A.G.Manfredotti, C.Guastini, A.Musatti, M.Nardelli
Gazz. Chim. Ital., **101,** 815, 1971

83.C   **Iron 1,8 - bis(fluoroboro) - 2,7,9,14,15,20 - hexaoxa - 3,6,10,13,16,19 - hexa - aza - 4,5,11,12,17,18 - hexamethylbicyclo(6.6.6)eicosa - 3,5,10,12,16,18 - hexaene cyclohexane**
$C_{12}H_{18}B_2F_2FeN_6O_6$, $0.5C_6H_{12}$
For complete entry see 61.5

83.39   **hexakis(Methylcyanide) iron(ii) tetrachloroferrate(iii)**
$C_{12}H_{18}FeN_6^{2+}$, $2Cl_4Fe^-$
B.A.Stork-Blaisse, G.C.Verschoor, C.Romers
Acta Cryst. (B), **28**, 2445, 1972

83.40   **hexakis(Methylcyanide) iron(ii) tetrachloroferrate(iii)**
$C_{12}H_{18}FeN_6^{2+}$, $2Cl_4Fe^-$
G.Constant, J.-C.Daran, Y.Jeannin   *J. Organometal. Chem.*, **44**, 353, 1972

83.41   **(1 - Diacetyl - monoximatoimino - 3 - diacetylmonoximeimino - propane)methyl - aquo - cobalt(iii) perchlorate**
$C_{12}H_{24}CoN_4O_3^+$, ClO_4^-
S.Bruckner, M.Calligaris, G.Nardin, L.Randaccio
Inorg. Chim. Acta, **3**, 278, 1969

83.C   **bis(Acetato) - bis(morpholine) copper(ii) dihydrate**
$C_{12}H_{24}CuN_2O_6$
For complete entry see 81.41

83.42   **(1,5 - Diazacyclo - octane) nickel(ii) perchlorate dihydrate**
$C_{12}H_{24}N_4Ni^{2+}$, $2ClO_4^-$, $2H_2O$
D.J.Royer, V.H.Schievelbein, A.R.Kalyanaraman, J.A.Bertrand
Inorg. Chim. Acta, **6**, 307, 1972

83.43   **Dichloro - (1,7,10,16 - tetra - oxa - 4,13 - diazacyclo - octadecane) copper(ii)**
$C_{12}H_{26}Cl_2CuN_2O_4$
M.Herceg, R.Weiss   *Acta Cryst. (B)*, **29**, 542, 1973

83.44   **bis(2 - Imino - 4 - amino - 4 - methyl - pentane) nickel(ii) bromide**
$C_{12}H_{28}N_4Ni^{2+}$, $2Br^-$
F.Hanic, F.Pavelcik, D.Gyepesova   *Kristallografija*, **17**, 10, 1972

83.C   **1 - (trans - (Dichloro(diethylamine) - platinio)) - 2 - diethylammoniumethane**
$C_{12}H_{30}Cl_2N_2Pt$
For complete entry see 71.23

83.C   **Trichloromethyl - mercury chloride - 1,10 - phenanthroline**
$C_{13}H_8Cl_4HgN_2$
For complete entry see 71.25

83.C   **Tetracarbonyl - 2 - (dimethylaminomethyl) phenyl - manganese**
$C_{13}H_{12}MnNO_4$
For complete entry see 71.26

83.C **(Pyridine - 2,6 - dicarboxylic acid)(pyridine - 2,6 - dicarboxylato) copper(ii) hydrate**
$C_{14}H_8CuN_2O_8$, xH_2O
For complete entry see 81.42

83.C **bis(Hydrogen pyridine - 2,6 - dicarboxylato) copper(ii) trihydrate**
$C_{14}H_8CuN_2O_8$, $3H_2O$
For complete entry see 81.43

83.C **bis(Hydrogen pyridine - 2,6 - dicarboxylato) nickel(ii) trihydrate**
$C_{14}H_8N_2NiO_8$, $3H_2O$
For complete entry see 81.44

83.C **bis(Hydrogen pyridine - 2,6 - dicarboxylato) nickel(ii) trihydrate**
$C_{14}H_8N_2NiO_8$, $3H_2O$
For complete entry see 81.45

83.C **Copper(ii) 2 - pyridylacetate dihydrate**
$C_{14}H_{12}CuN_2O_4$, $2H_2O$
For complete entry see 81.46

83.45 **bis(1 - Methyl - 3 - o - chlorophenyl - triazene 1 - oxide) nickel(ii)**
$C_{14}H_{14}N_6NiO_2$
R.C.Srivastava, G.L.Dwivedi, N.Ram *Acta Cryst. (A)*, **28**, S89, 1972

83.C **bis(Picolinato)bis(thiourea) copper(ii)**
$C_{14}H_{16}CuN_6O_4S_2$
For complete entry see 81.47

83.C **Chloro - (1,6 - bis(2' - pyridyl) - 2,5 - diazahexane) copper(ii) chloride hydrate**
$C_{14}H_{18}ClCuN_4{}^+$, Cl^- , xH_2O
For complete entry see 76.45

83.46 **Bromo - bis(2 - (2 - aminoethyl)pyridine) copper(ii) bromide**
$C_{14}H_{20}BrCuN_4{}^+$, Br^-
P.Singh, V.C.Copeland, W.E.Hatfield, D.J.Hodgson
J. Phys. Chem., **76**, 2887, 1972

83.47 **bis(2 - (2 - Aminoethyl)pyridine) copper(ii) perchlorate**
$C_{14}H_{20}CuN_4{}^{2+}$, $2ClO_4{}^-$
D.L.Lewis, W.E.Hatfield, D.J.Hodgson
Amer. Cryst. Assoc., Abstr. Papers (Winter Meeting), 71, 1973

83.C **bis((Acetato)aquo(pyridine)) nickel(ii)**
$C_{14}H_{20}N_2NiO_6$
For complete entry see 81.48

83.48   Dichloro - (2,3,9,10 - tetramethyl - 1,4,8,11 - tetra - aza - cyclotetradeca - 1,3,8,10 - tetraene) cobalt(iii) hexafluorophosphate
$C_{14}H_{24}Cl_2CoN_4^+$, F_6P^-
H.W.Smith, G.W.Svetich, E.C.Lingafelter
Amer. Cryst. Assoc., Abstr. Papers (Winter Meeting), 72, 1973

83.49   trans - bis(1 - Acetyl - 3 - t - butyl - 1,2,4 - triazabutadiene) platinum
$C_{14}H_{26}N_6O_2Pt$
R.Mason, K.M.Thomas, A.R.Galbraith, B.L.Shaw, C.M.Elson
J. C. S. Chem. Comm., 297, 1973

83.50   Chloro - (2,2′,2′ - terpyridine) palladium(ii) chloride dihydrate
$C_{15}H_{11}ClN_3Pd^+$, Cl^- , $2H_2O$
G.M.Intille, C.E.Pfluger, W.A.Baker Junior
J. Cryst. Mol. Struct., **3**, 47, 1973

83.51   Chloro - (2,2′,2″ - terpyridine) palladium(ii) tetrachloropalladate(ii)
$2C_{15}H_{11}ClN_3Pd^+$, Cl_4Pd^{2-}
G.M.Intille, C.E.Pfluger, W.A.Baker Junior
Cryst. Struct. Comm., **2**, 217, 1973

83.52   Dinitrato - tris(pyridine) cadmium(ii)
$C_{15}H_{15}CdN_5O_6$
A.F.Cameron, D.W.Taylor, R.H.Nuttall   *J. C. S. Dalton*, 1608, 1972

83.53   trans - Trichloro - tri(pyridine) - molybdenum(iii)
$C_{15}H_{15}Cl_3MoN_3$
J.V.Brencic   *Abstr. Ital.-Yug. Congr*, 161, 1973

83.54   Trichloro - tripyridine - titanium(iii) pyridine solvate
$C_{15}H_{15}Cl_3N_3Ti$, C_5H_5N
R.K.Collins, M.G.B.Drew   *Inorg. Nucl. Chem. Letters*, **8**, 975, 1972

83.55   Dinitrato - tris(pyridine) cobalt(ii)
$C_{15}H_{15}CoN_5O_6$
A.F.Cameron, D.W.Taylor, R.H.Nuttall   *J. C. S. Dalton*, 1603, 1972

83.56   Dinitrato - tris(pyridine) copper(ii)
$C_{15}H_{15}CuN_5O_6$
A.F.Cameron, D.W.Taylor, R.H.Nuttall   *J. C. S. Dalton*, 1603, 1972

83.57   Dinitrato - tris(pyridine) zinc(ii)
$C_{15}H_{15}N_5O_6Zn$
A.F.Cameron, D.W.Taylor, R.H.Nuttall   *J. C. S. Dalton*, 1603, 1972

83.58   Chloro - penta(propionitrile) iron(ii) tetrachloroferrate(iii)
$C_{15}H_{25}ClFeN_5^{2+}$, $2Cl_4Fe^-$
G.Constant, J.-C.Daran, Y.Jeannin   *Eur. Cryst. Meeting*, 1973

83.59   **Tri(histamine) nickel(ii) perchlorate**
$C_{15}H_{27}N_9Ni^{2+}$, $2ClO_4^-$
J.-J.Bonnet, Y.Jeannin   *Bull. Soc. Fr. Mineral. Cristallogr.*, **95**, 61, 1972

83.60   **(+)$_{546}$ - tris - (R,R - 2,4 - Diaminopentane) cobalt(iii) chloride monohydrate (absolute configuration)**
$C_{15}H_{42}CoN_6^{3+}$, $3Cl^-$, H_2O
A.Kobayashi, F.Marumo, Y.Saito   *Acta Cryst. (B)*, **28**, 3591, 1972

83.61   **(N,N' - Ethylene - bis - (2 - amino - 5 - chlorobenzylideneiminato)) cobalt(ii)**
$C_{16}H_{14}Cl_2CoN_2$
R.Karlsson, L.M.Engelhardt, M.Green   *J. C. S. Dalton*, 2463, 1972

83.62   **Structure III**
$C_{16}H_{15}N_4NiS_2^+$, ClO_4^- , H_2O
N.W.Alcock, P.A.Tasker   *J. C. S. Chem. Comm.*, 1239, 1972
Residue 1 also classified in 85

83.63   **Cesium tetrakis(succinimidato) copper(ii) dihydrate**
$C_{16}H_{16}CuN_4O_8^{2-}$, $2Cs^+$, $2H_2O$
T.Tsukihara, Y.Katsube, K.Fujimori, T.Ito
Bull. Chem. Soc. Jap., **45**, 2959, 1972

83.64   **Isothiocyanato - (1,7 - bis(2 - pyridyl) - 2,6 - diazaheptane) copper(ii) thiocyanate (α form)**
$C_{16}H_{20}CuN_5S^+$, CNS^-
N.A.Bailey, E.D.McKenzie   *J. C. S. Dalton*, 1566, 1972

83.65   **Isothiocyanato - (1,7 - bis(2 - pyridyl) - 2,6 - diazaheptane) copper(ii) thiocyanate (γ form)**
$C_{16}H_{20}CuN_5S^+$, CNS^-
N.A.Bailey, E.D.McKenzie   *J. C. S. Dalton*, 1566, 1972

83.C   **Chloro - (bis(2 - ((2 - pyridylmethyl)amino)ethyl)disulfide) nickel(ii) perchlorate**
$C_{16}H_{22}ClN_4NiS_2^+$, ClO_4^-
For complete entry see 85.24

83.66   **Condensation product between copper(ii) ions, oxalodihydrazide and acetaldehyde**
$C_{16}H_{26}CuN_8O_7^{2-}$, $3H_4N^+$, ClO_4^- , $6H_2O$
G.R.Clark, B.W.Skelton, T.N.Waters   *J. C. S. Chem. Comm.*, 1163, 1972

83.67   **racemic - 5,7,7,12,14,14 - Hexamethyl - 1,4,8,11 - tetra - azacyclotetradeca - 4,11 - diene - nickel(ii) thiocyanate monohydrate**
$C_{16}H_{32}N_4Ni^{2+}$, $2CNS^-$, H_2O
F.Hanic, D.Miklas   *J. Cryst. Mol. Struct.*, **2**, 115, 1972

83.C   **Tetracarbonyl - 2 - (N - phenylformimidoyl)phenyl - manganese**
$C_{17}H_{10}MnNO_4$
For complete entry see 71.40

83.C π - Methallyl - (2 - (R,S) - α - phenylethylimino - 3 - penten - 4 - olato) palladium(ii)
$C_{17}H_{23}NOPd$
For complete entry see 72.21

83.68 (Fluoroboro - tris(2 - aldoximo - 6 - pyridyl) phosphine) nickel(ii) tetrafluoroborate
$C_{18}H_9BFN_6NiO_3P^+$, BF_4^-
M.R.Churchill, A.H.Reis Junior *Inorg. Chem.*, **11**, 1811, 1972

83.69 μ - N,N' - Dehydrosemidinato - bis(tricarbonyl iron)
$C_{18}H_{10}Fe_2N_2O_6$
P.E.Baikie, O.S.Mills *Inorg. Chim. Acta*, **1**, 55, 1967

83.70 (Fluoro - boro - tris(2 - aldoximo - 6 - pyridyl)phosphine) iron(ii) tetrafluoroborate dichloromethane solvate
$C_{18}H_{12}BFFeN_6O_3P^+$, BF_4^- , CH_2Cl_2
M.R.Churchill, A.H.Reis Junior *Inorg. Chem.*, **11**, 2299, 1972

83.C N,N' - Ethylene - bis(acetylacetoniminato) - methyl - pyridino - cobalt(iii)
$C_{18}H_{26}CoN_3O_2$
For complete entry see 71.42

83.71 bis(N - t - Butylpyrrole - 2 - carbaldimino) copper(ii) (triclinic form)
$C_{18}H_{26}CuN_4$
C.H.Wei *Inorg. Chem.*, **11**, 2315, 1972

83.72 bis(N - t - Butylpyrrole - 2 - carbaldimino) copper(ii) (tetragonal form)
$C_{18}H_{26}CuN_4$
C.H.Wei *Inorg. Chem.*, **11**, 2315, 1972

83.73 bis(N - t - Butylpyrrole - 2 - carbaldimino) nickel(ii)
$C_{18}H_{26}N_4Ni$
C.H.Wei, J.R.Einstein *Acta Cryst. (B)*, **28**, 2591, 1972

83.74 bis(2 - (2 - Dimethylaminoethyl)pyridine - hydroxy - copper(ii)) perchlorate
$C_{18}H_{30}Cu_2N_4O_2^{2+}$, $2ClO_4^-$
D.L.Lewis, W.E.Hatfield, D.J.Hodgson
Amer. Cryst. Assoc., Abstr. Papers (Winter Meeting), 71, 1973

83.75 Di - μ - hydroxo - bis(2 - (2 - ethylaminoethyl)pyridine) dicopper(ii) perchlorate
$C_{18}H_{30}Cu_2N_4O_2^{2+}$, $2ClO_4^-$
D.L.Lewis, W.E.Hatfield, D.J.Hodgson *Inorg. Chem.*, **11**, 2216, 1972

83.76 2,3,9,10 - Tetramethyl - 1,4,8,11 - tetra - aza - cyclotetradeca - 1,3,8,10 - tetraene - bis(acetonitrile) iron(ii) hexafluorophosphate
$C_{18}H_{30}FeN_6^{2+}$, $2F_6P^-$
H.W.Smith, G.W.Svetich, E.C.Lingafelter
Amer. Cryst. Assoc., Abstr. Papers (Winter Meeting), 72, 1973

83.C   **tris(Hexamethyldisilylamido) iron(iii)**
$C_{18}H_{54}FeN_3Si_6$
For complete entry see 63.6

83.77   **Dichloro - 2 - (2' - pyridyl)3 - (N - 2 - picolylimino) - 4 - oxo - 1,2,3,4 - tetrahydro - quinazoline - copper(ii)**
$C_{19}H_{15}Cl_2CuN_5O$
A.Mangia, C.Pelizzi, G.Pelizzi *Abstr. Ital.-Yug. Congr,* 189, 1973

83.78   **bis(Tropolonato) - dioxo - pyridine - uranium(vi)**
$C_{19}H_{15}NO_6U$
G.Bombieri, S.Degetto, G.Marangoni, R.Graziani, E.Forsellini
Inorg. Nucl. Chem. Letters, **9,** 233, 1973
Also classified in 84

83.C   **Aquo - 2 - benzothiazolin - 2 - ylideneaminomethylpyridine(picolinato) copper(ii) perchlorate**
$C_{19}H_{17}CuN_4O_3S^+$, ClO_4^-
For complete entry see 81.53

83.C   **π - Cyclopentadienyl - azotoluene - 2 - yl - nickel**
$C_{19}H_{18}N_2Ni$
For complete entry see 71.45

83.C   **(Dihydro - bis(3,5 - dimethyl - 1 - pyrazolyl)borate) - (trihapto - cycloheptatrienyl)dicarbonyl - molybdenum**
$C_{19}H_{23}BMoN_4O_2$
For complete entry see 75.31

83.C   **cis - bis(Hexafluoroacetylacetonato) - bis(pyridine) copper(ii)**
$C_{20}H_{12}CuF_{12}N_2O_4$
For complete entry see 77.6

83.C   **cis - bis(Hexafluoroacetylacetonato) - bis(pyridine) zinc(ii)**
$C_{20}H_{12}F_{12}N_2O_4Zn$
For complete entry see 77.7

83.79   **bis(2,2' - Bipyridine) silver(ii) nitrate monohydrate**
$C_{20}H_{16}AgN_4^{2+}$, $2NO_3^-$, H_2O
J.L.Atwood, M.L.Simms, D.A.Zatko *Cryst. Struct. Comm.,* **2,** 279, 1973

83.80   **Dioxo - fluoro - (dipyridyl) vanadium(v)**
$C_{20}H_{16}F_2N_4O_4V$
A.J.Edwards, D.R.Slim, J.Sala-Pala, J.-E.Guerchais
C. R. Acad. Sci., Fr., C, **276,** 1377, 1973

83.81   **Trinitrato - bis(bipyridyl) lanthanum(iii)**
$C_{20}H_{16}LaN_7O_9$
A.R.Al-Karaghouli, J.S.Wood *Inorg. Chem.,* **11,** 2293, 1972

83.82 bis(α,α' - Dipyridyl) lanthanum(iii) nitrate (α form)
$C_{20}H_{16}LaN_7O_9$
V.B.Kravchenko *Zh. Strukt. Khim.*, **13**, 345, 1972

83.83 (2,4,6 - Trimethyl - 3,5 - di(trifluoromethyl) - Dewar - pyridine) - (2,4,6 - trimethyl - 3,5 - di(trifluoromethyl) - pyridine) - dichloropalladium
$C_{20}H_{18}Cl_2F_{12}N_2Pd$
Y.Kobayashi, A.Ohsawa, Y.Iitaka *Tetrahedron Letters*, 2643, 1973

83.84 Di(iodo - bis(2,2' - bipyridylamine) copper(ii)) iodide perchlorate
$2C_{20}H_{18}CuIN_6^+$, I^- , ClO_4^-
J.E.Johnson, R.A.Jacobson *J. C. S. Dalton*, 580, 1973

83.C Dicarbonyl - (2,2' - bipyridyl) - (pyridine) - (π - allyl) - molybdenum tetrafluoroborate
$C_{20}H_{18}MoN_3O_2^+$, BF_4^-
For complete entry see 72.27

83.85 2,11 - Dimethyldibenzo(6,7,13,14) - 1,5,8,12 - tetra - aza - cyclotetradeca - 2,4,6,8,10,13 - hexaene - nickel(ii)
$C_{20}H_{18}N_4Ni$
F.Hanic, M.Handlovic, O.Lindgren
Collect. Czechosl. Chem. Communic., **37**, 2119, 1972

83.86 Aquo - bis(2,2' - bipyridyl) palladium dinitrate
$C_{20}H_{18}N_4OPd^{2+}$, $2NO_3^-$
P.C.Chieh *J. C. S. Dalton*, 1643, 1972

83.87 μ_4 - Oxo - hexa - μ - chloro - tetrakis(pyridine copper(ii))
$C_{20}H_{20}Cl_6Cu_4N_4O$
B.T.Kilbourn, J.D.Dunitz *Inorg. Chim. Acta*, **1**, 209, 1967

83.C bis(Picolinato)bis(allylthiourea) copper(ii)
$C_{20}H_{24}CuN_6O_4S_2$
For complete entry see 81.54

83.88 Tetra - μ - adenine - diaquo - dicopper(ii) perchlorate dihydrate
$C_{20}H_{24}Cu_2N_{20}O_2^{4+}$, $4ClO_4^-$, $2H_2O$
A.Terzis, A.L.Beauchamp, R.Rivest *Inorg. Chem.*, **12**, 1166, 1973

83.89 bis(Aniline) - bis(dimethylglyoximato) cobalt(iii) chloride
$C_{20}H_{28}CoN_6O_4^+$, Cl^-
L.Battaglia, A.B.Corradi, C.G.Palmieri, M.Nardelli, M.E.V.Tani
Abstr. Ital.-Yug. Congr, 197, 1973

83.90 cis - bis(Dimethylphenylphosphine)bis - (5 - methyltetrazolato) palladium(ii)
$C_{20}H_{28}N_8P_2Pd$
G.B.Ansell *J. C. S. Dalton*, 371, 1973
Also classified in 86

83.91   **Carbonato - tetrakis(pyridine) cobalt(iii) perchlorate monohydrate**
$C_{21}H_{20}CoN_4O_3{}^+$, $ClO_4{}^-$, H_2O
K.Kaas, A.M.Sorensen   *Acta Cryst. (B),* **29,** 113, 1973

83.C   **Acetylacetonato - C - meso - (5,5,7,12,12,14 - hexamethyl - 1,4,8,11 - tetra - aza - cyclotetradecane) nickel(ii) perchlorate**
$C_{21}H_{43}N_4NiO_2{}^+$, $ClO_4{}^-$
For complete entry see 77.12

83.92   **bis(ω - Nitroacetophenonato) copper(ii) α - picoline**
$C_{22}H_{19}CuN_3O_6$
M.Bonamico, G.Dessy, V.Fares, L.Scaramuzza
J. C. S. Dalton, 2477, 1972
Also classified in 84

83.93   **N,N' - Ethylene - bis(benzoylacetonimide) cobalt(ii)**
$C_{22}H_{22}CoN_2O_2$
W.T.Robinson, G.A.Rodley   *Acta Cryst. (A),* **28,** S52, 1972
Also classified in 84

83.C   **N,N' - Ethylene - bis(benzoylacetoniminato) - nitrosyl - cobalt(ii)**
$C_{22}H_{22}CoN_3O_3$
For complete entry see 76.49

83.94   **Nickel(ii) α,β - dioximino - butyric anisidide (α form)**
$C_{22}H_{24}N_6NiO_8$
E.Hadicke, G.Henning, H.-C.Mez   *Eur. Cryst. Meeting,* 1973

83.95   **Nickel(ii) α,β - dioximino - butyric anisidide (β form)**
$C_{22}H_{24}N_6NiO_8$
E.Hadicke, G.Henning, H.-C.Mez   *Eur. Cryst. Meeting,* 1973

83.C   **bis(Isothiocyanato) - (N - (diphenylphosphinoethyl) - N' - diethylethylenediamine) cobalt(ii)**
$C_{22}H_{28}CoN_4PS_2$
For complete entry see 86.28

83.C   **((2 - ((2 - (Diethylamino)ethyl)amino)ethyl)diphenylphosphine oxide) di - isothiocyanato - cobalt(ii)**
$C_{22}H_{29}CoN_4OPS_2$
For complete entry see 76.50

83.96   **1 - (p - Methylphenyl) - ethyldioxy - bis(dimethylglyoximato) - pyridine - cobalt**
$C_{22}H_{30}CoN_5O_6$
A.Chiaroni, C.Pascard-Billy   *Bull. Soc. Chim. Fr.,* 781, 1973
Also classified in 84

83.97 **bis - (N,N - Diethylnicotinamide) zinc isothiocyanate dihydrate**
$C_{22}H_{32}N_6O_4S_2Zn$
F.Bigoli, A.Braibanti, M.A.Pellinghelli, A.Tiripicchio
Abstr. Ital.-Yug. Congr, 191, 1973

83.C **N,N' - Ethylene - bis(salicylideneiminato) - pyridine - vinyl - cobalt(iii)**
$C_{23}H_{22}CoN_3O_2$
For complete entry see 71.57

83.C **N,N' - (1,2 - Dimethyl - ethylenediamine) - bis(3 - methoxy - salicylideneiminato) - cobalt(ii) pyridine**
$C_{23}H_{23}CoN_3O_2$
For complete entry see 78.16

83.98 **1,1,1 - tris(Pyridine - 2 - aldiminomethyl)ethane - iron(ii) perchlorate**
$C_{23}H_{24}FeN_6^{2+}$, $2ClO_4^-$
E.B.Fleischer, A.E.Gebala, D.R.Swift, P.A.Tasker
Inorg. Chem., **11**, 2775, 1972

83.99 **1,1,1 - tris(Pyridine - 2 - aldiminomethyl)ethane - zinc(ii) perchlorate**
$C_{23}H_{24}N_6Zn^{2+}$, $2ClO_4^-$
E.B.Fleischer, A.E.Gebala, D.R.Swift, P.A.Tasker
Inorg. Chem., **11**, 2775, 1972

83.100 **Cyclo(bis(o - phenylene) - bis(acetylacetoniminate)) - carbonyl - iron hydrazine**
$C_{23}H_{26}FeN_6O$
V.L.Goedken, J.Molin-Case, Y.-A.Whang
J. C. S. Chem. Comm., 337, 1973

83.C **bis(Isothiocyanato) - (N - methyl - N - (diphenylphosphinoethyl) - N' - diethyl - ethylenediamine) cobalt(ii)**
$C_{23}H_{32}CoN_4PS_2$
For complete entry see 86.32

83.C **Di - μ - propionato - bis(propionato - p - toluidine copper(ii))**
$C_{23}H_{38}Cu_2N_2O_8$
For complete entry see 81.57

83.101 **Di - μ - hydroxo - bis(di(1,10 - phenanthroline) chromium(iii)) chloride hexahydrate**
$C_{24}H_{18}Cr_2N_4O_2^{4+}$, $4Cl^-$, $6H_2O$
J.T.Veal, W.E.Hatfield, D.J.Hodgson   *Acta Cryst. (B),* **29,** 12, 1973

83.C **bis(6 - Mercapto - 9 - benzylpurine) palladium(ii) dimethylacetamide solvate**
$C_{24}H_{18}N_8PdS_2$, $2C_4H_9NO$
For complete entry see 85.28

83.102   cis,cis - 1,3,5 - tris(Pyridine - 2 - aldimino)cyclohexane - nickel(ii) perchlorate
$C_{24}H_{24}N_6Ni^{2+}$, $2ClO_4^-$
E.B.Fleischer, A.E.Gebala, D.R.Swift, P.A.Tasker
Inorg. Chem., **11,** 2775, 1972

83.103   N,N - Diethylnicotinamide zinc isothiocyanate dimer
$C_{24}H_{28}N_8O_2S_4Zn$
F.Bigoli, A.B.F.Braibanti, M.A.Pellinghelli, A.Tiripicchio
Abstr. Ital.-Yug. Congr, 191, 1973
Also classified in 84

83.104   hexakis(2 - Methylimidazole) cadmium(ii) tetrafluoroborate
$C_{24}H_{36}CdN_{12}^{2+}$, $2BF_4^-$
J.Reedijk, G.C.Verschoor   *Acta Cryst. (B)*, **29,** 721, 1973

83.105   Chloro - (dodeca(dimethylamino)cyclohexaphosphazene) cobalt(ii) di - μ - chloro - bis(dichlorocobaltate(ii)) chloroform solvate
$2C_{24}H_{72}ClCoN_{18}P_6^+$, $Cl_6Co_2^{2-}$, $2CHCl_3$
W.Harrison, J.Trotter   *J. C. S. Dalton*, 61, 1973

83.106 μ - (2,2'.6',2'' - Terpyridyl cadmium) - bis(pentacarbonyl manganese)
$C_{25}H_{11}CdMn_2N_3O_{10}$
W.Clegg, P.J.Wheatley   *J. C. S. Dalton*, 90, 1973

83.C   bis(Dimethylglyoximato)(tri - n - butylphosphine)(4 - pyridyl) cobalt(iii)
$C_{25}H_{45}CoN_5O_4P$
For complete entry see 71.66

83.107   bis(1 - (2 - Thiazolylazo) - 2 - naphthol) iron(ii) chloroform ethanol solvate monohydrate
$C_{26}H_{16}FeN_6O_2S_2$, $CHCl_3$, C_2H_6O , H_2O
M.Kurahashi, A.Kawase, K.Hirotsu, M.Fukuyo, A.Shimada
Bull. Chem. Soc. Jap., **45,** 1940, 1972
Residue 1 also classified in 84

83.C   Di - μ - chloro - dicarbonyl - rhodium(i) - bis(phenylazophenyl - 2C,N') rhodium(iii)
$C_{26}H_{18}Cl_2N_4O_2Rh_2$
For complete entry see 71.68

83.C   Acetato - bis(phenylazophenyl - 2C,N') rhodium(ii)
$C_{26}H_{21}N_4O_2Rh$
For complete entry see 71.69

83.C   Zinc(ii) dithizonate
$C_{26}H_{22}N_8S_2Zn$
For complete entry see 85.29

83.C **bis(Dithiophenylacetato) nickel(ii) bis - pyridine**
$C_{26}H_{24}N_2NiS_4$
For complete entry see 85.30

83.C **Chloro - bis(dimethylglyoximato) - triphenylstibine - rhodium(iii)**
$C_{26}H_{29}ClN_4O_4RhSb$
For complete entry see 86.37

83.C **Chloro - (dimethylglyoximato) - (dimethylglyoxime) - (triphenylphosphine) - rhodium(iii) chloride methylene chloride solvate**
$C_{26}H_{30}N_4O_4PRh^+$, Cl^- , CH_2Cl_2
For complete entry see 86.39

83.C **Chloro(bis(2' - (diethylamino)ethyl) - 2 - (diphenylphosphino)ethylamine) cobalt(ii) perchlorate**
$C_{26}H_{42}ClCoN_2P^+$, ClO_4^-
For complete entry see 86.43

83.C **tris(Acetylacetonato)(1,10 - phenanthroline) europium(iii)**
$C_{27}H_{29}EuN_2O_6$
For complete entry see 77.14

83.C **Isothiocyanato - (bis(2 - (diethylamino)ethyl) - 2 - (diphenylarsino)ethylamine) nickel(ii) tetraphenylborate**
$C_{27}H_{42}AsN_4NiS^+$, $C_{24}H_{20}B^-$
For complete entry see 76.51

83.C **bis(ω - Nitroacetophenonato) copper(ii) γ - picoline**
$C_{28}H_{26}CuN_4O_6$
For complete entry see 84.25

83.108 **bis(Trimethylsilylmethyl) - bis(bipyridyl) chromium(iii) iodide**
$C_{28}H_{27}CrN_4Si_2^+$, I^-
F.Sanz *Eur. Cryst. Meeting,* 1973
Residue 1 also classified in 63

83.C **Carbonyl - chloro - bis(trimethylphosphine) - (1,2,3 - triphenylpropenylium - 1,3 - diyl) iridium tetrafluoroborate dichloromethane solvate**
$C_{28}H_{33}ClIrOP_2^+$, BF_4^- , CH_2Cl_2
For complete entry see 72.35

83.C **bis(Trimethylsilyl - methyl) - bis(2,2' - bipyridyl) chromium(iii) iodide**
$C_{28}H_{38}CrN_4Si_2^+$, I^-
For complete entry see 63.13

83.109 **bis(Terpyridine) copper(ii) nitrate**
$C_{30}H_{22}CuN_6^{2+}$, $2NO_3^-$
M.Mathew, G.J.Palenik *J. Coord. Chem.,* **1,** 243, 1971

83.110   tris(2,2' - Bipyridyl) copper(ii) perchlorate
$C_{30}H_{24}CuN_6{}^{2+}$, $2ClO_4{}^-$
O.P.Anderson   *J. C. S. Dalton*, 2597, 1972

83.C   Di - μ - methoxo - bis(salicylaldehyde anthraniloylhydrazonato) - dimanganese(iii) methanol solvate
$C_{30}H_{28}Mn_2N_6O_6$, $2CH_4O$
For complete entry see 78.22

83.111   Di - μ - imidazolato - octa(imidazole) - tricopper(ii) perchlorate
$C_{30}H_{30}Cu_3N_{20}{}^{4+}$, $4ClO_4{}^-$
G.Ivarsson, B.K.S.Lundberg, N.Ingri   *Acta Chem. Scand.*, **26,** 3005, 1972

83.C   Di - iodo - (N,N - bis(2 - diphenylarsinoethyl) - 2 - methoxyethylamino) nickel(ii)
$C_{31}H_{35}As_2I_2NNiO$
For complete entry see 86.49

83.C   Chloro - (N,N - bis(2 - diphenylphosphinoethyl) - 2 - methoxyethylamine) cobalt(ii) hexafluorophosphate
$C_{31}H_{35}ClCoNOP_2{}^+$, F_6P^-
For complete entry see 86.50

83.112   Dibromo - tetrakis(1,8 - naphthyridine) dinickel tetraphenylborate
$C_{32}H_{24}Br_2N_8Ni^+$, $C_{24}H_{20}B^-$
D.Gatteschi, C.Mealli, L.Sacconi   *J. Amer. Chem. Soc.*, **95,** 2736, 1973
Residue 2 classified in 62

83.113   tetrakis(1,8 - Naphthyridine) iron(ii) perchlorate
$C_{32}H_{24}FeN_8{}^{2+}$, $2ClO_4{}^-$
P.Singh, A.Clearfield, I.Bernal   *J. Coord. Chem.*, **1,** 29, 1971

83.C   cis - Diphenyl - bis(2,2' - bipyridyl) chromium(iii) iodide
$C_{32}H_{26}CrN_4{}^+$, I^-
For complete entry see 71.77

83.C   Dibromo - tetra(2 - dimethylaminophenyl) - hexacopper(i) benzene solvate
$C_{32}H_{40}Br_2Cu_6N_4$, $1.5C_6H_6$
For complete entry see 71.80

83.114 μ - Chloro - bis(copper(ii) - 5,5,7,12,12,14 - hexamethyl - 1,4,8,11 - tetra - aza - cyclotetradecane) perchlorate
$C_{32}H_{72}ClCu_2N_8{}^{3+}$, $3ClO_4{}^-$
R.A.Bauer, W.R.Robinson, D.W.Margerum
J. C. S. Chem. Comm., 289, 1973

83.C   cis - bis(- 2 - Methoxyphenyl) - bis(2,2' - bipyridyl) chromium(iii) iodide monohydrate
$C_{34}H_{30}CrN_4O_2{}^+$, I^- , H_2O
For complete entry see 71.83

83.115 **tris(1,10 - Phenanthroline) copper(ii) perchlorate**
$C_{36}H_{24}CuN_6^{2+}$, $2ClO_4^-$
O.P.Anderson *J. C. S. Dalton,* 1237, 1973

83.116 **tris(Diphenyltriazine) cobalt(iii)**
$C_{36}H_{30}CoN_9$
W.R.Krigbaum, B.Rubin *Acta Cryst. (B),* **29,** 749, 1973

83.117 **Chloro - (3,4' - bis(ethoxycarbonyl) - 5 - chloro - 3',4,5' - trimethyldipyrromethenyl) palladium(ii) (3,4' - bis(ethoxycarbonyl) - 5 - chloro - 3',4,5' - trimethyldipyrromethene)**
$C_{36}H_{41}Cl_3N_4O_8Pd$
F.C.March, J.E.Fergusson, W.T.Robinson *J. C. S. Dalton,* 2069, 1972

83.C **3 - Methylpyridine - tris(2,2,6,6 - tetramethyl - 3,5 - heptanedionato) lutetium(iii)**
$C_{39}H_{64}LuNO_6$
For complete entry see 77.20

83.C **(tris(2 - Diphenylphosphinoethyl)amine) - chlorocobalt(ii) hexafluorophosphate**
$C_{42}H_{42}ClCoNP_3^+$, F_6P^-
For complete entry see 86.64

83.118 **bis(8 - Quinolinolato) - bis(2,6 - di - isopropylphenoxo) titanium(iv)**
$C_{42}H_{46}N_2O_4Ti$
P.H.Bird, A.R.Fraser, C.F.Lau *Inorg. Chem.,* **12,** 1322, 1973
Also classified in 84

83.C **Carbonyl - chloro - (4 - fluorophenyl - di - imide - 2C,N')bis(triphenylphosphine) - iridium(iii) tetrafluoroborate acetone solvate**
$C_{43}H_{34}ClFIrN_2OP_2^+$, BF_4^- , C_3H_6O
For complete entry see 71.97

83.C **tris - (2,2,6,6 - Tetramethylheptane - 3,5 - dionato) europium(iii) di - pyridine**
$C_{43}H_{67}EuN_2O_6$
For complete entry see 77.22

83.119 **bis(2 - Methyl - 8 - quinolinolato) - bis(2,6 - di - isopropylphenoxo) titanium(iv)**
$C_{44}H_{50}N_2O_4Ti$
P.H.Bird, A.R.Fraser, C.F.Lau *Inorg. Chem.,* **12,** 1322, 1973
Also classified in 84

83.C **Tetra - μ - benzoato - bis(quinoline - cobalt(ii))**
$C_{46}H_{34}Co_2N_2O_8$
For complete entry see 81.61

83.C   bis(μ - Hydroxy - μ - acetato - (4 - hydroxy - bis(3,5 - (N - (2' - hydroxyphenyl) - formimidoyl)toluene) - cobalt(ii) - cobalt(iii))) dihydrate ethanol solvate

$C_{46}H_{38}Co_4N_4O_{12}$, $2C_2H_6O$, $2H_2O$

For complete entry see 81.62

83.120   1,3 - Diphenyltriazene - copper(ii)

$C_{48}H_{30}Cu_2N_{12}$

M.Corbett, B.F.Hoskins, N.J.McLeod, B.P.O'Day

Acta Cryst. (A), **28**, S76, 1972

83.121   1,3 - Diphenyltriazene - nickel(ii)

$C_{48}H_{30}N_{12}Ni_2$

M.Corbett, B.F.Hoskins, N.J.McLeod, B.P.O'Day

Acta Cryst. (A), **28**, S76, 1972

83.122   1,3 - Diphenyltriazene - palladium(ii)

$C_{48}H_{30}N_{12}Pd_2$

M.Corbett, B.F.Hoskins, N.J.McLeod, B.P.O'Day

Acta Cryst. (A), **28**, S76, 1972

83.123   Carbonyl - bis(triphenylphosphine)(1,4 - di - (p - fluorophenyl)tetrazene) iridium tetrafluoroborate

$C_{49}H_{38}F_2IrN_4OP_2^+$, BF_4^-

F.W.B.Einstein, D.Sutton   *Inorg. Chem.*, **11**, 2827, 1972

Residue 1 also classified in 86

83.C   bis(5 - Phenyltetrazolato) - bis(triphenylphosphine) palladium methylene dichloride solvate

$C_{50}H_{40}N_8P_2Pd$, CH_2Cl_2

For complete entry see 86.66

83.C   bis(1,5 - Diphenyl - 1,3,5 - pentanetrionato)tetrapyridine - dicobalt(ii) pyridine solvate

$C_{54}H_{44}Co_2N_4O_6$, $4C_5H_5N$

For complete entry see 77.24

83.124   Diaquo - (1',2',3',4' - tetra - acetyl - 3 - ethyl - riboflavin) - zinc perchlorate

$C_{54}H_{68}N_8O_{22}Zn^{2+}$, $2ClO_4^-$

P.Kierkegaard, M.Leijonmarck, P.-E.Werner

Acta Chem. Scand., **26**, 2980, 1972

Residue 1 also classified in 84

METAL COMPLEXES (OXYGEN LIGAND)

84.1 **Dichloro - bis(formamido) - iron(ii)**
$C_2H_6Cl_2FeN_2O_2$
G.Constant, J.C.Daran, Y.Jeannin *J. Inorg. Nucl. Chem.*, **33,** 4209, 1971

84.2 **Silver diethylphosphate**
$(C_4H_{10}AgO_4P)_n$
J.P.Hazel, R.L.Collin *Acta Cryst. (B),* **28,** 2951, 1972
Also classified in 46

84.3 **cis - Dioxo - bis(2 - hydroxyethanolato) molybdenum(vi)**
$C_4H_{10}MoO_6$
F.A.Schroder, J.Scherle, R.Hazell *Eur. Cryst. Meeting,* 1973

84.4 **bis(Imidazolidinone) mercury(ii) chloride**
bis(Ethyleneurea) mercury(ii) chloride
$C_6H_{12}Cl_2HgN_4O_2$
R.J.Majeste, L.M.Trefonas *Inorg. Chem.,* **11,** 1834, 1972

84.5 **bis(Ethyl carbamate) - dinitrato - dioxo - uranium(vi)**
$C_6H_{14}N_4O_{12}U$
R.Graziani, G.Bombieri, E.Forsellini, S.Degetto, G.Marangoni
J. C. S. Dalton, 451, 1973

84.6 **tris(1,2 - Ethanediol) copper(ii) sulfate**
$C_6H_{18}CuO_6^{2+}$, O_4S^{2-}
B.-M.Antti, B.K.S.Lundberg, N.Ingri *Acta Chem. Scand.,* **26,** 3984, 1972

84.7 **Erbium nitrate - dimethylsulfoxide**
$C_6H_{18}ErN_3O_{12}S_3$
L.A.Alsanov, L.I.Soleva, S.S.Goukhberg, M.A.Porai-Koshits
Zh. Strukt. Khim., **12,** 1113, 1971

84.8 **Ytterbium tris(dimethylsulfoxide)nitrate**
$C_6H_{18}N_3O_{12}S_3Yb$
K.K.Bhandary, H.Manohar, K.Venkatesan
Cryst. Struct. Comm., **2,** 99, 1973

84.C **(Pyrrole - 2 - (N - 3 - propanolato) - carboxaldimino) copper(ii)**
$C_8H_{10}CuN_2O$
For complete entry see 83.13

84.C   **Dichloro - bis(N,N - dimethylacetamide) zinc**
$C_8H_{18}Cl_2N_2O_2Zn$
For complete entry see 81.34

84.9   **Lanthanum(iii) nitrate tetra - dimethylsulfoxide**
$C_8H_{24}LaN_3O_{13}S_4$
K.K.Bhandary, H.Manohar *Acta Cryst. (B),* **29,** 1093, 1973

84.10   **Neodymium(iii) nitrate - dimethylsulfoxide complex**
$C_8H_{24}N_3NdO_{13}S_4$
L.A.Aslanov, L.I.Soleva, M.A.Porai-Koshits, S.S.Goukhberg
Zh. Strukt. Khim., **13,** 655, 1972

84.C   **Tricarbonyl(1 - 3 - η - hexen - 5 - one) iron hexafluorophosphate**
$C_9H_9FeO_4^+$, F_6P^-
For complete entry see 72.9

84.11   **Di - μ - (pyridine - 1 - oxide) - bis(dibromo - copper(ii))**
$C_{10}H_{10}Br_4Cu_2N_2O_2$
J.E.Whinnery, W.H.Watson *J. Coord. Chem.,* **1,** 207, 1971

84.C   **bis(Pyrazine - 2 - carboxamide) copper(ii) perchlorate**
$C_{10}H_{10}CuN_6O_2^{2+}$, $2ClO_4^-$
For complete entry see 83.23

84.12   **Diphenylsulfoxide - mercury(ii) chloride**
$(C_{12}H_{10}Cl_2HgOS)_n$
P.Biscarini, L.Fusina, G.D.Nivellini, A.Mangia, G.Pelizzi
J. C. S. Dalton, 159, 1973

84.C   **Cyclo - octatetraenyl(dichloro)(tetrahydrofuran) zirconium**
$C_{12}H_{16}Cl_2OZr$
For complete entry see 75.12

84.C   **tris(Methyl - vinyl - ketone) tungsten**
$C_{12}H_{18}O_3W$
For complete entry see 71.20

84.C   **bis(N,N - Diethyldithiocarbamato) - dioxo - trimethylamineoxide - uranium(vi)**
$C_{13}H_{29}N_3O_3S_4U$
For complete entry see 80.12

84.13   **Di - isoselenocyanato - nickel tetra(dimethylformamide)**
$C_{14}H_{24}N_6NiO_4Se_2$
G.V.Tsintsadze, M.A.Porai-Koshits, A.S.Antsyshkina
Zh. Strukt. Khim., **8,** 296, 1967

84.C   **Cyclo - octatetraenyl - (allyl) - t - butoxy - zirconium**
$C_{15}H_{22}OZr$
For complete entry see 72.19

84.14 **bis(ω - Nitroacetophenonato) copper(ii)**
$C_{16}H_{12}CuN_2O_6$
M.Bonamico, G.Dessy, V.Fares, L.Scaramuzza
J. C. S. Dalton, 2477, 1972

84.C **Uranyl N,N' - ethylene - bis(salicylideneiminato) methanol complex**
$C_{17}H_{18}N_2O_5U$
For complete entry see 78.4

84.C **π - Methallyl - (2 - (R,S) - α - phenylethylimino - 3 - penten - 4 - olato) palladium(ii)**
$C_{17}H_{23}NOPd$
For complete entry see 72.21

84.15 **hexakis(2 - Imidazolidinone) cadmium(ii) perchlorate**
hexakis(Ethyleneurea) cadmium(ii) perchlorate
$C_{18}H_{36}CdN_{12}O_6{}^{2+}$, $2ClO_4{}^-$
J.N.Brown, A.G.Pierce Junior, L.M.Trefonas   *Inorg. Chem.*, **11**, 1830, 1972

84.16 **Tetrachloro - di - μ_3 - oxo - tetra - μ - propoxo - tetraoxodipropoxo - tetramolybdenum**
$C_{18}H_{42}Cl_4Mo_4O_{12}$
J.A.Beaver, M.G.B.Drew   *J. C. S. Dalton,* 1376, 1973

84.C **bis(Tropolonato) - dioxo - pyridine - uranium(vi)**
$C_{19}H_{15}NO_6U$
For complete entry see 83.78

84.17 **Pyridine - 1 - oxide mercury(i) perchlorate**
$(C_{20}H_{20}Hg_2N_4O_4{}^{2+})_n$, $2nClO_4{}^-$
D.L.Kepert, D.Taylor, A.H.White   *J. C. S. Dalton,* 392, 1973

84.C **bis(N - 3 - Hydroxypropyl - salicylideneiminato) dicopper(ii)**
$C_{20}H_{22}Cu_2N_2O_4$
For complete entry see 78.11

84.C **bis(Cyclo - octatetraene)(tetrahydrofuran) zirconium (absolute configuration)**
$C_{20}H_{24}OZr$
For complete entry see 75.32

84.18 **Gadolinium(iii) tri(2,6 - dimethyl - 4 - pyrone) chloride**
$C_{21}H_{24}GdO_6{}^{3+}$, $3Cl^-$
V.Tazzoli, A.D.Giusta, M.Cola, A.Coda, C.C.Bisi
Acta Cryst. (A), **28**, S88, 1972

84.19 **Di - μ - 4 - phenylpyridine - 1 - oxide bis(dichloro - copper(ii)) monohydrate**
$C_{22}H_{18}Cl_4Cu_2N_2O_2$, H_2O
W.H.Watson, D.R.Johnson   *J. Coord. Chem.*, **1**, 145, 1971

84.C   bis(ω - Nitroacetophenonato) copper(ii) α - picoline
$C_{22}H_{19}CuN_3O_6$
For complete entry see 83.92

84.20   bis(10 - Methylisoalloxazine) copper(ii) perchlorate tetrahydrate
$C_{22}H_{20}CuN_8O_6{}^{2+}$, $2ClO_4{}^-$, $2H_2O$
C.J.Fritchie Junior   *J. C. S. Chem. Comm.*, 1220, 1972

84.C   N,N' - Ethylene - bis(benzoylacetonimide) cobalt(ii)
$C_{22}H_{22}CoN_2O_2$
For complete entry see 83.93

84.C   N,N' - Ethylene - bis(benzoylacetoniminato) - nitrosyl - cobalt(ii)
$C_{22}H_{22}CoN_3O_3$
For complete entry see 76.49

84.21   Di - (N,N - diethylnicotinamide) - di - isothiocyanato - manganese(ii)
$C_{22}H_{28}MnN_6O_2S_2$
F.Bigoli, A.Braibanti, M.A.Pellinghelli, A.Tiripicchio
Acta Cryst. (B), **29,** 39, 1973

84.C   ((2 - ((2 - (Diethylamino)ethyl)amino)ethyl)diphenylphosphine oxide) di - isothiocyanato - cobalt(ii)
$C_{22}H_{29}CoN_4OPS_2$
For complete entry see 76.50

84.C   1 - (p - Methylphenyl) - ethyldioxy - bis(dimethylglyoximato) - pyridine - cobalt
$C_{22}H_{30}CoN_5O_6$
For complete entry see 83.96

84.C   Di - μ - phenylmethoxo - bis(pentane - 2,4 - dionato) dicopper(ii)
$C_{24}H_{28}Cu_2O_6$
For complete entry see 77.13

84.C   N,N - Diethylnicotinamide zinc isothiocyanate dimer
$C_{24}H_{28}N_8O_2S_4Zn$
For complete entry see 83.103

84.22   hexakis(1,4 - Dithiane monosulfoxide) manganese(ii) perchlorate
$C_{24}H_{48}MnO_6S_{12}{}^{2+}$, $2ClO_4{}^-$
A.L.Spek   *Cryst. Struct. Comm.*, **2,** 331, 1973

84.C   bis(1 - (2 - Thiazolylazo) - 2 - naphthol) iron(ii) chloroform ethanol solvate monohydrate
$C_{26}H_{16}FeN_6O_2S_2$, $CHCl_3$, C_2H_6O , H_2O
For complete entry see 83.107

84.C **1,1 - bis(Phenyl - diethyl - phosphine) - 3,3,4,4 - tetra(trifluoromethyl) - 1 - pallada - 2,5 - dioxa - cyclopentane**
$C_{26}H_{30}F_{12}O_2P_2Pd$
For complete entry see 86.38

84.23 **tris(Octamethyl - methylene - diphosphonic diamide) copper(ii) perchlorate**
$C_{27}H_{78}CuN_{12}O_6P_6{}^{2+}$, $2ClO_4{}^-$
P.T.Miller, P.G.Lenhert, M.D.Joesten *Inorg. Chem.,* **11,** 2221, 1972

84.24 **Di - μ - tropolonato - bis(aquo(tropolonato) nickel(ii))**
$C_{28}H_{24}Ni_2O_{10}$
R.J.Irving, M.L.Post, D.C.Povey *J. C. S. Dalton,* 697, 1973

84.25 **bis(ω - Nitroacetophenonato) copper(ii) γ - picoline**
$C_{28}H_{26}CuN_4O_6$
M.Bonamico, G.Dessy, V.Fares, L.Scaramuzza
J. C. S. Dalton, 2477, 1972
Also classified in 83

84.C **Di - μ - methoxo - bis(salicylaldehyde anthraniloylhydrazonato) - dimanganese(iii) methanol solvate**
$C_{30}H_{28}Mn_2N_6O_6$, $2CH_4O$
For complete entry see 78.22

84.26 **hexakis(Pyridine - 1 - oxide) mercury(ii) perchlorate**
$C_{30}H_{30}HgN_6O_6{}^{2+}$, $2ClO_4{}^-$
D.L.Kepert, D.Taylor, A.H.White *J. C. S. Dalton,* 670, 1973

84.C **Di - iodo - (N,N - bis(2 - diphenylarsinoethyl) - 2 - methoxyethylamino) nickel(ii)**
$C_{31}H_{35}As_2I_2NNiO$
For complete entry see 86.49

84.C **Chloro - (N,N - bis(2 - diphenylphosphinoethyl) - 2 - methoxyethylamine) cobalt(ii) hexafluorophosphate**
$C_{31}H_{35}ClCoNOP_2{}^+$, F_6P^-
For complete entry see 86.50

84.C **Triphenylphosphine - (1 - (di(trifluoromethyl) - hydroxymethyl) - cyclopentadienyl) - (1,2 - di(carboxymethyl)ethylene - 1 - yl) - ruthenium**
$C_{32}H_{27}F_6O_5PRu$
For complete entry see 71.78

84.C **bis(Acetylacetonato) - bis(2,6 - di - isopropylphenoxo) titanium(iv)**
$C_{34}H_{48}O_6Ti$
For complete entry see 77.17

84.27 **Dioxo - dinitrato - bis(triphenylarsine oxide) uranium(vi)**
$C_{36}H_{30}As_2N_2O_{10}U$
C.Panattoni, R.Graziani, U.Croatto, B.Zarli, G.Bombieri
Inorg. Chim. Acta, **2,** 43, 1968

84.28   bis(N - Benzyldiphenylphosphinic amide) - dichlorocobalt(ii)
$C_{38}H_{36}Cl_2CoN_2O_2P_2$
R.M.Roy, J.W.Jeffery   *Acta Cryst. (B)*, **29**, 1083, 1973

84.C   tris(Dipivalomethanato) europium(iii) 3,3 - dimethylthietane - 1 - oxide
$C_{38}H_{67}EuO_7S$
For complete entry see 77.19

84.C   bis(Triphenylphosphine)hexafluoroacetone nickel(0)
3,3 - Di(trifluoromethyl) - 1 - nickela - 2 - oxa - cyclopropane
bis(triphenylphosphine)
$C_{39}H_{30}F_6NiOP_2$
For complete entry see 71.92

84.29   tris(Tetra - isopropyl - methylenediphosphonate) copper(ii) perchlorate
$C_{39}H_{90}CuO_{18}P_6^{2+}$, $2ClO_4^-$
P.T.Miller, P.G.Lenhert, M.D.Joesten   *Inorg. Chem.*, **12**, 218, 1973

84.30 μ - bis(2(1H) - Tetrahydropyrimidinone) - octakis(2(1H) -
tetrahydropyrimidinone) - dicobalt(ii) perchlorate
μ - bis(Propyleneurea) - octakis(propyleneurea) dicobalt(ii) perchlorate
$C_{40}H_{80}Co_2N_{20}O_{10}^{4+}$, $4ClO_4^-$
M.E.Brown, J.N.Brown, L.M.Trefonas   *Inorg. Chem.*, **11**, 1836, 1972

84.C   bis(8 - Quinolinolato) - bis(2,6 - di - isopropylphenoxo) titanium(iv)
$C_{42}H_{46}N_2O_4Ti$
For complete entry see 83.118

84.C   bis(2 - Methyl - 8 - quinolinolato) - bis(2,6 - di - isopropylphenoxo)
titanium(iv)
$C_{44}H_{50}N_2O_4Ti$
For complete entry see 83.119

84.C   bis(μ - Hydroxy - μ - acetato - (4 - hydroxy - bis(3,5 - (N - (2' -
hydroxyphenyl) - formimidoyl)toluene) - cobalt(ii) - cobalt(iii))) dihydrate
ethanol solvate
$C_{46}H_{38}Co_4N_4O_{12}$, $2C_2H_6O$, $2H_2O$
For complete entry see 81.62

84.C   Diaquo - (1',2',3',4' - tetra - acetyl - 3 - ethyl - riboflavin) - zinc perchlorate
$C_{54}H_{68}N_8O_{22}Zn^{2+}$, $2ClO_4^-$
For complete entry see 83.124

84.31   Lanthanum(iii) octa(2,6 - dimethyl - 4 - pyrone) perchlorate
$C_{56}H_{64}LaO_{16}^{3+}$, $3ClO_4^-$
V.Tazzoli, A.D.Giusta, M.Cola, A.Coda, C.C.Bisi
Acta Cryst. (A), **28**, S88, 1972

84.32   **Iodo - tetrakis(triphenylphosphine oxide) manganese(ii) di - iodo - tetracarbonyl - manganate(i)**
$C_{72}H_{60}IMnO_4P_4^+$, $C_4I_2MnO_4^-$
G.Ciani, M.Manassero, M.Sansoni   *J. Inorg. Nucl. Chem.*, **34,** 1760, 1972

METAL COMPLEXES
(SULPHUR OR SELENIUM LIGAND)

85.C   **Iron(ii) thioglycolate monohydrate**
$(C_2H_2FeO_2S)_n$, nH_2O
For complete entry see 81.2

85.C   **bis(2,4 - Dithiobiureto) nickel(ii) glycol solvate**
$C_4H_8N_6NiS_4$, $C_2H_6O_2$
For complete entry see 79.6

85.C   **Dichloro - (S - methyl - L - cysteine) palladium(ii) monohydrate**
$C_4H_9Cl_2NO_2PdS$, H_2O
For complete entry see 82.1

85.1   **Thiocyanato - (trimethylphosphine sulfide) copper(i)**
$(C_4H_9CuNPS_2)_n$
P.W.R.Corfield
Amer. Cryst. Assoc., Abstr. Papers (Winter Meeting), 92, 1973

85.2   **Reaction product of bis(S - amino - dithionitrito) nickel(ii) with ammonia, formaldehyde and methanol**
$C_4H_9N_5NiOS_4$
U.Thewalt, C.E.Bugg   *Chem. Ber.*, **105**, 1614, 1972
Also classified in 83

85.3   **Tetrabutylammonium bis(maleonitriledithiolato) copper(ii)**
$C_8CuN_4S_4{}^{2-}$, $2C_{16}H_{36}N^+$
C.R.Ollis, W.E.Hatfield, G.O.Carlisle   *J. Inorg. Nucl. Chem.*, **34**, 776, 1972
Residue 2 classified in 3

85.C   **Tetramethyl - p - phenylenediamine free radical bis(maleonitriledithiolato) nickel(ii)**
$C_8N_4NiS_4{}^{2-}$, $2C_{20}H_{16}N_2{}^+$
For complete entry see 12.11

85.4   **bis(Ethylene - 1,2 - dithiolene) palladium dimer**
$C_8H_8Pd_2S_8$
K.W.Browall, T.Bursh, L.V.Interrante, J.S.Kasper
Inorg. Chem., **11**, 1800, 1972

85.5 **bis(Ethylene - 1,2 - dithiolene) platinum dimer**
$C_8H_8Pt_2S_8$
K.W.Browall, T.Bursh, L.V.Interrante, J.S.Kasper
Inorg. Chem., **11**, 1800, 1972

85.6 μ,μ'',μ''' **- tris(Dithioacetato) -** μ_3 **- trithio - orthoacetato - triangulo - trinickel(ii)**
$C_8H_{12}Ni_3S_9$
M.Bonamico, G.Dessy, V.Fares, L.Scaramuzza
Acta Cryst. (A), **28**, S77, 1972

85.7 **bis(Tetra - n - butylammonium)** μ **- S,S' - (tetrakis(ethane - 1,2 - dithiolate) - di - iron(iii))**
$C_8H_{16}Fe_2S_8{}^{2-}$, $2C_{16}H_{36}N^+$
M.R.Snow, J.A.Ibers *Inorg. Chem.*, **12**, 249, 1973
Residue 2 classified in 3

85.C **Chloro - bis(acetone thiosemicarbazone) nickel(ii) chloride monohydrate**
$C_8H_{18}ClN_6NiS_2{}^+$, Cl^- , H_2O
For complete entry see 79.20

85.C **Nitrato - bis(acetone thiosemicarbazone) nickel(ii) nitrate monohydrate**
$C_8H_{18}N_7NiO_3S_2{}^+$, $NO_3{}^-$, H_2O
For complete entry see 79.21

85.8 **trans - Dichloro - tetrakis(thioacetamide) nickel(ii)**
$C_8H_{20}Cl_2N_4NiS_4$
R.L.Girling, J.E.O'Connor, E.L.Amma *Acta Cryst. (B)*, **28**, 2640, 1972

85.9 **Mercury dithiophosphate**
$(C_8H_{20}HgO_4PS_4)_n$
Y.Watanabe, H.Hagihara *Acta Cryst. (A)*, **28**, S89, 1972

85.10 **bis(Tetramethyldiphosphine disulfide) copper(i) fluoroborate**
$C_8H_{24}CuP_4S_4{}^+$, $BF_4{}^-$
P.W.R.Corfield
Amer. Cryst. Assoc., Abstr. Papers (Winter Meeting), 92, 1973

85.11 **Thiomorpholine - 3 - one tungsten pentacarbonyl**
$C_9H_7NO_6SW$
M.Cannas, G.Carta, G.Marongiu, E.F.Trogu
Abstr. Ital.-Yug. Congr, 166, 1973

85.12 **Tri -** μ **- methylmercapto - hexacarbonyl di - iron(ii) tetrakis(cis - 1,2 - di(perfluoromethyl)ethylene - 1,2 - dithiolato) di - iron**
$C_9H_9Fe_2O_6S_3{}^+$, $C_{16}F_{24}Fe_2S_8{}^-$
A.J.Schultz, R.Eisenberg *Inorg. Chem.*, **12**, 518, 1973
Residue 2 classified in 85

85.13 **Cyclo - tri - μ - (trimethylphosphine sulfide) - tris(chlorocopper(i))**
$C_9H_{27}Cl_3Cu_3P_3S_3$
J.A.Tiethof, J.K.Stalick, D.W.Meek   *Inorg. Chem.*, **12**, 1170, 1973

85.14 **bis(μ - (Trifluoromethyl)seleno - manganese tetracarbonyl) (at $-20°C$)**
$C_{10}F_6Mn_2O_8Se_2$
C.J.Marsden, G.M.Sheldrick   *J. Organometal. Chem.*, **40**, 175, 1972

85.C **bis(Pyrrolidine - carbodithioato) copper(ii)**
$C_{10}H_{16}CuN_2S_4$
For complete entry see 80.6

85.C **bis(Pyrrolidine - carbodithioato) nickel(ii)**
$C_{10}H_{16}N_2NiS_4$
For complete entry see 80.7

85.15 **bis(Dithiopivalato) nickel(ii)**
$C_{10}H_{18}NiS_4$
M.Bonamico, G.Dessy, V.Fares, L.Scaramuzza
Cryst. Struct. Comm., **2**, 201, 1973

85.16 **o - Aminothiophenol - di - iron hexacarbonyl**
$C_{12}H_5Fe_2NO_6S$
G.le Borgne, D.Grandjean   *Acta Cryst. (B)*, **29**, 1040, 1973
Also classified in 83

85.17 **Copper(i) hydrogen 2,5 - dithiahexane - 1,6 - dicarboxylate**
$C_{12}H_{19}CuO_8S_4$
H.van der Meer   *J. C. S. Dalton*, 1, 1973

85.C **bis - (μ - N,N - Dimethylthiocarbamoyl) - bis(chloro(trimethylphosphite) - palladium(ii))**
$C_{12}H_{21}Cl_2N_2O_6P_2Pd_2S_2$
For complete entry see 71.21

85.18 **catena - bis(μ - (O,O, - Di - isopropyldithiophosphato) - digold(i))**
$C_{12}H_{28}Au_2O_4P_2S_4$
S.L.Lawton, W.J.Rohrbaugh, G.T.Kokotailo   *Inorg. Chem.*, **11**, 2227, 1972

85.19 **Di - μ - (O,O' - di - isopropyldithiophosphato) - digold(i)**
$C_{12}H_{28}Au_2O_4S_4$
S.L.Lawton, W.J.Rohrbaugh, G.T.Kokotailo
Acta Cryst. (A), **28**, S78, 1972

85.20 **Hexa - μ - dithiocacodylato - tetrazinc - sulfide**
$C_{12}H_{36}As_6S_{13}Zn_4$
D.Johnstone, J.E.Fergusson, W.T.Robinson
Bull. Chem. Soc. Jap., **45**, 3721, 1972

85.C   **N,N - Di - n - butyldithiocarbamato - (1,2 - dicyanoethene - 1,2 - dithiolato) gold(iii)**
$C_{13}H_{18}AuN_3S_4$
For complete entry see 80.11

85.21   **bis(Dithiobenzoato) zinc(ii)**
$C_{14}H_{10}S_4Zn$
M.Bonamico, G.Dessy, V.Fares, L.Scaramuzza
J. C. S. Dalton, 2515, 1972

85.22   **μ - (1,2 - bis(Phenylthio)ethane) - bis(chlorogold(i))**
$C_{14}H_{14}Au_2Cl_2S_2$
M.G.B.Drew, M.J.Riedle   *J. C. S. Dalton*, 52, 1973

85.23   **Diaquo - bis(monothiobenzoate) zinc(ii)**
$C_{14}H_{14}O_4S_2Zn$
M.Bonamico, G.Dessy, V.Fares, L.Scaramuzza
Cryst. Struct. Comm., **2**, 255, 1973

85.C   **Di - μ - thioethyl - μ - dithio - bis(cyclopentadienyl - iron)**
$C_{14}H_{20}Fe_2S_4$
For complete entry see 73.21

85.C   **Tri - μ - methylmercapto - hexacarbonyl di - iron(ii) tetrakis(cis - 1,2 - di(perfluoromethyl)ethylene - 1,2 - dithiolato) di - iron**
$C_{16}F_{24}Fe_2S_8{}^-$, $C_9H_9Fe_2O_6S_3{}^+$
For complete entry see 85.12

85.C   **Benzene - 1,2 - dithiolato - di(π - cyclopentadienyl) - titanium(iv)**
$C_{16}H_{14}S_2Ti$
For complete entry see 73.27

85.C   **Structure III**
$C_{16}H_{15}N_4NiS_2{}^+$, $ClO_4{}^-$, H_2O
For complete entry see 83.62

85.24   **Chloro - (bis(2 - ((2 - pyridylmethyl)amino)ethyl)disulfide) nickel(ii) perchlorate**
$C_{16}H_{22}ClN_4NiS_2{}^+$, $ClO_4{}^-$
P.E.Riley, K.Seff   *Inorg. Chem.*, **11**, 2993, 1972
Residue 1 also classified in 83

85.C   **N,N' - Dimethyl - N,N' - bis(β - mercaptoethyl) ethylenediamine zinc(ii) chloride dihydrate**
$C_{16}H_{36}Cl_4N_4S_4Zn_4$, $2H_2O$
For complete entry see 76.47

85.25   **Di - μ - phenylthio - di(tricarbonyl iron)**
$C_{18}H_{10}Fe_2O_6S_2$
W.Henslee, R.E.Davis   *Cryst. Struct. Comm.*, **1**, 403, 1972

85.C   bis(Diethyldithiocarbamato) - bis(bis(cis - 1,2 - perfluoromethyl)ethene - 1,2 - dithiolato) - dinickel
$C_{18}H_{20}F_{12}N_2Ni_2S_8$
For complete entry see 80.23

85.C   Ethylmercapto - cyclopentadienyl - iron - acetonitrile dimer hexafluorophosphate
$C_{18}H_{26}Fe_2N_2S_2{}^{2+}$, $2F_6P^-$
For complete entry see 73.31

85.C   Dicyclopentadienyl - di - iron trisulfide dimer
$C_{20}H_{20}Fe_4S_6$
For complete entry see 73.37

85.26   Dithiocumato(trithioperoxycumato) nickel(ii)
$C_{20}H_{22}NiS_5$
J.P.Fackler Junior, J.A.Fetchin, D.C.Fries
J. Amer. Chem. Soc., **94,** 7323, 1972

85.27   bis(Trithioperoxycumato) zinc(ii)
$C_{20}H_{22}S_6Zn$
J.P.Fackler Junior, J.A.Fetchin, D.C.Fries
J. Amer. Chem. Soc., **94,** 7323, 1972

85.C   Dichloro - bis - μ - ethanethiolato - bis(tripropylphosphine) diplatinum(ii)
$C_{22}H_{52}Cl_2P_2Pt_2S_2$
For complete entry see 86.29

85.C   Monothiodibenzoylmethanato - (π - syn - 1 - t - butyl - 2 - methallyl) palladium(ii)
$C_{23}H_{26}OPdS$
For complete entry see 72.32

85.28   bis(6 - Mercapto - 9 - benzylpurine) palladium(ii) dimethylacetamide solvate
$C_{24}H_{18}N_8PdS_2$, $2C_4H_9NO$
H.I.Heitner, S.J.Lippard, H.R.Sunshine
J. Amer. Chem. Soc., **94,** 8936, 1972
Residue 1 also classified in 83

85.29   Zinc(ii) dithizonate
$C_{26}H_{22}N_8S_2Zn$
A.Mawby, H.M.N.H.Irving   *J. Inorg. Nucl. Chem.*, **34,** 109, 1972
Also classified in 83

85.30   bis(Dithiophenylacetato) nickel(ii) bis - pyridine
$C_{26}H_{24}N_2NiS_4$
M.Bonamico, G.Dessy, V.Fares, L.Scaramuzza
Abstr. Ital.-Yug. Congr, 170, 1973
Also classified in 83

85.C (O - Methyl - phosphorodithioato) - (1 - diphenylarsino - 2 - diphenylphosphino - ethane) - nickel(ii) benzene solvate
$C_{27}H_{27}AsNiO_2PS_2$, $0.5C_6H_6$
For complete entry see 86.44

85.31 Tetraethylammonium tetra(benzylmercapto - μ_3 - sulfido - iron)
$C_{28}H_{28}Fe_4S_8{}^{2-}$, $2C_8H_{20}N^+$
B.A.Averill, T.Herskovitz, R.H.Holm, J.A.Ibers
J. Amer. Chem. Soc., **95**, 3523, 1973
Residue 2 classified in 3

85.32 Tricarbonyl - bis(diphenylphosphino)methane - dithioformato - manganese
$C_{29}H_{23}MnO_3P_2S_2$
F.W.Einstein, E.Enwall, N.Flitcroft, J.M.Leach
J. Inorg. Nucl. Chem., **34**, 885, 1972
Also classified in 86

85.33 bis(cis - 1,2 - Diphenylethylene - 1,2 - dithiolato) palladium cyclohexa - 1,3 - diene adduct
$C_{34}H_{28}PdS_4$
G.R.Clark, J.M.Waters, K.R.Whittle   *J. C. S. Dalton*, 821, 1973

85.34 bis(Dithiocumato) platinum(ii) dimer
$C_{40}H_{44}Pt_2S_8$
J.M.Burke, J.P.Fackler Junior   *Inorg. Chem.*, **11**, 3000, 1972

85.35 bis(Triphenylphosphine) - (1,2 - bis(trifluoromethyl) - ethylene - 1,2 - dithiolato) ruthenium carbonyl (orange form)
$C_{41}H_{30}F_6OP_2RuS_2$
I.Bernal, A.Clearfield, E.F.Epstein, J.S.Ricci Junior, A.Balch, J.S.Miller
J. C. S. Chem. Comm., 39, 1973
Also classified in 86

85.36 bis(Triphenylphosphine) - (1,2 - bis(trifluoromethyl) - ethylene - 1,2 - diolato) ruthenium carbonyl (violet form)
$C_{41}H_{30}F_6OP_2RuS_2$
I.Bernal, A.Clearfield, E.F.Epstein, J.S.Ricci Junior, A.Balch, J.S.Miller
J. C. S. Chem. Comm., 39, 1973
Also classified in 86

85.37 bis(Tetraphenyl - dithioimido - diphosphinato) manganese(ii)
$C_{48}H_{40}MnN_2P_4S_4$
O.Siiman, M.Wrighton, H.B.Gray   *J. Coord. Chem.*, **2**, 159, 1972

METAL COMPLEXES (P, AS, SB LIGAND)

86.1   bis(Dimethylphosphonato) mercury(ii)
$C_4H_{12}HgO_6P_2$
G.G.Mather, A.Pidcock   *J. C. S. Dalton,* 560, 1973

86.2   trans - bis(1,4 - Diphospha - 2,6,7 - trioxabicyclo(2.2.2)octane) iron tricarbonyl
$C_9H_{12}FeO_9P_4$
D.A.Allison, J.Clardy, J.G.Verkade   *Inorg. Chem.,* **11,** 2804, 1972

86.3   (1,2 - bis(Dimethylarsino) - 1,1,2 - trifluoro - 2 - chloroethane) - tetracarbonyl - chromium
$C_{10}H_{12}As_2ClCrF_3O_4$
I.W.Nowell, S.Rettig, J.Trotter   *J. C. S. Dalton,* 2381, 1972

86.4   (1,2 - bis(Dimethylarsino) - 1,1,2 - trifluoroethane) - tetracarbonyl - chromium
$C_{10}H_{13}As_2CrF_3O_4$
I.W.Nowell, S.Rettig, J.Trotter   *J. C. S. Dalton,* 2381, 1972

86.5   (1,2 - bis(Dimethylarsino) - 1,1,2 - trifluoroethane) - tetracarbonyl - molybdenum (re - interpretation of data of Nowell and Trotter,J.Chem.Soc.(A),2922,1971)
$C_{10}H_{13}As_2F_3MoO_4$
I.W.Nowell, S.Rettig, J.Trotter   *J. C. S. Dalton,* 2381, 1972

86.6   (1,2 - bis(Dimethylphosphino) - 1,1,2 - trifluoroethane) - tetracarbonyl - molybdenum
$C_{10}H_{13}F_3MoO_4P_2$
I.W.Nowell, S.Rettig, J.Trotter   *J. C. S. Dalton,* 2381, 1972

86.7   (1,2 - bis(Dimethylarsino) - 1,1 - difluoroethane) - tetracarbonyl - chromium
$C_{10}H_{14}As_2CrF_2O_4$
I.W.Nowell, S.Rettig, J.Trotter   *J. C. S. Dalton,* 2381, 1972

86.C π - (Cyclopentadienyl) - trans - iododicarbonyl(trimethyl phosphite) molybdenum
$C_{10}H_{14}IMoO_5P$
For complete entry see 73.11

86.8   **fac - Chloro - (1,3 - bis(dimethylarsino)propane)tricarbonyl - manganese**
$C_{10}H_{18}As_2ClMnO_3$
C.A.Bear, J.Trotter   *J. C. S. Dalton,* 673, 1973

86.9   **(1,2 - bis(Dimethylarsino) - 1,1,2 - trifluoro - 2 - trifluoromethyl - ethane) -**
tetracarbonyl - molybdenum (re - interpretation of data of Roberts and
Trotter,J.Chem.Soc.(A),1501,1971)
$C_{11}H_{12}As_2F_6MoO_4$
I.W.Nowell, S.Rettig, J.Trotter   *J. C. S. Dalton,* 2381, 1972

86.10   **(1,3 - bis(Dimethylarsino) - 2 - chloro - 1,1,3,3 - tetrafluoropropane)**
tetracarbonyl - chromium
$C_{11}H_{13}As_2ClCrF_4O_4$
I.W.Nowell, J.Trotter   *J. C. S. Dalton,* 2378, 1972

86.C   **π - (Methylcyclopentadienyl) - trans - iododicarbonyl(trimethyl phosphite)**
molybdenum
$C_{11}H_{16}IMoO_5P$
For complete entry see 73.13

86.C   **trans - Chloromethyl - bis(trimethylarsine) - platinum(ii) hexafluorobut - 2 -**
yne
$C_{11}H_{21}As_2ClF_6Pt$
For complete entry see 71.16

86.11   **Pentacarbonyl manganese(dimethylarsenide) pentacarbonyl chromium**
$C_{12}H_6AsCrMnO_{10}$
H.Vahrenkamp   *Chem. Ber.,* **105,** 1486, 1972

86.12   **Tetracarbonyl - dimethylphenylarsine - manganese(0)**
$C_{12}H_{11}AsMnO_4$
M.Laing, T.Ashworth, P.Sommerville, E.Singleton, R.Reimann
J. C. S. Chem. Comm., 1251, 1972

86.C   **bis - (μ - N,N - Dimethylthiocarbamoyl) - bis(chloro(trimethylphosphite) -**
palladium(ii))
$C_{12}H_{21}Cl_2N_2O_6P_2Pd_2S_2$
For complete entry see 71.21

86.C   **1,2 - bis(Trifluoromethyl) - 1 - (trans - dimethyl - trimethylphosphino -**
gold(iii)) - 2 - (trimethylphosphino - gold(i)) - ethylene
$C_{12}H_{24}Au_2F_6P_2$
For complete entry see 71.22

86.13   **trans - Chloro - hydrido - bis(triethylphosphine) palladium**
$C_{12}H_{31}ClP_2Pd$
M.L.Schneider, H.M.M.Shearer   *J. C. S. Dalton,* 354, 1973

86.C   bis - (π - 2 - Methylallyl) - bis(trimethylphosphite) ruthenium (monoclinic form)

$C_{14}H_{32}O_6P_2Ru$

For complete entry see 72.17

86.14   Octacarbonyl - μ - (1,2 - bis(dimethylarsino) - 3,3,4,4 - tetrafluorocyclobutene) dimanganese

$C_{16}H_{12}As_2F_4Mn_2O_8$

L.Y.Y.Chan, F.W.B.Einstein   *J. C. S. Dalton*, 111, 1973

86.15   Di - μ - chloro - (dichloromercurio) - bis(dimethylphenylphosphine) platinum(ii)

$C_{16}H_{22}Cl_4HgP_2Pt$

R.W.Baker, M.J.Braithwaite, R.S.Nyholm   *J. C. S. Dalton*, 1924, 1972

86.16   (1,3 - Dimethylarsino - 2,2,4,4 - tetramethylcyclobutane) tetracarbonyl - chromium (at −150°C)

$C_{16}H_{26}As_2CrO_4$

F.W.B.Einstein, R.D.G.Jones   *Inorg. Chem.*, **12**, 1148, 1973

86.17   bis(Dimethyl - phenyl - phosphine) - π - borallyl - platinum

$C_{16}H_{28}B_3P_2Pt$

L.J.Guggenberger, A.R.Kane, E.L.Muetterties
J. Amer. Chem. Soc., **94**, 5665, 1972

86.18   bis(Dicarbonyl(triethylphosphine) - nitrosylferrio) mercury

$C_{16}H_{30}Fe_2HgN_2O_6P_2$

F.S.Stephens   *J. C. S. Dalton*, 2257, 1972

86.19   trans - Dibromo - bis(di(t - butyl)fluorophosphine) nickel(ii)

$C_{16}H_{36}Br_2F_2NiP_2$

W.S.Sheldrick, O.Stelzer   *J. C. S. Dalton*, 926, 1973

86.20   Tetracarbonyl - diphenylmethylphosphine - manganese(0)

$C_{17}H_{13}MnO_4P$

M.Laing, T.Ashworth, P.Sommerville, E.Singleton, R.Reimann
J. C. S. Chem. Comm., 1251, 1972

86.21   Trichloro(triphenylphosphine) gold(iii)

$C_{18}H_{15}AuCl_3P$

G.Bandoli, D.A.Clemente, G.Marangoni, L.Cattalini
J. C. S. Dalton, 886, 1973

86.C   1,1 - bis(Trimethylphosphite) - 2,3,4,5,6,7 - hexa(trifluoromethyl) - 1 - nickela - cyclohepta - cis,trans,cis - triene

$C_{18}H_{18}F_{18}NiO_6P_2$

For complete entry see 71.41

86.C   (Di - isopropyl - phenyl - phosphine) - π - pentenyl - σ - methyl - nickel

$C_{18}H_{31}NiP$

For complete entry see 71.43

86.C   **Dichloro - (diphenyl - methyl - phosphine) ruthenium benzene**
$C_{19}H_{19}Cl_2PRu$
For complete entry see 74.8

86.22   **Dicarbonyl - nitrosyl - triphenylarsine - cobalt(0)**
$C_{20}H_{15}AsCoNO_3$
M.Sacerdoti, G.Gilli   *Abstr. Ital.-Yug. Congr,* 156, 1973

86.23   **Dithiocyanato - triphenylarsine - mercury(ii)**
$C_{20}H_{15}AsHgN_2S_2$
R.C.Makhija, A.L.Beauchamp, R.Rivest
J. C. S. Chem. Comm., 1043, 1972

86.24   **Dicarbonyl - nitrosyl - triphenylphosphine - cobalt**
$C_{20}H_{15}CoNO_3P$
V.G.Albano, P.L.Bellon, G.Ciani   *J. Organometal. Chem.*, **38,** 155, 1972

86.C   **1,1 - bis(Phenyl - dimethyl - arsine) - 1 - nickela - octafluoro - cyclopentane**
$C_{20}H_{22}As_2F_8Ni$
For complete entry see 71.49

86.C   **cis - bis(Dimethylphenylphosphine)bis - (5 - methyltetrazolato) palladium(ii)**
$C_{20}H_{28}N_8P_2Pd$
For complete entry see 83.90

86.25   **Dichloro - bis(o - phenylenebis(dimethylarsine)) cobalt chloride**
$C_{20}H_{32}As_4Cl_2Co^+$, Cl^-
P.K.Bernstein, G.A.Rodley, R.Marsh, H.B.Gray
Inorg. Chem., **11,** 3040, 1972

86.26   **Chloro - bis(o - phenylenebis(dimethylarsine)) nickel chloride**
$C_{20}H_{32}As_4Cl_2Ni^+$, Cl^-
P.K.Bernstein, G.A.Rodley, R.Marsh, H.B.Gray
Inorg. Chem., **11,** 3040, 1972

86.C   **bis(Dimethyl - phenyl - phosphine) - butadiene - carbonyl - iridium tetrafluoroborate**
$C_{21}H_{26}IrOP_2^+$, BF_4^-
For complete entry see 72.28

86.C   **bis(Dimethyl - phenyl - phosphine) - butadiene - carbonyl - iridium perchlorate**
$C_{21}H_{26}IrOP_2^+$, ClO_4^-
For complete entry see 72.29

86.C   **trans - 1,3 - Diphenyl - 2 - (dichloro - triethylphosphine - platinum) - imidazolidine**
$C_{21}H_{29}Cl_2N_2PPt$
For complete entry see 71.50

86.C   **Ethylene - (dicyclohexyl - (3 - (ethyl - vinyl - boro)propyl) - phosphine) nickel**
$C_{21}H_{40}BNiP$
For complete entry see 72.30

86.27   **Tetracarbonyl(triphenylphosphine - aurio) cobalt**
$C_{22}H_{15}AuCoO_4P$
T.L.Blundell, H.M.Powell   *J. Chem. Soc. (A)*, 1685, 1971

86.28   **bis(Isothiocyanato) - (N - (diphenylphosphinoethyl) - N' - diethylethylenediamine) cobalt(ii)**
$C_{22}H_{28}CoN_4PS_2$
A.B.Orlandini, C.Calabresi, C.A.Ghilardi, P.L.Orioli, L.Sacconi
J. C. S. Dalton, 1383, 1973
Also classified in 76, 83

86.C   **bis(Ethylene)(tricyclohexylphosphine) nickel**
$C_{22}H_{41}NiP$
For complete entry see 72.31

86.29   **Dichloro - bis - μ - ethanethiolato - bis(tripropylphosphine) diplatinum(ii)**
$C_{22}H_{52}Cl_2P_2Pt_2S_2$
M.C.Hall, J.A.J.Jarvis, B.T.Kilbourn, P.G.Owston
J. C. S. Dalton, 1544, 1972
Also classified in 85

86.30   **Pentacarbonyl(triphenylphosphine) chromium(0)**
$C_{23}H_{15}CrO_5P$
H.J.Plastas, J.M.Stewart, S.O.Grim   *Inorg. Chem.,* **12,** 265, 1973

86.31   **Pentacarbonyl(triphenylphosphite) chromium(0)**
$C_{23}H_{15}CrO_8P$
H.J.Plastas, J.M.Stewart, S.O.Grim   *Inorg. Chem.,* **12,** 265, 1973

86.C   **Methyl - tetracarbonyl - triphenylphosphine - manganese(i)**
$C_{23}H_{18}MnO_4P$
For complete entry see 71.55

86.C   **Dichloro - (diphenyl - methyl - phosphine) ruthenium p - cymene**
$C_{23}H_{27}Cl_2PRu$
For complete entry see 74.9

86.32   **bis(Isothiocyanato) - (N - methyl - N - (diphenylphosphinoethyl) - N' - diethyl - ethylenediamine) cobalt(ii)**
$C_{23}H_{32}CoN_4PS_2$
A.B.Orlandini, C.Calabresi, C.A.Ghilardi, P.L.Orioli, L.Sacconi
J. C. S. Dalton, 1383, 1973
Also classified in 76, 83

86.C   **(Pentafluorophenyl)(triphenylphosphine) gold(i)**
$C_{24}H_{15}AuF_5P$
For complete entry see 71.60

86.C   o - (1 - **Manganese - tetracarbonyl - ethyl)phenyl - diphenylphosphine**
$C_{24}H_{18}MnO_4P$
For complete entry see 71.61

86.C   **bis(Triethylphosphine) - (hexakis(trifluoromethyl)benzene) platinum**
$C_{24}H_{30}F_{18}P_2Pt$
For complete entry see 74.10

86.33   **Tetrabromo - tris(dimethylphenylphosphine) molybdenum(iv)**
$C_{24}H_{33}Br_4MoP_3$
M.G.B.Drew, J.D.Wilkins, A.P.Wolters   *J. C. S. Chem. Comm.*, 1278, 1972

86.34   **tris(Dimethylphenylphosphine) - tetrachloro molybdenum(iv) ethanol solvate**
$C_{24}H_{33}Cl_4MoP_3$, C_2H_6O
L.Manojlovic-Muir   *Inorg. Nucl. Chem. Letters,* **9,** 59, 1973

86.C   **(Tricyclohexylphosphine)methyl - nickel(ii) 2,4 - pentanedionate**
$C_{24}H_{43}NiO_2P$
For complete entry see 71.62

86.C   **(2,2' - bis(Dimethylarsino) - octafluoro - 1,1' - bi(cyclobut - 1 - enyl)) - bis(dimethyl - arsino) - nonacarbonyl - tetracobalt - dihydride**
$C_{25}H_{26}As_4Co_4F_8O_9$
For complete entry see 75.35

86.35   **Chloro - carbonyl - tris(phenyldimethylphosphine) iridium(i)**
$C_{25}H_{33}ClIrOP_3$
J.-Y.Chen, J.Halpern, J.Molin-Case   *J. Coord. Chem.*, **2,** 239, 1973

86.C   **(Cyclo - octa - 1,5 - diene) - bis(dimethylphenylphosphine) methyl iridium(i)**
$C_{25}H_{37}IrP_2$
For complete entry see 75.36

86.C   **bis(Dimethylglyoximato)(tri - n - butylphosphine)(4 - pyridyl) cobalt(iii)**
$C_{25}H_{45}CoN_5O_4P$
For complete entry see 71.66

86.36   **Dichloro - nitrosyl - bis(methyl - diphenyl - phosphine) cobalt**
$C_{26}H_{26}Cl_2CoNOP_2$
C.P.Brock, J.P.Collman, G.Dolcetti, P.H.Farnham, J.A.Ibers, J.E.Lester, C.A.Reed   *Inorg. Chem.*, **12,** 1304, 1973

86.37   **Chloro - bis(dimethylglyoximato) - triphenylstibine - rhodium(iii)**
$C_{26}H_{29}ClN_4O_4RhSb$
A.C.Villa, A.G.Manfredotti, C.Guastini
Cryst. Struct. Comm., **2,** 129, 1973
Also classified in 83

86.38 1,1 - bis(Phenyl - diethyl - phosphine) - 3,3,4,4 - tetra(trifluoromethyl) - 1 - pallada - 2,5 - dioxa - cyclopentane
$C_{26}H_{30}F_{12}O_2P_2Pd$
R.Countryman, B.R.Penfold *Acta Cryst. (A)*, **28**, S83, 1972
Also classified in 84

86.39 Chloro - (dimethylglyoximato) - (dimethylglyoxime) - (triphenylphosphine) - rhodium(iii) chloride methylene chloride solvate
$C_{26}H_{30}N_4O_4PRh^+$, Cl^-, CH_2Cl_2
A.C.Villa, A.G.Manfredotti, C.Guastini
Cryst. Struct. Comm., **2**, 133, 1973
Residue 1 also classified in 83

86.40 Dichloro - dicarbonyl - tris(dimethylphenylphosphine) molybdenum(ii) methanol solvate
$C_{26}H_{33}Cl_2MoO_2P_3$, CH_4O
A.Mawby, G.E.Pringle *J. Inorg. Nucl. Chem.*, **34**, 517, 1972

86.41 trans - Dichloro - bis(4,4 - dimethoxy - 1 - phenylphosphorinan) nickel(ii)
$C_{26}H_{38}Cl_2NiO_4P_2$
A.T.McPhail, J.C.H.Steele *J. C. S. Dalton*, 2680, 1972

86.42 Dichloro - bis(1 - phenyl - 4,4 - dimethoxyphosphorinane) nickel(ii)
$C_{26}H_{38}Cl_2NiO_4P_2$
A.T.McPhail, J.J.Breen, J.C.H.Steele Junior, L.D.Quin
Phosphorus, **1**, 255, 1972

86.43 Chloro(bis(2' - (diethylamino)ethyl) - 2 - (diphenylphosphino)ethylamine) cobalt(ii) perchlorate
$C_{26}H_{42}ClCoN_2P^+$, ClO_4^-
P.Dapporto, G.Fallani *J. C. S. Dalton*, 1498, 1972
Residue 1 also classified in 83

86.C μ - (α,ω - Octadi - π - enyl)bis(bromo - tri - isopropylphosphine - nickel(ii))
$C_{26}H_{54}Br_2Ni_2P_2$
For complete entry see 72.34

86.C Nonacarbonyl - μ - dimethylarsino - μ - (2 - (diphenylphosphino)tetrafluorocyclobut - 1 - enyl) - tri - iron
$C_{27}H_{16}AsF_4Fe_3O_9P$
For complete entry see 71.72

86.44 (O - Methyl - phosphorodithioato) - (1 - diphenylarsino - 2 - diphenylphosphino - ethane) - nickel(ii) benzene solvate
$C_{27}H_{27}AsNiO_2PS_2$, $0.5C_6H_6$
L.Gastaldi, P.Porta, A.A.G.Tomlinson *Abstr. Ital.-Yug. Congr*, 181, 1973
Residue 1 also classified in 85

86.C (Dimenthenyl - methyl - phosphine) - π - pentenyl - σ - methyl - nickel
$C_{27}H_{49}NiP$
For complete entry see 71.73

86.45 **(bis(Diphenylphosphino)methane) tricarbonyl - iron**
$C_{28}H_{22}FeO_3P_2$
F.A.Cotton, K.I.Hardcastle, G.A.Rusholme *J. Coord. Chem.*, **2**, 217, 1973

86.C **bis(Acetylacetonato) - triphenylphosphine - palladium benzene solvate**
$C_{28}H_{29}O_4PPd$, $0.5C_6H_6$
For complete entry see 77.15

86.46 **bis(9 - Phenyl - 9 - phosphabicyclo(3.3.1)nonane) nickel chloride**
$C_{28}H_{38}Cl_2NiP_2$
A.E.Smith *Inorg. Chem.*, **11**, 3017, 1972

86.C **bis - μ - Dicyclohexylphosphido - bis(π - ethylene nickel)**
$C_{28}H_{52}Ni_2P_2$
For complete entry see 72.36

86.C **Tricarbonyl - bis(diphenylphosphino)methane - dithioformato - manganese**
$C_{29}H_{23}MnO_3P_2S_2$
For complete entry see 85.32

86.47 **bis(Thiocyanato) - (1,3 - bis(diphenylphosphino) - 2 - trifluoromethyl - propene) palladium(ii) methylene dichloride**
$C_{30}H_{23}F_3N_2P_2PdS_2$, CH_2Cl_2
R.T.Simpson, S.Jacobson, A.J.Carty, M.Mathew, G.J.Palenik
J. C. S. Chem. Comm., 388, 1973

86.48 **Dicarbonyl - bis(diphenylethylphosphine) platinum(0)**
$C_{30}H_{30}O_2P_2Pt$
V.G.Albano, P.L.Bellon, M.Manassero
J. Organometal. Chem., **35**, 423, 1972

86.C **1,2,3,4 - Tetra(trifluoromethyl)butadienyl - (cyclopentadienyl) - triphenyl - phosphine - ruthenium**
$C_{31}H_{21}F_{12}PRu$
For complete entry see 71.76

86.49 **Di - iodo - (N,N - bis(2 - diphenylarsinoethyl) - 2 - methoxyethylamino) nickel(ii)**
$C_{31}H_{35}As_2I_2NNiO$
L.Sacconi, J.J.van der Zee, K.G.Shields, C.H.L.Kennard
Cryst. Struct. Comm., **2**, 229, 1973
Also classified in 83, 84

86.50 **Chloro - (N,N - bis(2 - diphenylphosphinoethyl) - 2 - methoxyethylamine) cobalt(ii) hexafluorophosphate**
$C_{31}H_{35}ClCoNOP_2^+$, F_6P^-
P.Dapporto, G.Fallani, L.Sacconi *J. Coord. Chem.*, **1**, 269, 1971
Residue 1 also classified in 83, 84

86.C **Triphenylphosphine - (1 - (di(trifluoromethyl) - hydroxymethyl) - cyclopentadienyl) - (1,2 - di(carboxymethyl)ethylene - 1 - yl) - ruthenium**
$C_{32}H_{27}F_6O_5PRu$
For complete entry see 71.78

86.C **1,2 - bis(Dicyclohexylphosphino)ethane - (tetramethylethylene) - nickel**
$C_{32}H_{60}NiP_2$
For complete entry see 72.37

86.C **Structure III**
$C_{34}H_{25}FeNiO_3P$
For complete entry see 71.82

86.C **Iodo - (N,N - bis(2 - diphenylphosphinoethyl) - N - (2 - diethylaminoethyl)amine) cobalt(ii) iodide**
$C_{34}H_{42}CoIN_2P_2^+$, I^-
For complete entry see 76.52

86.C **Iodo - (N,N - bis(2 - diphenylphosphinoethyl) - N - (2 - diethylaminoethyl)amine) nickel(ii) iodide**
$C_{34}H_{42}IN_2NiP_2^+$, I^-
For complete entry see 76.53

86.51 **di - μ - Chloro - bis(di - t - butyl - p - tolylphosphine) tetracarbonyl - diruthenium(i)**
$C_{34}H_{50}Cl_2O_4P_2Ru_2$
R.Mason, K.M.Thomas, D.F.Gill, B.L.Shaw
J. Organometal. Chem., **40**, C67, 1972

86.C **(Cyclo - octa - 1,5 - diene)(1,2 - bis(diphenylphosphino)ethane) methyl iridium(i)**
$C_{35}H_{39}IrP_2$
For complete entry see 71.84

86.52 **Bromo - bis(triphenylphosphine) copper(i) benzene solvate**
$C_{36}H_{30}BrCuP_2$, $0.5C_6H_6$
P.H.Davis, R.L.Belford, I.C.Paul *Inorg. Chem.*, **12,** 213, 1973

86.53 **Dichloro - dinitrosyl - bis(triphenylphosphine) molybdenum**
$C_{36}H_{30}Cl_2MoN_2O_2P_2$
M.O.Visscher, K.G.Caulton *J. Amer. Chem. Soc.*, **94,** 5923, 1972

86.54 **Di - iodo - bis(triphenylphosphine) palladium(ii) dichloromethane solvate**
$C_{36}H_{30}I_2P_2Pd$, $2CH_2Cl_2$
T.Debaerdemaker, A.Kutoglu, G.Schmid, L.Weber
Acta Cryst. (B), **29,** 1283, 1973

86.C **(Cyclo - octa - 1,5 - diene)(1,3 - bis(diphenylphosphino)propane)methyl iridium(i)**
$C_{36}H_{41}IrP_2$
For complete entry see 71.85

86.C **(Di - (o - tolyl) - t - butylphosphine) - dibenzyl - platinum - o,o' - (t - butylphosphine)**
$C_{36}H_{44}P_2Pt$
For complete entry see 71.86

86.C **(±) - cis - Di(o - ((o - tolyl)(t - butyl)phosphino)benzyl) platinum**
$C_{36}H_{44}P_2Pt$
For complete entry see 71.87

86.C **(±) - trans - Di(o - ((o - tolyl)(t - butyl)phosphino)benzyl) platinum**
$C_{36}H_{44}P_2Pt$
For complete entry see 71.88

86.55 **trans - Dihydrido - tetrakis(diethyl - phenylphosphonite) ruthenium(ii)**
$C_{36}H_{62}O_8P_4Ru$
L.J.Guggenberger *Inorg. Chem.*, **12**, 1317, 1973

86.56 **Carbonyl - nitrosyl - bis(triphenylphosphine) cobalt**
$C_{37}H_{30}CoNO_2P_2$
V.G.Albano, P.L.Bellon, G.Ciani *J. Organometal. Chem.*, **38**, 155, 1972

86.57 **Nitrosyl - carbonyl - bis(triphenylphosphine) iridium**
$C_{37}H_{30}IrNO_2P_2$
C.P.Brock, J.A.Ibers *Inorg. Chem.*, **11**, 2812, 1972

86.C **Iodo - methyl - bis(triphenylphosphine) platinum - sulfur dioxide**
$C_{37}H_{33}IP_2Pt$, O_2S
For complete entry see 71.89

86.C **2,5 - Di(trifluoromethyl) - 3 - (tricarbonyl - diphenylphosphino - iron) - 4 - diphenylphosphino - 1 - tricarbonyl - 1 - ferra - cyclopenta - 2,4 - diene dicarbonyl iron benzene solvate**
$C_{38}H_{20}F_6Fe_3O_8P_2$, $0.5C_6H_6$
For complete entry see 71.90

86.C **Chloro - (2,2' - bis(diphenylphosphino)stilbene) - rhodium**
$C_{38}H_{30}ClP_2Rh$
For complete entry see 72.40

86.C **Dichloro - carbonyl - bis(triphenylphosphine) - difluoromethyl - iridium**
$C_{38}H_{31}Cl_2F_2IrP_2$
For complete entry see 71.91

86.C **bis(Triphenylphosphine) - (ethylene) platinum**
$C_{38}H_{34}P_2Pt$
For complete entry see 72.42

86.58 **trans - Dithiocyanato - bis((3,3 - dimethylbutynyl)diphenylphosphine) - palladium(ii)**
$C_{38}H_{38}N_2P_2Pd_2S_2$
G.Beran, A.J.Carty, P.C.Chieh, H.A.Patel *J. C. S. Dalton*, 488, 1973

86.59   **Dicarbonyl(thiocarbonyl) - bis(triphenylphosphine) iridium(i) hexafluorophosphate acetone solvate**
$C_{39}H_{24}IrO_2P_2S^+$, F_6P^- , C_3H_6O
J.S.Field, P.J.Wheatley   *J. C. S. Dalton*, 2269, 1972

86.60   **Chloro - cyano - (N - thiocyanato) - carbonyl - bis(triphenylphosphine) - iridium(iii)**
$C_{39}H_{30}ClIrN_2OP_2S$
J.A.Ibers, D.S.Hamilton, W.H.Baddley   *Inorg. Chem.*, **12,** 229, 1973

86.C   **bis(Triphenylphosphine)hexafluoroacetone nickel(0)**
3,3 - Di(trifluoromethyl) - 1 - nickela - 2 - oxa - cyclopropane
bis(triphenylphosphine)
$C_{39}H_{30}F_6NiOP_2$
For complete entry see 71.92

86.C   **bis(Triphenylphosphine) - hexafluoroacetone - nickel(0)**
$C_{39}H_{30}F_{12}NiOP_2$
For complete entry see 72.43

86.61   **Nitrato - tris(diphenylmethylphosphine) copper(i)**
$C_{39}H_{39}CuNO_3P_3$
M.Mathew, G.J.Palenik, A.J.Carty   *Canad. J. Chem.*, **49,** 4119, 1971

86.C   **Chloro - carbonyl - bis(triphenylphosphine) - difluoromethyl - chlorodifluoroacetato - iridium benzene solvate**
$C_{40}H_{31}Cl_2F_4IrO_3P_2$, C_6H_6
For complete entry see 71.93

86.C   **1 - (Diphenylphosphino) - 2 - (dimethylarsino)tetrafluorocyclobut - 1 - ene - μ - (1 - (diphenylphosphino) - 2 - (dimethylarsino)tetrafluorocyclobut - 1 - ene) - tetracarbonyl - di - iron**
$C_{40}H_{32}As_2F_8Fe_2O_4P_2$
For complete entry see 75.48

86.C   **Acetato - bis(triphenylphosphine)dicarbonyl - manganese(i)**
$C_{40}H_{33}MnO_4P_2$
For complete entry see 81.60

86.C   **bis(μ - Diphenylphosphido - μ' - carbonyl - π - methylcyclopentadienyl - carbonyl - iron) rhodium hexafluorophosphate**
$C_{40}H_{34}F_6Fe_2O_4P_3Rh$
For complete entry see 73.46

86.C   **bis(Triphenylphosphine) - (N - methyl - hexafluoroacetonimide) platinum(0)**
$C_{40}H_{36}F_{12}NO_2Pt$
For complete entry see 72.44

86.62   **trans - Dihydrido - tetra(diethoxyphenylphosphine) - ruthenium**
$C_{40}H_{62}O_8P_4Ru$
L.J.Guggenberger   *Acta Cryst. (A)*, **28,** S84, 1972

86.C   bis(Triphenylphosphine) - (1,2 - bis(trifluoromethyl) - ethylene - 1,2 - dithiolato) ruthenium carbonyl (orange form)

$C_{41}H_{30}F_6OP_2RuS_2$

For complete entry see 85.35

86.C   bis(Triphenylphosphine) - (1,2 - bis(trifluoromethyl) - ethylene - 1,2 - diolato) ruthenium carbonyl (violet form)

$C_{41}H_{30}F_6OP_2RuS_2$

For complete entry see 85.36

86.C   (1,4 - bis(o - Diphenylphosphino - phenyl)buta - 1,3 - diene) ruthenium(0) carbonyl

$C_{41}H_{32}OP_2Ru$

For complete entry see 72.45

86.63   Iodo - (1,1,1 - tris(diphenylphosphinomethyl) - ethane) nickel(i)

$C_{41}H_{39}INiP_3$

P.Dapporto, G.Fallani, S.Midollini, L.Sacconi

J. C. S. Chem. Comm., 1161, 1972

86.C   (1,4 - bis(o - Diphenylphosphino - phenyl) - buta - diyl) ruthenium(ii) dicarbonyl

$C_{42}H_{34}O_2P_2Ru$

For complete entry see 71.94

86.C   bis(Triphenylphosphine) - bis(π - allyl) ruthenium toluene solvate

$C_{42}H_{40}P_2Ru$, C_7H_8

For complete entry see 72.46

86.64   (tris(2 - Diphenylphosphinoethyl)amine) - chlorocobalt(ii) hexafluorophosphate

$C_{42}H_{42}ClCoNP_3^+$, F_6P^-

M.di Vaira, A.B.Orlandini *Inorg. Chem.*, **12**, 1292, 1973

Residue 1 also classified in 83

86.C   Osmium compound X

$C_{43}H_{28}O_7Os_3P_2$

For complete entry see 71.95

86.C   Carbonyl - chloro - (tetracyanoethylene)bis(triphenylarsine) iridium

$C_{43}H_{30}As_2ClIrN_4O$

For complete entry see 72.47

86.C   Osmium compound VI

$C_{43}H_{30}O_7Os_3P_2$

For complete entry see 71.96

86.C   **Carbonyl - chloro - (4 - fluorophenyl - di - imide - 2C,N')bis(triphenylphosphine) - iridium(iii) tetrafluoroborate acetone solvate**
$C_{43}H_{34}ClFIrN_2OP_2{}^+$, $BF_4{}^-$, C_3H_6O
For complete entry see 71.97

86.C   **Tetrafluoroethylene - (1,1,1 - tris(diphenylphosphinomethyl)ethane) nickel**
$C_{43}H_{39}F_4NiP_3$
For complete entry see 72.48

86.C   **Manganese compound 4**
$C_{44}H_{28}Mn_2O_8P_2$
For complete entry see 71.98

86.C   **Osmium compound IV**
$C_{44}H_{30}O_8Os_3P_2$
For complete entry see 71.99

86.C   **1,1 - bis(Triphenylphosphine) - 1 - platina - benzocyclopentenedione (monoclinic form)**
$C_{44}H_{34}O_2P_2Pt$
For complete entry see 71.100

86.C   **1,1 - bis(Triphenylphosphine) - 1 - platina - benzocyclopentenedione (triclinic form)**
$C_{44}H_{34}O_2P_2Pt$
For complete entry see 71.101

86.C   **Ethylene - bis(tri - o - tolyl - phosphite) nickel**
$C_{44}H_{46}NiO_6P_2$
For complete entry see 72.49

86.C   **(σ - Dicyanovinyl)(carbonyl)(π - dicyanoacetylene) - bis(triphenylphosphine) iridium(i)**
$C_{45}H_{31}IrN_4OP_2$
For complete entry see 71.102

86.C   **Acrylonitrile - bis(tri - o - phosphite) nickel**
$C_{45}H_{45}NNiO_6P_2$
For complete entry see 72.50

86.C   **trans - bis(Triphenylphosphine) - chloro - carbonyl - (2 - dicarbadodecaborane(12) - acetylide) - (trans - 2 - dicarbadodecaborane(12) - vinyl) - iridium dichloromethane solvate**
$C_{45}H_{54}B_{20}ClIrOP_2$, $2.5CH_2Cl_2$
For complete entry see 71.103

86.C   **Compound I**
$C_{46}H_{39}Au_2FeP_2{}^+$, $BF_4{}^-$
For complete entry see 71.104

86.65 Tetra - μ - chloro - bis(di - μ - diphenylphosphino - bis(nitrosyl - ruthenium))
$C_{48}H_{40}Cl_4N_4O_4P_4Ru_4$
R.Eisenberg, A.P.Gaughan Junior, C.G.Pierpont, J.Reed, A.J.Schultz
J. Amer. Chem. Soc., **94**, 6240, 1972

86.C Bromo - bis(triphenylphosphine) - (1,2,3,4 - tetrakis(methoxycarbonyl) - buta - 1,3 - dienyl) palladium
$C_{48}H_{43}BrO_8P_2Pd$
For complete entry see 71.105

86.C Carbonyl - bis(triphenylphosphine)(1,4 - di - (p - fluorophenyl)tetrazene) iridium tetrafluoroborate
$C_{49}H_{38}F_2IrN_4OP_2{}^+$, $BF_4{}^-$
For complete entry see 83.123

86.C bis(Triphenylphosphine) - cyclopentadienyl - ruthenium - phenylacetylide chloro - copper(i) acetone solvate
$C_{49}H_{40}ClCuP_2Ru$, C_3H_6O
For complete entry see 71.107

86.C μ - Diphenylacetylene - bis(triphenylphosphine - bis(trifluorophosphine) rhodium) diethyl ether solvate
$C_{50}H_{40}F_{12}P_6Rh_2$, $C_4H_{10}O$
For complete entry see 72.51

86.66 bis(5 - Phenyltetrazolato) - bis(triphenylphosphine) palladium methylene dichloride solvate
$C_{50}H_{40}N_8P_2Pd$, CH_2Cl_2
P.Kreutzer, C.Weis, H.Boehme, T.Kemmerich, W.Beck, C.Spencer, R.Mason *Z. Naturforsch., B*, **27**, 745, 1972
Residue 1 also classified in 83

86.67 Copper(i) iodide - bis(diphenylphosphino)methane
$C_{50}H_{41}Cu_4I_4P_4$
N.Marsich, G.Nardin, L.Randaccio *Abstr. Ital.-Yug. Congr*, 194, 1973

86.68 Di - μ - diphenylphosphino - bis(diphenylmethylphosphine - nitrosyl - ruthenium)
$C_{50}H_{46}N_2O_2P_4Ru_2$
R.Eisenberg, A.P.Gaughan Junior, C.G.Pierpont, J.Reed, A.J.Schultz
J. Amer. Chem. Soc., **94**, 6240, 1972

86.C Ruthenium complex with triphenyl phosphite
$C_{51}H_{39}O_{12}P_3Ru_2$
For complete entry see 74.12

86.69 Dioxo - bis(1,2 - bis(diphenylarsino)ethane) rhodium tetrafluoroborate
$C_{52}H_{28}As_4O_2Rh^+$, $BF_4{}^-$
B.L.Kindberg, M.S.Weininger, E.L.Amma
Amer. Cryst. Assoc., Abstr. Papers (Winter Meeting), 40, 1973

86.70   Dioxo - bis(1,2 - bis(diphenylphosphino)ethane) iridium tetrafluoroborate
$C_{52}H_{28}IrO_2P_4^+$, BF_4^-
A.G.Gash, N.W.Terry III, E.L.Amma
Amer. Cryst. Assoc., Abstr. Papers (Winter Meeting), 40, 1973

86.71   Dioxo - bis(1,2 - bis(diphenylphosphino)ethane) rhodium tetrafluoroborate
$C_{52}H_{28}O_2P_4Rh^+$, BF_4^-
A.G.Gash, N.W.Terry III, E.L.Amma
Amer. Cryst. Assoc., Abstr. Papers (Winter Meeting), 40, 1973

86.72   Chloro - bis(1,2 - bis(diphenylphosphino)ethane) cobalt(ii) trichlorostannate(ii)
$C_{52}H_{48}ClCoP_4^+$, Cl_3Sn^-
J.K.Stalick, P.W.R.Corfield, D.W.Meek
J. Amer. Chem. Soc., **94**, 6194, 1972

86.73   Chloro - bis(1,2 - bis(diphenylphosphino)ethane) cobalt(ii) trichlorostannate(ii) chlorobenzene solvate
$C_{52}H_{48}ClCoP_4^+$, Cl_3Sn^- , C_6H_5Cl
J.K.Stalick, P.W.R.Corfield, D.W.Meek
J. Amer. Chem. Soc., **94**, 6194, 1972

86.74   Dioxo - bis(1,2 - bis(diphenylphosphino)ethane) cobalt tetrafluoroborate
$C_{52}H_{48}CoO_2P_4^+$, BF_4^-
A.G.Gash, N.W.Terry III, E.L.Amma
Amer. Cryst. Assoc., Abstr. Papers (Winter Meeting), 40, 1973

86.75   Nitrosyl - bis(1,2 - bis(diphenylphosphino)ethane) ruthenium tetraphenylborate acetone solvate
$C_{52}H_{48}NOP_4Ru^+$, $C_{24}H_{20}B^-$, C_3H_6O
C.G.Pierpont, R.Eisenberg *Inorg. Chem.*, **12**, 199, 1973
Residue 2 classified in 62

86.76   Trihydrido - bis(1,2 - bis(diphenylphosphino)ethane) rhenium(iii)
$C_{52}H_{51}P_4Re$
V.G.Albano, P.L.Bellon *J. Organometal. Chem.*, **37**, 151, 1972

86.77   Tetra(diphenylmethylphosphine) molybdenum tetrahydride
$C_{52}H_{56}MoP_4$
P.Meakin, L.J.Guggenberger, W.G.Peet, E.L.Muetterties, J.P.Jesson
J. Amer. Chem. Soc., **95**, 1467, 1973

86.78   Chloro - bis(2 - (diphenoxyphosphino - oxy)phenyl) (triphenylphosphite) iridium(iii)
$C_{54}H_{43}ClIrO_9P_3$
J.M.Guss, R.Mason *J. C. S. Dalton*, 2193, 1972

86.79   Tetracarbonyl - iron - carbonyltris(triphenylphosphite) diplatinum
$C_{59}H_{45}FeO_9P_2Pt_2$
V.G.Albano, G.Ciani, M.I.Bruce, G.Shaw, F.G.A.Stone
J. Organometal. Chem., **42**, C99, 1972

86.C   cis - Triphenylphosphine - phenyl - platinum - bis(triphenyl - lead)
$C_{60}H_{50}PPb_2Pt$
For complete entry see 69.28

86.C μ - (α,ω - Octadi - π - enyl)bis(bromo - bis(diphenylphosphino)ethane - nickel(ii)) chloroform solvate
$C_{60}H_{62}Br_2Ni_2P_4$, $2CHCl_3$
For complete entry see 72.53

86.C μ - (α,ω - Octadi - π - enyl) - bis(bromo - (bis(diphenylphosphino)ethane) nickel) chloroform solvate
$C_{60}H_{62}Br_2Ni_2P_4$, $8CHCl_3$
For complete entry see 72.54

86.C   Hexa - μ - acetato - μ_3 - oxo - tri(triphenylphosphine - ruthenium)
$C_{66}H_{63}O_{13}P_3Ru_3$
For complete entry see 81.63

86.80   Dihydrido - tetrakis(triphenylphosphine) ruthenium(ii)
$C_{72}H_{62}P_4Ru$
A.Immirzi, A.Luccarelli   *Cryst. Struct. Comm.*, **1**, 317, 1972

86.81   Thiocyanato - bis(triphenylphosphine) silver(i) dimer
$C_{74}H_{60}Ag_2N_2P_4S_2$
J.Howatson, B.Morosin   *Cryst. Struct. Comm.*, **2**, 51, 1973

86.82   tetrakis(Triphenylphosphine)di - μ - carbonyl - dirhodium(0) dichloromethane solvate
$C_{74}H_{60}O_2P_4Rh_2$, $2CH_2Cl_2$
P.Singh, C.B.Dammann, D.J.Hodgson   *Inorg. Chem.*, **12**, 1335, 1973

86.83   Dichloro - tris(1,2 - bis(diphenylphosphine)ethane) dicopper(i) acetone solvate
$C_{78}H_{72}Cl_2Cu_2P_6$, $2C_3H_6O$
V.G.Albano, P.L.Bellon, G.Ciani   *J. C. S. Dalton*, 1938, 1972

86.84   Triphenylphosphine - copper(i) hydride hexamer dimethylformamide solvate
$C_{108}H_{96}Cu_6P_6$, C_3H_7NO
M.R.Churchill, S.A.Bezman, J.A.Osborn, J.Wormald
Inorg. Chem., **11**, 1818, 1972

86.85   Tri - iodo - heptakis(tri - p - fluorophenylphosphine) undecagold (absolute configuration)
$C_{126}H_{84}Au_{11}F_{21}I_3P_7$
P.Bellon, M.Manassero, M.Sansoni   *J. C. S. Dalton*, 1481, 1972

86.86   hexakis(tris - p - Tolylphosphine - gold) tetraphenylborate
$C_{126}H_{126}Au_6P_6^{2+}$, $2C_{24}H_{30}B^-$
P.L.Bellon, M.Manassero, L.Naldini, M.Sansoni
J. C. S. Chem. Comm., 1035, 1972
Residue 2 classified in 62

C_1

C_2

C_4

$(C_4H_9CuNPS_2)_n$	85.1	5
C_4H_9GeN	69.6	2
$C_4H_9InO_2$	68.2	5
C_4H_9NO , $3CH_6N_3^+$, $3Cl^-$	60.16	2
C_4H_9NO , C_8H_5I	60.59	2
$C_4H_9NO_2$	1.66+	1
$C_4H_9NO_2$	48.32	1
$C_4H_9NO_2$	48.17	5
$C_4H_9NO_2$	48.18	5
$C_4H_9NO_3$	48.33	1
$C_4H_9NO_3$	48.19	5
$C_4H_9NO_3$	48.20	5
$C_4H_9NO_3S$	48.34	1
C_4H_9NSSn	69.9	3
C_4H_9NSn	69.8	2
$C_4H_9N_2O_3^+$, Cl^- , H_2O	48.35	1
$C_4H_9N_2O_3^+$, Cl^- , H_2O	48.21	5
$C_4H_9N_2O_3^+$, NO_3^-	48.22	5
$C_4H_9N_2O_3^+$, $H_2O_4P^-$, H_2O	48.23	5
$C_4H_9N_3O$	9.3	5
$C_4H_9N_3O_2$, H_2O	48.36+	1
$C_4H_9N_3O_2$, H_2O	48.24	5
$C_4H_9N_5NiOS_4$	85.2	5
$C_4H_9O_2P$	64.8	3
$(C_4H_{10}AgO_4P)_n$	84.2	5
$C_4H_{10}Ag_2N_2O_4^{2+}$, $2NO_3^-$	82.4+	4
$C_4H_{10}Ag_2N_2O_4^{2+}$, $2NO_3^-$	82.2	5
$C_4H_{10}Br_4O_2Ti^-$, $C_8H_{20}Br_2O_4Ti^+$	84.27	2
$C_4H_{10}CdCl_2N_2O_2$	83.28	2
$C_4H_{10}CdCl_2N_6O_4$	84.8	2
$C_4H_{10}CdN_4O_4S_2$	79.20	2
$(C_4H_{10}CdO_4)_n$	81.17	4
$C_4H_{10}ClCuS_2$	85.20	2
$C_4H_{10}ClHgO_3P$	86.1	3
$C_4H_{10}ClHgS^+$, Cl^- , Cl_2Hg	85.21	2
$C_4H_{10}Cl_2N_4Pd$	71.1	4
$C_4H_{10}Cl_2N_6O_4Zn$	84.9	2
$C_4H_{10}Cl_3NPt$	72.5	3
$C_4H_{10}Cl_3NPt$	72.6	3
$(C_4H_{10}Cl_3NTi)_n$	83.9	4
$C_4H_{10}Cl_6O_2S_2W_2$	84.2	4
$C_4H_{10}Cl_8O_2Sb_2$	66.8	2
$C_4H_{10}CuN_6O_4^{2+}$, $2Cl^-$	84.10	2
$C_4H_{10}Fe_2N_4O_4S_2$	85.22	2
$C_4H_{10}HgS_2$	85.23	2
$4C_4H_{10}LiO^+$, $C_{12}H_{14}Cr_2^{4-}$	71.12	4
$C_4H_{10}MoO_6$	84.3	5
$C_4H_{10}NO^+$, $C_{12}H_4N_4^-$	40.5	5
$2C_4H_{10}NO^+$, $2C_{12}H_4N_4^-$, $C_{12}H_4N_4$	40.4	4
$C_4H_{10}NO_2^+$, Cl^-	48.25	5
$C_4H_{10}NO_3^+$, Br^-	48.26	5
$C_4H_{10}NO_6P$	48.12	3
$C_4H_{10}NO_6P$	48.12	4
$C_4H_{10}N_2$, $6H_2O$	33.13	1
$0.5C_4H_{10}N_2$, Ag^+ , I^-	60.60	2
$C_4H_{10}N_4NiS_2$	85.4	3
$C_4H_{10}N_4O_2PbS_2$, H_2O	69.6	5
$C_4H_{10}N_6NiS_4^{2+}$, $2ClO_4^-$, C_2H_6O	79.8	5
$C_4H_{10}N_6NiS_4^{2+}$, $2ClO_4^-$, C_2H_6O	79.9	5
$C_4H_{10}N_8NiS_4$	83.30	2
$C_4H_{10}NiO_2S_2$	85.24	2
$C_4H_{10}O$	5.13	5
$C_4H_{10}O$, $CHBrCl_2$	60.61	2
$4C_4H_{10}O$, $Br_6Mg_4O_7$	67.17	2
$C_4H_{10}O_4$	45.1+	1
$C_4H_{10}O_4P^-$, $C_4H_{12}N_3^+$	8.13	5
$C_4H_{10}O_4P^-$, $C_6H_{15}N_4O_2^+$	48.51	5
$2C_4H_{10}O_4P^-$, Ba^{2+}	46.6	1
$2C_4H_{10}O_4P^-$, $C_4H_{14}N_2^{2+}$	3.14	5
$C_4H_{10}O_6S_3$	11.29	1
$C_4H_{10}O_6Zn$	81.29	2
$C_4H_{10}O_8Zn$	81.30	2
$C_4H_{11}Cl_2NPt$	72.5	2
$C_4H_{11}N$, $8.67H_2O$	61.4	2
$C_4H_{11}N$, $9.75H_2O$	61.5	2
$xC_4H_{11}N$, $C_{40}H_{37}EuO_8$	61.10	5
$(C_4H_{11}NNdO_7^+)_n$, nCl^-	81.18	4
$C_4H_{11}NO_2$, $C_{14}H_9I_3O_4$	60.26	5
$C_4H_{11}NO_3S_2$	4.8	1
$C_4H_{11}N_2O_2^+$, Br^-	1.68	1
$C_4H_{11}N_2O_2^+$, Cl^-	48.13+	4
$C_4H_{11}N_3O_4S$	8.12	5
$C_4H_{12}As^+$, Br^-	65.7	2
$C_4H_{12}As_2S_2$	65.8	2
$(C_4H_{12}BLi)_n$	62.3	3
$C_4H_{12}BN$	62.4	3
$C_4H_{12}B_2Cl_2N_2$	62.20	2
$C_4H_{12}B_2F_4N_2$	62.21	2
$C_4H_{12}B_2N_8NiO_8$, $4H_2O$	83.10	4
$C_4H_{12}B_4$	62.22	2
$C_4H_{12}B_6$	62.23	2
$C_4H_{12}B_{10}Br_4$	62.24	2
$C_4H_{12}Br_4Pd_3S_2$	85.26	2
$C_4H_{12}Cl_2CuN_6O_2$	83.10	3
$C_4H_{12}Cl_2CuO_2S_2$	84.3	3
$C_4H_{12}Cl_2N_2Pd$	76.2	4
$C_4H_{12}Cl_2O_2PdS_2$	85.28+	2
$C_4H_{12}Cl_3N_2Pt^+$, Cl^- , $0.5H_2O$	72.3	4
$C_4H_{12}Cl_4O_6Re_2$	81.7	3
$C_4H_{12}CuN_2O_4$	81.31+	2
$C_4H_{12}Cu_2O_{10}$, $2H_2O$	81.33	2
$C_4H_{12}F_4N_4P_4$	64.9	3
$C_4H_{12}Ge_2S_6$	69.7	5
$C_4H_{12}HgO_2P_2$	86.1	5
$C_4H_{12}In^-$, Cs^+	68.3	5
$C_4H_{12}In^-$, K^+	68.4	5
$C_4H_{12}Li_4$	67.5	3
$C_4H_{12}N^+$, $Ag_2I_3^-$	3.19	1
$C_4H_{12}N^+$, $AlO_8Si_2^-$, H_2O	3.14	3
$C_4H_{12}N^+$, $AlO_{12}Si_3^-$	3.15	3
$C_4H_{12}N^+$, Br_3Hg^-	3.20	1
$C_4H_{12}N^+$, Br_3Ni^-	3.21	1
$C_4H_{12}N^+$, $CeCl_6^-$	3.22	1
$C_4H_{12}N^+$, Cl^- , $5Cl_2OSe$	3.23	1
$C_4H_{12}N^+$, ClO_4^-	3.24	1
$C_4H_{12}N^+$, Cl_2I^-	3.25	1
$C_4H_{12}N^+$, Cl_3Mn^-	3.26	1
$C_4H_{12}N^+$, Cl_3Ni^-	3.27	1
$C_4H_{12}N^+$, Cs^+ , $H_{13}B_{11}^{2-}$	3.28	1
$C_4H_{12}N^+$, F^- , $4H_2O$	61.6	2
$C_4H_{12}N^+$, HgI_3^-	3.29	1
$C_4H_{12}N^+$, HCl_2^-	3.16+	3
$C_4H_{12}N^+$, HO^- , $5H_2O$	61.7	2
$C_4H_{12}N^+$, $H_4Cl_4O_2Rh^-$	3.9	5
$C_4H_{12}N^+$, $C_4H_8B_3CrO_4^-$	3.18	3
$C_4H_{12}N^+$, $C_4H_{16}B_{18}Br_6Co^-$	3.31	1
$C_4H_{12}N^+$, $C_{14}H_{10}MnO_3^-$	71.14	5
$C_4H_{12}N^+$, $C_{14}H_{21}B_{10}^-$	62.12	5
$C_4H_{12}N^+$, $C_{15}H_{30}N_5O_2S_6U^-$	80.31	2
$C_4H_{12}N^+$, $C_{18}H_{13}O_4Si^-$	3.32	1
$C_4H_{12}N^+$, $C_{18}H_{19}O_6Pb^-$	69.30	4
$2C_4H_{12}N^+$, $13Ag^+$, $15I^-$	3.19	3

C_5

$C_5H_{11}AgIN$	83.16	3
$C_5H_{11}BO_3$	62.40	2
$C_5H_{11}B_9O_3Re^-$, Cs^+	62.41	2
$C_5H_{11}Cl_2NO_2PdS$	82.1	3
$C_5H_{11}Cl_2NO_2PtS$	82.2	3
$C_5H_{11}Cl_2NO_2PtS$	82.3	3
$C_5H_{11}CuO_4^+$, $C_6N_3O_7^-$	77.1	2
$C_5H_{11}NO_2$	48.45	1
$C_5H_{11}NO_2$	48.16	3
$C_5H_{11}NO_2S$	41.11	1
$C_5H_{11}NO_2S$	48.46+	1
$C_5H_{11}NO_2S$	48.32	5
$C_5H_{11}N_2O_2^+$, ClO_4^-	48.48	1
$C_5H_{11}N_2O_3^+$, Cl^-	48.16	4
$C_5H_{11}N_2O_3^+$, Cl^-	48.17	4
$C_5H_{11}N_2O_3^+$, Cl^- , H_2O	48.49	1
$C_5H_{11}N_3^{2+}$, $2Cl^-$	59.1	1
$C_5H_{11}N_3^{2+}$, Cl_4Co^{2-}	32.8	4
$C_5H_{11}N_3^{2+}$, $2H_2O_4P^-$, H_2O	32.18+	3
$2C_5H_{11}N_3O_2$, $C_{10}H_{18}CoN_6O_4^-$, K^+	79.10	4
$C_5H_{11}O_4P$, H_2O	46.3	3
C_5H_{12}	5.17	1
$C_5H_{12}ClNS_2Sn$	69.11	3
$C_5H_{12}Cl_3NPt$	72.3	5
$C_5H_{12}Cl_3NPt$	72.1+	5
$C_5H_{12}CuN_5S^+$, CNS^-	83.12	4
$C_5H_{12}LiN_2O_4^+$, Br^- , H_2O	67.6	3
$C_5H_{12}N^+$, Cl^-	33.34	1
$C_5H_{12}N^+$, Cl^-	33.20	5
$C_5H_{12}N^+$, ClO_4^-	3.44	1
$C_5H_{12}N^+$, $C_7H_4BrO_2^-$	33.10	4
$C_5H_{12}N^+$, $C_7H_4ClO_2^-$	33.11	4
$C_5H_{12}N^+$, $C_{40}H_{36}EuO_8^-$	77.21	5
$2C_5H_{12}N^+$, $As_4S_6^{2-}$	33.12	4
$2C_5H_{12}N^+$, $BiBr_5^{2-}$	33.35	1
$2C_5H_{12}N^+$, Br_5Sb^{2-}	33.21	5
$C_5H_{12}NO_2^+$, Br^-	48.50	1
$C_5H_{12}NO_2^+$, Cl^-	48.51	1
$C_5H_{12}NO_2^+$, Cl^-	48.52	1
$C_5H_{12}NO_2^+$, Cl^-	48.17	3
$C_5H_{12}NO_2^+$, Cl^-	48.33	5
$C_5H_{12}NO_2^+$, Cl^- , H_2O	48.18	3
$C_5H_{12}NO_2S^+$, Cl^- , H_2O	48.18	4
$C_5H_{12}NO_2S^+$, CH_4N_2O , Cl^-	60.68	2
$C_5H_{12}N_2O_3S$	48.19	3
$C_5H_{12}N_3O_2^+$, Br^-	1.16	5
$C_5H_{12}N_3O_2^+$, Cl^-	8.14	5
$C_5H_{12}O_2$	5.3	3
$C_5H_{12}O_4$	45.11+	1
$C_5H_{12}O_5$	45.14	1
$C_5H_{12}O_5$	45.15	1
$C_5H_{12}O_5$	45.16	1
$C_5H_{12}S_4$	11.31	1
$C_5H_{13}NO_4P^-$, Ca^{2+} , Cl^-	59.1	5
$C_5H_{13}NO_5S$	59.2	1
$C_5H_{13}N_2O_2^+$, Br^-	48.20	3
$C_5H_{13}N_2O_2^+$, Br^-	48.19	4
$C_5H_{13}N_2O_2^+$, Cl^-	48.56	1
$C_5H_{13}N_2O_2^+$, Cl^-	48.21	3
$C_5H_{13}N_2O_3P$	64.6	4
$C_5H_{13}N_4NiS_2^+$, Cl^-	72.8	3
$(C_5H_{14}CuN_3O_2^+)_n$, $nCHO_2^-$	76.20	3
$2C_5H_{14}N^+$, $C_4H_6Cl_6Pt_2^{2-}$	72.3	3
$C_5H_{14}NO^+$, Cl^-	59.3	1
$C_5H_{14}NO^+$, Cl^-	59.1	4
$C_5H_{14}NO^+$, $C_4H_6CrN_6S_4^-$	59.4	1

$C_5H_{14}NO_5P$, H_2O	46.3	5
$C_5H_{14}N_3O_3P$, $0.5CH_6N_3^+$, $0.5Cl^-$	8.9	4
$C_5H_{14}N_8^{2+}$, $2Cl^-$, H_2O	8.43+	1
$C_5H_{15}As_5$	65.9	2
$C_5H_{16}ClN_5NiS$	76.27	2
$C_5H_{16}CoN_5O_3S_2$, $2H_2O$	76.28	2
$C_5H_{16}N_5NiS^+$, I^-	76.21	3
$C_5H_{16}N_6NiO_2S$	76.29	2
$(C_5H_{20}Cu_2N_{10}S_5^{2+})_n$, nO_4S^{2-} , $2nH_2O$	79.10	3
$C_5H_{21}B_{18}CoS_2$	71.2	4
$C_5H_{25}ClCoN_5^{2+}$, $2NO_3^-$	83.13	4

C_6

C_6BrCl_5	19.1	1
$C_6Br_2Cl_4$	19.2	1
$C_6Br_4O_2$	18.1	1
C_6Br_5Cl	19.3	1
C_6Br_6 , $C_6H_2Br_4$	60.69	2
$C_6Cl_2O_4^{2-}$, $2H_4N^+$, H_2O	6.5	1
$C_6Cl_4O_2$	18.2+	1
$C_6Cl_4O_2$	18.1	4
$C_6Cl_4O_2$, $2C_9H_7NO$	60.107	2
$C_6Cl_4O_2$, $C_{10}H_{16}N_2$	60.122	2
$C_6Cl_4O_2$, $C_{12}H_{10}$	60.19	5
$C_6Cl_4O_2$, $C_{12}H_{18}$	60.71	2
$C_6Cl_4O_2$, $C_{15}H_{12}$	60.27	5
$C_6Cl_4O_2$, $2C_{16}H_{20}N_2$	60.33	4
$C_6Cl_4O_2$, $C_{18}H_{12}N_2O_2Pd$	60.70	2
C_6Cl_6	19.4	1
$C_6CoO_{12}^{3-}$, $C_{36}H_{24}N_6Ni^{2+}$, K^+ , $2H_2O$		
	83.142	3
$C_6CrO_{12}^{3-}$, $3H_4N^+$, $2H_2O$	81.37	2
$C_6F_4O_2$, $C_{20}H_{12}$	60.152	2
C_6F_5NO , $2C_{12}F_{10}N_2O_2$	10.1	5
C_6F_6 , C_9H_{12}	60.21	4
C_6F_6 , $C_{12}H_{18}$	60.24+	5
C_6I_6	19.1+	3
$C_6MoN_6S_6^{3-}$, $3K^+$, $C_2H_4O_2$, H_2O	83.38	2
$C_6N_2O_8^{2-}$, $2H_4N^+$	6.6	1
$C_6N_2O_8^{2-}$, $2H_5O_2^+$, $2H_2O$	6.1	3
$C_6N_3O_7^-$, $C_5H_{11}CuO_4^+$	77.1	2
C_6N_4	7.21	1
C_6N_4	7.5	4
C_6N_4 , $C_{10}H_8$	60.72	2
C_6N_4 , $C_{10}H_{10}Fe$	60.119	2
C_6N_4 , $C_{16}H_{10}$	60.142	2
C_6N_4 , $C_{16}H_{10}$	60.44	3
C_6N_4 , $C_{18}H_{20}$	60.37	4
C_6N_4 , $2C_{18}H_{20}$	60.35	5
C_6N_4 , $8C_{20}H_{10}N_4$, $2CH_2Cl_2$	60.52	3
C_6N_4 , $C_{20}H_{12}$	60.53	3
C_6N_4O	38.2+	4
$C_6N_4O_4$	21.1	3
$C_6N_6O_3$, $C_6H_{15}OP$	60.73	2
$C_6N_6O_3$, $C_6H_{15}O_4P$	60.15+	4
$C_6N_6O_6$	40.2+	1
$C_6N_6O_6$, $C_{12}H_{12}S_2$	60.74	2
$C_6N_6O_{12}$	15.1	1
$C_6N_9^{3-}$, $3Na^+$, $3H_2O$	33.36	1
$C_6NbO_{13}^{3-}$, $3H_4N^+$, H_2O	81.19	4
$C_6O_{12}Rh^{3-}$, $3K^+$, $4.5H_2O$	81.20	4
$C_6O_{12}Sb^{3-}$, $3K^+$, $4H_2O$	66.2	3

$C_6H_{24}Cu_2N_{12}S_6^{2+}$, $2ClO_4^-$	79.7	4
$C_6H_{24}Cu_4N_{12}S_6^{4+}$, $4NO_3^-$, $4H_2O$	79.15	5
$C_6H_{24}N_4NiO_2^{2+}$, $2Br^-$	83.10	5
$C_6H_{24}N_4NiO_2^{2+}$, $2ClO_4^-$	76.43	3
$C_6H_{24}N_4NiO_2^{2+}$, $2NO_3^-$	83.51	2
$C_6H_{24}N_4NiO_2^{2+}$, $C_9H_{30}N_6Ni^{2+}$, $4Cl^-$, H_2O	83.33	4
$C_6H_{24}N_6Ni^{2+}$, K^+, $3CNSe^-$	76.41	2
$C_6H_{24}N_6Ni^{2+}$, $2NO_3^-$	76.42	2
$C_6H_{24}N_6Ni^{2+}$, O_4S^{2-}	76.44	3
$3C_6H_{24}N_6Ni^{2+}$, $O_{15}Si_6^{6-}$, $26H_2O$	76.43	2
$C_6H_{24}N_6Ru^{3+}$, $3Cl^-$, $3.5H_2O$	76.45	3
$C_6H_{24}N_6Zn^{2+}$, $O_3S_2^{2-}$	76.19	5
$C_6H_{24}N_{12}NiS_6^{2+}$, $2Br^-$	79.28	2
$C_6H_{24}N_{12}O_6Ti^{3+}$, $3ClO_4^-$	79.8	4
$C_6H_{24}N_{12}O_6Ti^{3+}$, $3ClO_4^-$	79.16	5
$C_6H_{24}N_{12}O_6Ti^{3+}$, $3I^-$	79.29	2
$C_6H_{24}N_{12}O_6Ti^{3+}$, $3I^-$	79.19	3
$C_6H_{24}N_{12}O_6V^{3+}$, $3I^-$	79.17+	5
$C_6H_{24}N_{12}O_6Zn^{2+}$, $2NO_3^-$	84.4	4
$C_6H_{24}N_{12}PbS_6^{2+}$, $2ClO_4^-$	69.13	4
$C_6H_{24}N_{12}STe_2^{4+}$, $4HF_2^-$	70.13	2
$C_6H_{26}N_{12}O_{12}U^{2+}$, I_4^{2-}	79.30	2
$C_6H_{32}B_{26}Co_2^{2-}$, $2Cs^+$, H_2O	62.12	3

C_7

$C_7HF_2MnO_5$	71.10	2
$C_7HF_5O_2$	13.1	5
$C_7HF_6O_4Rh$	77.2	2
$C_7H_2Br_3N$	7.9	4
$C_7H_2Cl_3N$	7.23	1
$C_7H_2Cl_3N$	7.10	4
$C_7H_2N_4$	20.7	5
$C_7H_2N_5O_{10}^-$, Rb^+	12.7	4
$C_7H_3Br_2N_3$	9.19	1
$C_7H_3Br_5$	19.5	4
$C_7H_3Cl_5$	19.1	5
$C_7H_3F_2NO_6Ti^{2-}$, $2K^+$, $2H_2O$	81.23	5
$C_7H_3FeNO_4$	72.13	2
$C_7H_3NO_4^{2-}$, Ca^{2+}, $3H_2O$	33.51	1
$C_7H_4BrN_3$	9.4	4
$C_7H_4ClMnO_6$	71.3	5
C_7H_4ClNOS	41.15	1
$C_7H_4ClNO_4$	13.1	1
C_7H_4ClNS	41.6	4
$C_7H_4ClN_2O_4^-$, K^+	12.8	4
$C_7H_4ClN_3$	9.20	1
$C_7H_4ClO_2^-$, $C_7H_5ClO_2$, K^+	14.1	5
$C_7H_4CrO_3S$	75.1	2
$C_7H_4CuN_3$	83.14	4
C_7H_4IN	7.24	1
$C_7H_4NO_4^-$, $C_7H_5NO_4$, K^+	14.3	1
$C_7H_4NO_4^-$, $C_7H_5NO_4$, Rb^+	14.2	1
$C_7H_4N_2O_6$, $C_{12}H_9NS$	60.130	2
$C_7H_4O_5^{2-}$, Ca^{2+}, $3H_2O$	38.23	1
$C_7H_5AlCl_4O$	68.14	2
$C_7H_5BrO_2$	13.2	1
$C_7H_5BrO_2$	13.3	1
$C_7H_5BrO_2$	13.2	5
$C_7H_5Br_2NO_2$	13.4+	1
$C_7H_5Br_3FeO_2Sn$	73.3	3
$C_7H_5ClCoHgO_2^+$, Cl^-, $2Cl_2Hg$	73.1	5
C_7H_5ClO	22.1	4
$C_7H_5ClO_2$	13.6	1
$C_7H_5ClO_2$	13.7	1
$C_7H_5ClO_2$, $C_7H_4ClO_2^-$, K^+	14.1	5
$C_7H_5ClO_2^-$, $C_2H_6N_3S^+$	8.32	1
$C_7H_5ClO_3$, $C_7H_8N_4O_2$	60.24	3
$C_7H_5ClO_3$, $C_8H_{10}N_4O_2$	60.104	2
$2C_7H_5ClO_3$, $C_7H_8N_5O_2$	60.17	4
$C_7H_5Cl_2CoHgO_2$	73.2	5
$C_7H_5Cl_2NO_2$	13.8	1
$C_7H_5Cl_3FeO_2Sn$	73.4	3
$C_7H_5Cl_4O_2Sb$	66.4	3
$C_7H_5Cl_6OSb$	66.2	5
$C_7H_5FO_2$	13.9	1
$C_7H_5IO_3$	42.8	1
C_7H_5NOS	41.2	3
C_7H_5NOS	40.7	5
$C_7H_5NOS_2$	39.41	1
$C_7H_5NO_2$	40.8	5
$C_7H_5NO_3$	15.20+	1
$C_7H_5NO_3S$	41.16+	1
$C_7H_5NO_4$	13.10	1
$C_7H_5NO_4$	13.11+	1
$C_7H_5NO_4$	13.1	4
$C_7H_5NO_4$	33.27	5
$C_7H_5NO_4$	33.28	5
$C_7H_5NO_4$	33.29	5
$C_7H_5NO_4$	33.30	5
$C_7H_5NO_4$, H_2O	33.30	5
$C_7H_5NO_4$, $C_7H_4NO_4^-$, K^+	14.3	1
$C_7H_5NO_4$, $C_7H_4NO_4^-$, Rb^+	14.2	1
$C_7H_5NO_5$	13.13	1
$C_7H_5NO_5$	13.1	3
$C_7H_5NO_8V^-$, H_4N^+, $1.3H_2O$	81.24	5
$C_7H_5NS_2$	41.18	1
$C_7H_5NS_2$	41.7	4
$C_7H_5N_2O_4^-$, K^+	12.5	3
$C_7H_5N_3O$	22.2	4
$C_7H_5N_5O_8$	16.24	1
$C_7H_5O_2^-$, Na^+	22.2	1
$C_7H_5O_2^-$, $4CH_4N_2S$, Tl^+	60.98	2
$C_7H_5O_2^-$, $C_7H_6O_2$, K^+	14.4	1
$C_7H_5O_2^-$, $C_{16}H_{36}N^+$, $39.5H_2O$	61.14	2
$C_7H_5O_3$, $C_{10}H_{15}N_2$	60.22	4
$C_7H_5O_3^-$, $C_7H_6O_3$, K^+, H_2O	14.6	1
$C_7H_5O_3^-$, $C_7H_6O_3$, H_4N^+, H_2O	14.5	1
$C_7H_5O_5MnS^-$, H_4N^+	13.15	1
$C_7H_5O_6S^-$, $H_7O_3^+$	13.2	3
$C_7H_5O_6S^-$, $H_7O_3^+$	13.2	4
$C_7H_6BF_3FeO_5S$	72.9	3
$C_7H_6BrMnN_2O_3$	71.4	5
C_7H_6BrNO	13.16	1
C_7H_6BrNO	13.17	1
$C_7H_6BrN_3O$	9.5+	4
$C_7H_6BrN_4O$, $C_7H_8BrN_5$	44.26	3
C_7H_6ClNO	19.21	1
C_7H_6ClNO	19.22	1
C_7H_6ClNO	10.3	3
C_7H_6ClNO	13.3+	5
$C_7H_6ClNO_2$	18.17	1
$C_7H_6ClNO_2$	19.23	1
C_7H_6FNO	13.18	1
C_7H_6FNO	13.19	1
$C_7H_6FeO_3$	72.14	2
C_7H_6INO	13.5	5
$C_7H_6NO_2^-$, K^+, $3H_2O$	14.7	1
$C_7H_6NO_2^-$, Na^+, $5H_2O$	14.8	1

C₈

Formula	Page	
$(C_8H_{16}N_8NiPd)_n$	76.49	3
$2C_8H_{16}NaO_2^+$, $C_{24}H_{28}Al_2^{2-}$	68.21	3
$2C_8H_{16}NaO_2^+$, $C_{32}H_{32}Al_2^{2-}$	68.31	5
$C_8H_{16}Ni_2S_6$	85.49+	2
$C_8H_{16}O_2$	20.21	1
$C_8H_{16}O_4$	38.30	1
$C_8H_{16}O_5S$	45.52	1
$C_8H_{16}O_6S$	45.13	4
$C_8H_{16}O_{10}Rh_2$	81.28	4
$C_8H_{16}P_2S_2$	64.24	2
$C_8H_{17}CoN_5O_4$	83.26	3
$C_8H_{17}GeNO_3$	69.17	3
$C_8H_{17}NO_2Si$	63.2	3
$C_8H_{17}N_2O_3^+$, Br^-	48.34	3
$C_8H_{17}O_2P$, H_2O	64.25	2
$C_8H_{17}O_3P$	64.17	3
C_8H_{18}	5.22	1
$C_8H_{18}BF_4N_4NiO^+$, BF_4^-	76.23	5
$C_8H_{18}ClN_6NiS_2^+$, Cl^-, H_2O	79.20	5
$C_8H_{18}Cl_2Cu_2N_4O_8$	82.11	5
$C_8H_{18}Cl_2N_2O_2Zn$	81.34	5
$C_8H_{18}Cl_2N_2P_2$	64.10	4
$C_8H_{18}Cl_2PdSe_2$	85.10	3
$C_8H_{18}Cr_2N_4O_{10}$	82.12	5
$C_8H_{18}CuO_8$	81.66	2
$C_8H_{18}CuO_8$	81.22	3
$C_8H_{18}HgS_2$	85.51	2
$C_8H_{18}NO^+$, Br^-	58.3	1
$C_8H_{18}NO^+$, $CH_3O_3S^-$	32.33	3
$C_8H_{18}NOS^+$, I^-	3.34	3
$C_8H_{18}NO_2^+$, Br^-	3.17	5
$C_8H_{18}NO_2^+$, I^-	59.13	1
$C_8H_{18}NO_2^+$, I^-	59.14	1
$C_8H_{18}NO_2^+$, I^-	38.3	3
$C_8H_{18}N_2^+$, $C_4H_{22}B_{18}Ni^{2-}$	37.5	3
$C_8H_{18}N_2O_{11}Zn$	82.32	2
$C_8H_{18}N_2S_2Sn$	69.12	5
$C_8H_{18}N_4O_4S_6Te$	70.4+	5
$C_8H_{18}N_6NiS_2$	76.47	2
$C_8H_{18}N_6NiS_2$	76.50	3
$C_8H_{18}N_6NiS_2$	76.51	3
$C_8H_{18}N_6Pt^{2+}$, $2Cl^-$, $2H_2O$	83.29	4
$C_8H_{18}N_7NiO_6S_2^+$, NO_3^-, H_2O	79.21	5
$C_8H_{18}O_2$, $4H_2O$	5.14	5
$C_8H_{18}O_5Sn_2$	69.20	2
$C_8H_{19}AsO_2$	65.3	3
$C_8H_{19}B_2^-$, $C_{16}H_{36}N^+$	61.4	5
$C_8H_{19}N_2O_2P$	64.11	4
$C_8H_{20}AlLi$	68.15	2
$C_8H_{20}Au_2Br_2$	71.16	2
$C_8H_{20}B_2F_4N_2$	3.35	3
$(C_8H_{20}BrCuP_2)_n$	86.3	3
$C_8H_{20}Br_2MgO_2$	67.3	2
$C_8H_{20}Br_2O_4Ti^+$, $C_4H_{10}Br_4O_2Ti^-$	84.27	2
$C_8H_{20}Br_4Pt_2S_2$	85.52	2
$C_8H_{20}ClCuN_6S_4$	85.53	2
$C_8H_{20}Cl_2N_4NiO_4$	83.30	4
$C_8H_{20}Cl_2N_4NiS_4$	85.8	5
$C_8H_{20}Cl_2PdSe_2$	85.11	3
$C_8H_{20}Cl_4O_4Ti_2$	84.28	2
$C_8H_{20}Cl_5Re_2S_4$	85.54	2
$C_8H_{20}CoN_6O_4^+$, NO_3^-	83.61	2
$2C_8H_{20}CoN_6S_2^+$, $C_8H_4O_{12}Sb_2^{2-}$, $4H_2O$	76.52	3
$C_8H_{20}CoS_4^{2+}$, $2ClO_4^-$	85.55	2
$C_8H_{20}CuN_2O_6$	83.62	2
$C_8H_{20}CuN_3O_2^+$, ClO_4^-	81.35	5
$C_8H_{20}CuN_4^{2+}$, $2NO_3^-$	76.53	3
$C_8H_{20}CuN_6S_2$	76.24	5
$(C_8H_{20}HgO_4PS_4)_n$	85.9	5
$C_8H_{20}I_2N_4Rh^+$, I^-	83.28	3
$C_8H_{20}Li_4$	67.4	2
$C_8H_{20}N^+$, Br_6Sb^-	3.36	3
$C_8H_{20}N^+$, Cl_4In^-	3.60	1
$C_8H_{20}N^+$, I_3^-	3.61+	1
$C_8H_{20}N^+$, $H_2Br_4O_2Re^-$	3.63	1
$C_8H_{20}N^+$, $C_4Br_3O_4W^-$	3.12	4
$C_8H_{20}N^+$, $C_4HCr_2O_4^-$	3.64	1
$C_8H_{20}N^+$, $C_4H_{18}B_{14}Co^-$	62.9	4
$C_8H_{20}N^+$, $C_8Br_2Co_2InO_8^-$	3.13	4
$C_8H_{20}N^+$, $C_8B_{40}B_{40}Co^-$	62.6	5
$C_8H_{20}N^+$, $C_9H_{15}O_3PbS_6^-$	69.15	4
$C_8H_{20}N^+$, $C_9H_{26}B_{17}CoN^-$	62.8	5
$C_8H_{20}N^+$, $C_{10}HCr_2O_{10}^-$	3.37	3
$C_8H_{20}N^+$, $C_{16}H_{16}N_2NiS_2^-$	85.27	3
$C_8H_{20}N^+$, $C_{20}H_{40}EuN_4S_8^-$	80.12	3
$C_8H_{20}N^+$, $C_{20}H_{40}N_4NpS_8^-$	80.13+	3
$C_8H_{20}N^+$, $C_{28}H_{20}O_7V^{2-}$, Na^+, $2C_3H_6O$	81.52	3
$2C_8H_{20}N^+$, $Br_6Pt_2^{2-}$	3.65	1
$2C_8H_{20}N^+$, ClO_4Ti^{2-}	3.66	1
$2C_8H_{20}N^+$, Cl_4Ni^{2-}	3.67	1
$2C_8H_{20}N^+$, Cl_5In^{2-}	3.68	1
$2C_8H_{20}N^+$, Cl_5OPa^{2-}	3.14	4
$2C_8H_{20}N^+$, $C_4H_{22}B_{18}Cu^{2-}$	62.34	2
$2C_8H_{20}N^+$, $C_8FeN_5OS_4^{2-}$	85.42	2
$2C_8H_{20}N^+$, $C_{14}Mo_2Ni_4O_{14}^{2-}$	3.15	4
$2C_8H_{20}N^+$, $C_{16}Ni_3O_{16}W_2^{2-}$	3.16	4
$2C_8H_{20}N^+$, $C_{28}H_{28}Fe_4S_8^{2-}$	85.31	5
$3C_8H_{20}N^+$, $C_8H_{42}B_{34}Co_3^{3-}$	62.15	3
$3C_8H_{20}N^+$, $C_{12}InN_6S_6^{3-}$	68.7	4
$4C_8H_{20}N^+$, $C_8N_8S_8U^{4-}$	3.17	4
$C_8H_{20}NO_4P$	46.13	1
$C_8H_{20}NO_4P$, $CdCl_2$, $3H_2O$	46.14	1
$C_8H_{20}N_2NiO_6$, $2H_2O$	82.33	2
$C_8H_{20}N_2NiS_2$	85.56	2
$C_8H_{20}N_4NiS_4^{2+}$, $2Br^-$	85.10	4
$C_8H_{20}NiO_4P_2S_4$	85.57+	2
$C_8H_{20}NiP_2S_4$	85.60	2
$C_8H_{20}Ni_2S_8$	85.12	3
$C_8H_{20}O_4P_2PbS_4$	69.14	4
$C_8H_{20}O_4P_2S_4Zn$	85.61	2
$C_8H_{20}P_2S_2$	64.26	2
$C_8H_{20}P_2S_4Se_2$	64.27	2
$C_8H_{20}P_2S_4Se_2Te$	64.18	3
$C_8H_{20}P_2Se_5$	64.19	3
$C_8H_{22}B_{18}Fe_2O_4^{2-}$, $2Cs^+$, C_3H_6O, H_2O	62.13	3
$C_8H_{22}Be_2^{2-}$, $2C_4H_{10}NaO^+$	67.5	2
$C_8H_{22}Cl_2CoN_4^+$, NO_3^-	76.25	5
$C_8H_{22}CoN_5O_2^{2+}$, $2Cl^-$, H_2O	76.17	4
$2C_8H_{22}CoN_5O_2^{2+}$, $4I^-$, H_2O	76.18	4
$C_8H_{22}CoN_6O_4^+$, Br^-	76.19	4
$C_8H_{22}CoN_6O_4^+$, Br^-	83.31	4
$C_8H_{22}CoN_6O_4^+$, ClO_4^-	76.54	3
$C_8H_{22}CoN_6O_4^+$, ClO_4^-	76.20	4
$C_8H_{22}CoN_6O_4^+$, ClO_4^-	76.21	4
$C_8H_{22}CuN_{10}O_4^{2+}$, $2Cl^-$	83.29	3
$C_8H_{22}N_{10}NiO_4^{2+}$, $2Cl^-$, $2H_2O$	83.30	3
$C_8H_{23}ClCoN_5^{2+}$, Cl^-, ClO_4^-	76.55	3
$C_8H_{23}ClCoN_5^{2+}$, $2ClO_4^-$	76.49	3
$C_8H_{23}ClCoN_5^{2+}$, $2ClO_4^-$	76.26	5

C_9

C_{12}

$C_{12}H_{36}Be_3N_6$	67.13	3
$C_{12}H_{36}Cl_4Pt_4$	71.29	2
$C_{12}H_{36}Co_3N_6O_6^{2+}$, $2C_2H_3O_2^-$	83.65	3
$C_{12}H_{36}CrN_4OSi_3$	83.61	4
$C_{12}H_{36}Cs_4O_4Si_4$	63.4	3
$C_{12}H_{36}CuN_4O_2^{2+}$, $2NO_3^-$	76.78	3
$C_{12}H_{36}K_4O_4Si_4$	63.5	3
$C_{12}H_{36}N_6W$	83.123	2
$2C_{12}H_{36}N_9P_3^+$, $Mo_6O_{19}^{2-}$	64.17	5
$C_{12}H_{36}O_4Rb_4Si_4$	63.6	3
$C_{12}H_{36}O_6Si_8$	63.18	2
$C_{12}H_{40}O_4Pt$	71.30+	2
$C_{12}H_{41}B_9P_2PtS$	86.10	3
$C_{12}H_{42}N_6Pt_2^{2+}$, $2I^-$	76.80	2
$C_{12}H_{54}Co_4N_{12}O_6^{6+}$, $3O_6S_2^{2-}$, $8H_2O$	76.39	4
$C_{12}H_{54}Cr_4N_{12}O_6^{6+}$, $6N_3^-$, $4H_2O$	76.81	2

C_{13}

$C_{13}Cl_{11}$	12.11	4
$C_{13}H_3Co_2F_9O_4$	71.24	5
$C_{13}H_3N_6^-$, H_4N^+	7.6	3
$C_{13}H_4BrN_3O_7$, $C_{14}H_{10}$	60.42	3
$C_{13}H_4BrN_3O_7$, $C_{20}H_{16}$	60.55	3
$C_{13}H_4BrN_3O_7$, $C_{26}H_{16}$	60.156	2
$C_{13}H_5Co_2O_4Pt$	83.124	2
$C_{13}H_5MnMoO_8$	73.41	2
$C_{13}H_5MoN_4S_4^-$, $C_{24}H_{20}P^+$	64.52	3
$C_{13}H_5N_3O_7$	28.2	5
$C_{13}H_6Cl_2N_2O_2$	31.11	4
$C_{13}H_7BrN_2$	9.7	3
$C_{13}H_7BrO$	28.2	3
$C_{13}H_8BrN_3OS$	41.23	5
$C_{13}H_8Br_2O$	19.15	5
$C_{13}H_8Br_4GeN_2O_3W$	83.66	3
$C_{13}H_8Cl_4HgN_2$	71.25	5
$C_{13}H_8Co_2O_6$	75.14	5
$C_{13}H_8CrO_3$	74.13	2
$C_{13}H_8CrO_5S$	71.14	4
$C_{13}H_8CrO_6$	71.32	2
$C_{13}H_8FeO_5$	75.10	3
$C_{13}H_8Fe_2O_5$	75.26	2
$C_{13}H_8Fe_2O_6$	75.6	4
$C_{13}H_8N_2$	28.15	1
$C_{13}H_8N_2O_3$	36.21	1
$C_{13}H_8N_4O_4$	15.9	5
$C_{13}H_8O$	28.3	4
$C_{13}H_8O_2$	28.3	5
$C_{13}H_9BrN_2O_4$	35.13	3
$C_{13}H_9Br_3O$	20.29	1
$C_{13}H_9ClO_2$	13.49	1
$C_{13}H_9ClO_2$	13.50	1
$C_{13}H_9Cl_2N$	16.12	4
$C_{13}H_9Cl_2N$	16.13	4
$C_{13}H_9CrNO_3$	74.14	2
$C_{13}H_9FeO_4$	72.21	3
$C_{13}H_9Fe_3NO_{10}Si$	63.5	4
$C_{13}H_9IO_2$	13.14	5
$C_{13}H_9IO_2$	13.15	5
$C_{13}H_9N$	36.22	1
$C_{13}H_9N$	36.5	3
$C_{13}H_9N$	36.12	5
$C_{13}H_9N$, $C_4H_5N_3O$, H_2O	60.132	2

$C_{13}H_9NO$	36.6	3
$C_{13}H_9NOS$	41.13	3
$C_{13}H_9NO_2$	33.69	1
$C_{13}H_9N_3O$	35.19	5
$C_{13}H_9N_3S$	9.38	1
$C_{13}H_{10}$	28.16	1
$C_{13}H_{10}BrN$	33.42	3
$C_{13}H_{10}BrN$	16.14	4
$C_{13}H_{10}BrNO$	16.48	1
$C_{13}H_{10}BrNO$	16.49	1
$C_{13}H_{10}Br_2O_4S_2$	11.10	5
$C_{13}H_{10}Br_2S_2$	11.21	3
$C_{13}H_{10}ClNO$	16.50	1
$C_{13}H_{10}ClNO$	16.51	1
$C_{13}H_{10}ClNO$	16.52+	1
$C_{13}H_{10}Cl_2FeO_2Sn$	69.22	3
$C_{13}H_{10}Cl_2HgN_4S$	85.13	4
$C_{13}H_{10}Cl_2O_2$	17.38	1
$C_{13}H_{10}FeN_2$	73.10	3
$C_{13}H_{10}FeO_3$	72.41	2
$C_{13}H_{10}N_2O_4$	50.6	1
$C_{13}H_{10}N_2O_4$	35.14+	3
$C_{13}H_{10}N_2O_4$	35.15	4
$C_{13}H_{10}N_2O_4$	35.16	4
$C_{13}H_{10}N_4S$	41.35	1
$C_{13}H_{10}N_4S$	32.32	3
$C_{13}H_{10}O$	19.42+	1
$C_{13}H_{10}O$, $C_{12}H_{11}N$	60.26	4
$C_{13}H_{10}O_2S$	39.74	1
$C_{13}H_{10}O_2S$, H_2O	17.39	1
$C_{13}H_{10}O_3Rh_2$	73.42	2
$C_{13}H_{11}Cl_2N_2O^+$, I^- , H_2O	33.43	3
$C_{13}H_{11}Cl_2N_3O_3$	35.20	5
$C_{13}H_{11}FeN$	73.10	4
$C_{13}H_{11}I_2IrN_2O_3$	71.33	2
$C_{13}H_{11}N$	16.9	3
$C_{13}H_{11}NO_2$	37.8	4
$C_{13}H_{11}NO_2$	34.4	5
$C_{13}H_{11}NO_2S_2$	13.51	1
$C_{13}H_{11}NO_3$	10.29	1
$C_{13}H_{11}NS$	39.34	3
$C_{13}H_{11}NS$	41.19	4
$C_{13}H_{11}NS$, $C_{12}H_4N_4$	60.24	4
$C_{13}H_{11}N_2^+$, $C_{12}H_4N_4^-$	36.23	1
$C_{13}H_{11}O_2P$	64.36	2
$C_{13}H_{11}PS_4$	64.18	5
$C_{13}H_{12}BrNO_2$	33.70	1
$C_{13}H_{12}BrNO_2S$	16.54	1
$C_{13}H_{12}Br_2N_2SZn$	83.62	4
$C_{13}H_{12}ClNO_2$	1.33	5
$C_{13}H_{12}ClNO_3S$	39.16	4
$C_{13}H_{12}ClN_3O_2S$	41.20	4
$C_{13}H_{12}Cl_2CuN_2S$	85.14	4
$C_{13}H_{12}CrO_3$	75.11	3
$C_{13}H_{12}Cu_3S_6$	85.69	2
$C_{13}H_{12}FeO$	73.43	2
$C_{13}H_{12}FeO_2$	75.12	3
$C_{13}H_{12}FeO_3$	75.13	3
$C_{13}H_{12}FeO_3$	75.15	5
$C_{13}H_{12}FeO_3$	75.16	5
$C_{13}H_{12}MnNO_4$	71.26	5
$C_{13}H_{12}N_2O_4$	32.19	5
$C_{13}H_{12}N_2O_4S$	9.8	3
$C_{13}H_{12}N_3O_2^+$, Cl^- , $2H_2O$	36.13	5
$C_{13}H_{12}N_4O_2$, $2C_{10}H_8O_2$	60.27	4
$C_{13}H_{12}N_4O_2$, $2C_{10}H_8O_2$	60.28	4

Formula		
$C_{14}H_{10}PdS_4$	85.72	2
$C_{14}H_{10}S_4Zn$	85.21	5
$C_{14}H_{10}S_4Zn$	85.19	4
$C_{14}H_{11}BrO$	19.46	1
$C_{14}H_{11}ClO$	19.47	1
$C_{14}H_{11}Cl_3MoN_2O_3Sn$	69.23	3
$C_{14}H_{11}FO$	19.48	1
$C_{14}H_{11}NO_2$	16.10	3
$C_{14}H_{12}$	19.49	1
$C_{14}H_{12}$	29.1	1
$C_{14}H_{12}$	22.9	4
$C_{14}H_{12}$, $4Cl_3Sb$	60.136	2
$C_{14}H_{12}Ag_4O^{4+}$, $4ClO_4^-$	74.6	3
$C_{14}H_{12}As_2Co_2F_4O_6$	86.11	3
$C_{14}H_{12}As_2F_4Fe_2O_6$	75.33	2
$C_{14}H_{12}BrClO_2$	27.8	3
$C_{14}H_{12}BrNO_3$	38.17	4
$C_{14}H_{12}Br_2OS$	11.9	4
$C_{14}H_{12}ClN_3O$	35.17	3
$2C_{14}H_{12}Cl_2N_2Ni$, $2CHCl_3$	83.128	2
$C_{14}H_{12}Cl_2N_2Zn$	83.68	3
$C_{14}H_{12}Cl_2O_2$	17.40	1
$C_{14}H_{12}CoS_4^-$, $C_{19}H_{18}As^+$, $0.5C_2H_6O$	85.73	2
$C_{14}H_{12}CrO_3S$	75.17	5
$C_{14}H_{12}CuN_2O_2$	78.7	2
$C_{14}H_{12}CuN_2O_4$	78.8	2
$C_{14}H_{12}CuN_2O_4$, $2H_2O$	81.46	5
$C_{14}H_{12}N^+$, I^-	36.9	3
$C_{14}H_{12}N_2$	9.9	3
$C_{14}H_{12}N_2NiO_2$	78.9	2
$C_{14}H_{12}N_2NiO_4$	78.10	2
$C_{14}H_{12}N_2O$	40.28	5
$C_{14}H_{12}N_2OSe$	8.49	1
$C_{14}H_{12}N_2O_2$	36.26	1
$C_{14}H_{12}N_2O_2$	9.10	3
$C_{14}H_{12}N_2O_2$	16.11	3
$C_{14}H_{12}N_2O_3$	16.20	5
$C_{14}H_{12}N_2O_4Pd$	78.1	3
$C_{14}H_{12}N_2O_5U$	83.65	4
$C_{14}H_{12}N_2S_4$	11.10	4
$C_{14}H_{12}N_4O_4Pd$	83.66	4
$C_{14}H_{12}N_4S_3$	39.35	3
$C_{14}H_{12}OS$	39.75	1
$C_{14}H_{12}O_2$	31.12	4
$C_{14}H_{12}O_3$	59.6	3
$C_{14}H_{12}O_4$	38.50	1
$C_{14}H_{12}O_5$	59.9	5
$C_{14}H_{12}O_7$	25.20	1
$C_{14}H_{12}S_2$	39.20	4
$C_{14}H_{12}S_2$	39.25	5
$C_{14}H_{13}BrN_4O_2$, H_2O	36.18	5
$C_{14}H_{13}BrN_4O_3$, H_2O	43.31	1
$C_{14}H_{13}BrO_2$	52.23	1
$C_{14}H_{13}BrO_2$	31.16	5
$C_{14}H_{13}ClO$	31.16	1
$C_{14}H_{13}Cl_2NO_3$	50.5	3
$C_{14}H_{13}HgNO_3$	71.22	4
$C_{14}H_{13}MnO_2$	75.34	2
$C_{14}H_{13}MoO_2$	71.29	5
$C_{14}H_{13}NO$	1.141	1
$C_{14}H_{13}N_4O_2^+$, I^-, H_2O	36.11	3
$C_{14}H_{14}$	19.50	1
$C_{14}H_{14}$	19.51+	1
$C_{14}H_{14}$	23.1	5
$C_{14}H_{14}$, $2Cl_3Sb$	60.137	2
$C_{14}H_{14}$, $4Cl_3Sb$	60.138	2
$C_{14}H_{14}Ag^+$, ClO_4^-	74.7	3
$C_{14}H_{14}As_2$	65.24	2
$C_{14}H_{14}As_2Br_2$	65.25	2
$C_{14}H_{14}As_2I_2$	65.26	2
$C_{14}H_{14}Au_2Cl_2S_2$	85.22	5
$C_{14}H_{14}BrNO$	37.6	5
$C_{14}H_{14}BrNO_2S$	16.55	1
$C_{14}H_{14}BrNO_2S$	16.56	1
$C_{14}H_{14}Br_2O_4$, $5C_4H_8O_2$	31.16	3
$C_{14}H_{14}Br_2Se$	11.63	1
$C_{14}H_{14}ClN_2O_4Tl$	68.22	2
$C_{14}H_{14}Cl_2CoN_2$	83.130+	2
$C_{14}H_{14}Cl_2CoN_6O_2$	83.67	4
$C_{14}H_{14}Cl_2CuN_2$	83.132	2
$C_{14}H_{14}Cl_2CuN_2O$	83.69	3
$C_{14}H_{14}Cl_2HgN_2O_3$, $0.5H_2O$	84.37	2
$C_{14}H_{14}Cl_2Se$	11.64	1
$C_{14}H_{14}Cl_4$	20.30	1
$C_{14}H_{14}CrO_3$	75.9	4
$C_{14}H_{14}CuO_8$, $2H_2O$	81.78	2
$C_{14}H_{14}FeO$	73.55	2
$C_{14}H_{14}FeO_2$	73.14	3
$C_{14}H_{14}FeO_3$	75.18	5
$C_{14}H_{14}GdN_7O_9$	83.70	3
$C_{14}H_{14}Hg$	71.15	3
$C_{14}H_{14}NO_3$	17.41	1
$C_{14}H_{14}N_2$	9.40+	1
$C_{14}H_{14}N_2$, $C_{12}H_4N_4$	60.17+	5
$C_{14}H_{14}N_2O_3$	9.11+	3
$C_{14}H_{14}N_2S$	16.12	3
$C_{14}H_{14}N_2Tl^+$, ClO_4^-	68.17	5
$C_{14}H_{14}N_3O_3S^-$, Na^+, H_2O, C_2H_6O	9.13	5
$C_{14}H_{14}N_4O_2$	36.13	4
$C_{14}H_{14}N_4S$	8.17	5
$C_{14}H_{14}N_6Ni$	83.71	3
$C_{14}H_{14}N_6NiO_2$	83.45	5
$C_{14}H_{14}NiO_6$	78.11	2
$C_{14}H_{14}O_2$	17.42	1
$C_{14}H_{14}O_2$	24.5	5
$C_{14}H_{14}O_2PS_2^-$, K^+	64.19	5
$C_{14}H_{14}O_2Ru$	73.57	2
$C_{14}H_{14}O_3$	38.18	4
$C_{14}H_{14}O_3$	38.15	5
$C_{14}H_{14}O_4$	31.17	5
$C_{14}H_{14}O_4S_2$	39.21	4
$C_{14}H_{14}O_4S_2Zn$	85.23	5
$C_{14}H_{14}O_4S_4Te$	70.24	2
$C_{14}H_{14}O_4Se_2$	39.22	4
$C_{14}H_{14}O_8Zn$	78.12	2
$C_{14}H_{14}S$	11.65	1
$C_{14}H_{14}S$, I_2	60.139	2
$C_{14}H_{14}S_2$	11.66	1
$C_{14}H_{14}S_2$	11.25	3
$C_{14}H_{14}S_2$	11.23+	3
$C_{14}H_{14}S_2$	11.11	4
$C_{14}H_{14}Se$	11.67	1
$C_{14}H_{14}Te$	70.25	2
$C_{14}H_{15}BrO_3$	53.1	1
$C_{14}H_{15}BrO_3$	59.7	3
$C_{14}H_{15}BrO_3S$	31.17	1
$C_{14}H_{15}BrO_7$	59.25	1
$C_{14}H_{15}ClN_4O_5$	44.25	4
$C_{14}H_{15}Cl_2N_3O_3$	35.24	5
$C_{14}H_{15}NO_2$	21.14	3
$C_{14}H_{15}NO_4S$	11.11	5
$C_{14}H_{15}NS$	11.12	5

Formula	Reference
$C_{16}H_{11}Cl_2NO$	35.19 4
$C_{16}H_{11}FeO_4P$	86.12 3
$C_{16}H_{11}N_3O_3$	9.44 1
$C_{16}H_{11}N_4^+$, Cl^-, H_2O	36.16 3
$C_{16}H_{12}$	31.22 1
$C_{16}H_{12}$	30.1 3
$C_{16}H_{12}$	31.20 3
$C_{16}H_{12}$	24.6 5
$C_{16}H_{12}$	31.21 5
$0.5C_{16}H_{12}$, $C_{19}H_{12}FeO_3$	72.25 5
$C_{16}H_{12}As_2F_4I_2Mn_2O_8$	86.3 4
$C_{16}H_{12}As_2F_4Mn_2O_8$	86.14 5
$C_{16}H_{12}Br_2$	29.4 4
$C_{16}H_{12}Cl_2$	26.3 4
$C_{16}H_{12}Cl_2CuN_4$	83.81 3
$C_{16}H_{12}CrO_3$	75.46 2
$C_{16}H_{12}CuN_2O_6$	84.14 5
$C_{16}H_{12}Fe_2O_6$	71.37 5
$C_{16}H_{12}Fe_2O_6$	72.20 5
$C_{16}H_{12}Fe_2O_6$	75.25 5
$C_{16}H_{12}Fe_2O_8$	75.18 3
$C_{16}H_{12}N^+$, ClO_4^-	12.5 5
$C_{16}H_{12}NO^+$, HO_4Se^-	50.4 4
$C_{16}H_{12}N_4OS$	41.37 1
$C_{16}H_{12}OS_2$	39.27 4
$C_{16}H_{12}O_2$	19.10 3
$C_{16}H_{12}O_2$	26.4 4
$C_{16}H_{12}O_2$	27.14 4
$C_{16}H_{12}O_6$	50.6 3
$C_{16}H_{12}O_{12}Re_4S_4$	85.21 4
$C_{16}H_{13}BiN_2O_2S_2$	66.7 4
$C_{16}H_{13}Br$	26.37 1
$C_{16}H_{13}Br$	28.6 3
$C_{16}H_{13}Cl$	29.4 5
$C_{16}H_{13}ClN_2O$	35.20 4
$C_{16}H_{13}ClN_4O_4$	15.26 1
$C_{16}H_{13}MoN_3O_2S$	72.27 3
$C_{16}H_{13}NO_3$	1.33 3
$C_{16}H_{13}N_2O_2^+$, Br^-, $2H_2O$	58.12 3
$0.5C_{16}H_{14}$, $C_{19}H_{14}FeO_3$	60.51 3
$C_{16}H_{14}BrNO$	31.23 1
$C_{16}H_{14}BrN_5O_3$	37.23 1
$C_{16}H_{14}BrO_6$	38.21 4
$C_{16}H_{14}ClFeN_2O_2$	78.13 2
$C_{16}H_{14}ClFeN_2O_2$, CH_3NO_2	78.14 2
$C_{16}H_{14}Cl_2CoN_2$	83.61 5
$C_{16}H_{14}Cl_2Cu_2N_2O_2$	78.2 3
$C_{16}H_{14}Cl_2N_2$	35.21 4
$C_{16}H_{14}Cl_2N_2O_2Ti$, C_4H_8O	78.2 2
$C_{16}H_{14}Cl_2O$	31.21 3
$C_{16}H_{14}Cl_2S_2$	11.69 1
$C_{16}H_{14}Cl_4O_4$	28.4 5
$C_{16}H_{14}CoN_2O_2$, $CHCl_3$	78.15 2
$C_{16}H_{14}CrO_5$	74.11 3
$C_{16}H_{14}CuN_2O_2$	78.16 2
$C_{16}H_{14}CuN_2O_2$, $CHCl_3$	78.17 2
$C_{16}H_{14}CuN_2O_2$, $CHCl_3$	60.1 3
$C_{16}H_{14}CuN_2O_2$, $C_6H_5NO_3$	78.18 2
$C_{16}H_{14}CuN_2O_2$, $C_6H_5NO_3$	60.16 3
$2C_{16}H_{14}CuN_2O_2$, Na^+, ClO_4^-, C_8H_{10}	60.145 2
$C_{16}H_{14}CuO_{10}$	81.80 2
$C_{16}H_{14}F_2$	5.30 1
$C_{16}H_{14}MoS_2$	73.26 3
$C_{16}H_{14}N_2NiO_2$	78.3 4
$C_{16}H_{14}N_2OS$	41.35 5
$C_{16}H_{14}N_2O_2S$	39.31 5
$C_{16}H_{14}O$	38.55 1
$C_{16}H_{14}O_2$	17.6 3
$C_{16}H_{14}O_2$	6.2 4
$C_{16}H_{14}O_2S$	1.34 3
$C_{16}H_{14}S_2Ti$	73.27 5
$C_{16}H_{14}S_3$	39.38 3
$C_{16}H_{15}Br$	5.31 1
$C_{16}H_{15}Br_2N_3O_2Si$, C_3H_6O	63.19 2
$C_{16}H_{15}Co$	73.67 2
$C_{16}H_{15}Co_3OS$	73.28 5
$C_{16}H_{15}Co_3O_2$	73.27 3
$C_{16}H_{15}N$	37.11 3
$C_{16}H_{15}N_2O^+$, Br^-	32.45 1
$C_{16}H_{15}N_4NiS_2^+$, ClO_4^-, H_2O	83.62 5
$C_{16}H_{16}$	31.24 1
$C_{16}H_{16}$	31.25+ 1
$C_{16}H_{16}$	19.11 3
$C_{16}H_{16}$	23.8+ 3
$C_{16}H_{16}$	31.18 4
$C_{16}H_{16}$	31.22 5
$C_{16}H_{16}Ag^+$, NO_3^-	75.47 2
$C_{16}H_{16}Br_2N_2$	28.7 3
$C_{16}H_{16}Br_2O_2$	17.18 5
$C_{16}H_{16}Ce^-$, $C_6H_{14}KO_3^+$	75.26 5
$C_{16}H_{16}Cl_2Cu_2N_2O_2$	78.4 4
$C_{16}H_{16}Cl_4CuN_2O_4$	81.40 4
$C_{16}H_{16}Cl_4Pd_2$	72.47 2
$C_{16}H_{16}Co_2O_4$	75.27 5
$C_{16}H_{16}CuN_2O_2$	78.21 2
$C_{16}H_{16}CuN_2O_2$	78.19+ 2
$C_{16}H_{16}CuN_2O_2$	78.5 4
$C_{16}H_{16}CuN_2O_2$	78.3 5
$C_{16}H_{16}CuN_2S_4$	80.21 5
$C_{16}H_{16}CuN_4O_8^{2-}$, $2Cs^+$, $2H_2O$	83.63 5
$C_{16}H_{16}CuN_4O_8^{2-}$, $2K^+$	83.79 4
$C_{16}H_{16}CuN_4O_8^{2-}$, $2Li^+$	83.80 4
$C_{16}H_{16}Fe$	75.48 2
$C_{16}H_{16}Fe_2O_4Pb$	73.68 2
$C_{16}H_{16}Fe_2O_4Sn$	73.69 2
$C_{16}H_{16}HgN_2O_2S_2$	85.75 2
$C_{16}H_{16}HgO_2$	71.39 2
$C_{16}H_{16}IN$	37.13 4
$C_{16}H_{16}MoO_4S_2Ti$	73.28 3
$C_{16}H_{16}N_2^{2+}$, $2Br^-$, H_2O	31.23 5
$C_{16}H_{16}N_2NiO_2$	78.22+ 2
$C_{16}H_{16}N_2NiO_6$	83.144 2
$C_{16}H_{16}N_2NiS_2^-$, $C_8H_{20}N^+$	85.27 3
$C_{16}H_{16}N_2NiS_4$	80.22 5
$C_{16}H_{16}N_2O_2$	9.45 1
$C_{16}H_{16}N_2O_2Zn$	78.24 2
$C_{16}H_{16}N_2O_3Zn$	78.25 2
$C_{16}H_{16}N_2O_4S$, C_2H_6O	50.6 4
$C_{16}H_{16}N_2O_5S$	50.7 3
$C_{16}H_{16}N_4S$	41.36 5
$C_{16}H_{16}NiO_4$	72.48 2
$C_{16}H_{16}Ni_2$	75.28 5
$C_{16}H_{16}OS_2$	39.39 3
$C_{16}H_{16}O_3$	17.46 1
$C_{16}H_{16}O_4$	27.14 4
$C_{16}H_{16}Th$	75.14 4
$C_{16}H_{16}Ti$	75.19 3
$C_{16}H_{16}U$	75.15 4
$C_{16}H_{17}BCo_2NO_{10}$	71.40 2
$C_{16}H_{17}BrN_4O_3$	36.17 4
$C_{16}H_{17}Mo_2O_4P$	73.70 2

C_{19}

C_{21}

C_{25}

C_{26}

C_{27}

C_{28}

$C_{28}H_{30}NP_2{}^+$, I^-	64.76	2
$C_{28}H_{31}BrO_5$	51.35	1
$C_{28}H_{31}ClP_2Pt$	86.50	2
$C_{28}H_{31}IO_6$	56.1	1
$C_{28}H_{32}Cl_4Cu_4$	75.77	2
$C_{28}H_{32}P^+$, Br^-	64.41	5
$C_{28}H_{33}BrO_5S$	50.35	1
$C_{28}H_{33}ClIrOP_2{}^+$, $BF_4{}^-$, CH_2Cl_2	72.35	5
$C_{28}H_{33}IO_9$	56.2	1
$C_{28}H_{34}O_3$	17.11	3
$C_{28}H_{35}AsN_2O_3S_4U$	80.19	3
$C_{28}H_{35}N_2O_3Se_4U$	85.32	4
$C_{28}H_{35}N_2O_3PS_4U$	80.20	3
$C_{28}H_{36}Cl_2Ga_2N_6O_2{}^{2+}$, $2Cl^-$, H_2O	68.30	5
$C_{28}H_{36}NO_4{}^+$, Br^-	58.110	1
$C_{28}H_{36}N_4Ni^{2+}$, $2ClO_4{}^-$	83.198	2
$C_{28}H_{36}N_4Ni^{2+}$, $2ClO_4{}^-$	83.199	2
$C_{28}H_{36}N_8P_4$	64.37	4
$C_{28}H_{36}O_2$	30.7	4
$C_{28}H_{36}O_3$, $0.5H_2O$	56.2	5
$C_{28}H_{37}BrNO_5{}^+$, Br^- , C_2H_6O	58.42	5
$C_{28}H_{37}Cl_3NOP_2Re$	83.200	2
$C_{28}H_{38}BrN_5O_8$, $0.75C_4H_8O_2$, H_2O	48.57	4
$C_{28}H_{38}Cl_2NiP_2$	86.46	3
$C_{28}H_{38}CoN_7O_2$	78.20	5
$C_{28}H_{38}CrN_4Si_2{}^+$, I^-	63.13	5
$C_{28}H_{38}NO_{12}$, $1.5H_2O$	52.6	4
$C_{28}H_{38}O_7$, $2H_2O$	51.49	4
$C_{28}H_{39}BrO_5$	55.2	1
$C_{28}H_{39}BrO_9$	54.12	3
$C_{28}H_{39}BrO_{16}$	52.31	1
$C_{28}H_{40}Al_2Ti_2$	73.111	2
$C_{28}H_{40}KO_{10}{}^+$, I^-	67.25	4
$C_{28}H_{40}KO_{10}{}^+$, I^-	38.31	5
$C_{28}H_{40}NiP_2$	86.51+	2
$C_{28}H_{40}O_{10}$	38.26	4
$C_{28}H_{41}BrO_4$	54.20	1
$C_{28}H_{42}AsN_5NiS_2$	76.86	3
$C_{28}H_{42}Br_2O_9$	54.13	3
$C_{28}H_{43}IO_4$	51.40	1
$C_{28}H_{44}Br_2O$	51.37+	3
$C_{28}H_{44}Cu_4N_4O_8$	77.25	3
$C_{28}H_{44}O_2$	51.43	5
$C_{28}H_{46}Co_4O_{16}$	77.16	5
$C_{28}H_{46}NO_3{}^+$, Br^-	58.111	1
$C_{28}H_{46}NO_3{}^+$, Br^-	58.60	3
$C_{28}H_{48}Cd_2N_4S_8$	80.28	5
$C_{28}H_{48}N_4S_8Zn_2$	80.29	5
$C_{28}H_{48}N_8NiS_4{}^{2+}$, $2I^-$	79.12	4
$C_{28}H_{49}NiP$	71.65	4
$C_{28}H_{52}$, Ag^+ , $NO_3{}^-$	23.6	4
$C_{28}H_{52}Ni_2P_2$	72.36	5
$C_{28}H_{55}BrO$	51.50	4
$C_{28}H_{55}BrO$	51.51	4
$C_{28}H_{56}Mo_2N_4S_8$	71.66	4
$C_{28}H_{58}Cl_6Mg_4O_6$	67.26	4
$C_{28}H_{64}N_6Ni_2O_2{}^{2+}$, $2ClO_4{}^-$	76.87	3
$C_{28}H_{72}O_{16}Ti_4$	84.58	2
$C_{28}H_{72}O_{16}Ti_4$	84.22	3

C_{29}

$C_{29}H_{15}Fe_3O_{11}P$, $C_{29}H_{15}Fe_3O_{11}P$	86.53	2

$C_{29}H_{18}Br_2O_2S_2$	39.46	4
$C_{29}H_{18}Fe_2N_2O_6$	83.137	3
$C_{29}H_{20}Br$	20.12	4
$C_{29}H_{20}F_5NiP$	73.112	2
$C_{29}H_{20}F_6FeN_2O_2P_2$	86.45	3
$C_{29}H_{21}BrO_{11}$	59.60	1
$C_{29}H_{22}$	20.18	3
$C_{29}H_{22}MoO_4P_2$	86.46	3
$C_{29}H_{23}MnO_3P_2S_2$	85.32	5
$C_{29}H_{24}Br_2MoO_3P_2$, C_3H_6O	86.23	4
$C_{29}H_{25}NiP$	73.113	2
$C_{29}H_{25}O_3P$	64.38	4
$C_{29}H_{28}Br_2O_6$	53.39	1
$C_{29}H_{30}BrNO_{11}$, C_3H_6O	50.11	5
$C_{29}H_{30}ClIrO_3P_2$	86.24	4
$C_{29}H_{32}BrClO_6S$	50.18	3
$C_{29}H_{32}BrNO_9$	59.61	1
$C_{29}H_{33}BrClO_6S$	50.10	4
$C_{29}H_{33}Cl_3NP_2Re$	86.47	3
$C_{29}H_{34}BrFO_5$	51.39	3
$C_{29}H_{34}CoN_7$	49.2	3
$C_{29}H_{34}N_4$	49.3	3
$C_{29}H_{37}AgNO_5{}^+$, $BF_4{}^-$, $2C_3H_6O$	50.19	3
$C_{29}H_{39}BrN_2O_5S$	51.36	1
$C_{29}H_{39}NO_9{}^+$, I^-	58.35	4
$C_{29}H_{43}IO_4$	51.37	1
$C_{29}H_{44}Br_2O_4$	51.38	1
$C_{29}H_{45}BrO_4$	51.39	1
$C_{29}H_{46}FeN_9O_{14}$, H_2O	82.28	5
$C_{29}H_{47}NO_5$, CH_4O	58.61	3
$C_{29}H_{48}O_2S$	51.41	1
$C_{29}H_{50}$	56.3	5
$C_{29}H_{50}NO_3{}^+$, Br^- , H_2O	58.43	5

C_{30-34}

$C_{30}Co_8O_{24}$	71.59	3
$C_{30}Co_8O_{24}$, $0.5C_6H_8$	71.60	3
$C_{30}H_{14}$	30.23	1
$C_{30}H_{14}O_2$	30.24	1
$C_{30}H_{14}O_2$	30.25	1
$C_{30}H_{15}AuClF_{10}P$	71.52	2
$C_{30}H_{16}$	30.26	1
$C_{30}H_{16}$	30.8	4
$C_{30}H_{16}Fe_4O_{10}$, $C_2H_4Cl_2$	75.43	3
$C_{30}H_{18}Br_2O_7$	59.62	1
$C_{30}H_{18}Cl_2$	29.28	1
$C_{30}H_{18}Fe_2O_6$	72.69	2
$C_{30}H_{18}O_2$	30.27	1
$C_{30}H_{20}$	31.37+	3
$C_{30}H_{20}Br_2$	31.43	5
$C_{30}H_{20}Cl_2Cu_2Fe_2O_4$	71.74	5
$C_{30}H_{20}F_{12}O_2S$	11.16	5
$C_{30}H_{20}Ni_2O_6P_2$	86.55	2
$C_{30}H_{20}O_2$	31.48	1
$C_{30}H_{20}S_2$	39.47	4
$C_{30}H_{22}Br_2O_{10}$, $2CH_4O$, H_2O	59.29	3
$C_{30}H_{22}Cl_2$	20.13	4
$C_{30}H_{22}CuN_6{}^{2+}$, $2NO_3{}^-$	83.109	5
$C_{30}H_{22}CuO_4$	77.59	2
$C_{30}H_{22}CuO_4$	77.26	3
$C_{30}H_{22}O_2PdS_2$	85.92	2
$C_{30}H_{22}O_4Pd$	77.60	2

$C_{36}H_{42}N_{12}Ni_3O_{12}$, C_6H_6	83.215	2	$C_{38}H_{20}F_6Fe_3O_8P_2$, $0.5C_6H_6$	71.90 5
$C_{36}H_{44}Al_4Mo_2$	68.34	5	$C_{38}H_{24}O_8Os_3P_2$	71.73 4
$C_{36}H_{44}Cl_2N_4Sn$	49.9	3	$C_{38}H_{26}Cl_5F_5NiP_2$	71.74 4
$C_{36}H_{44}N_4Ni$	49.2+	5	$C_{38}H_{26}F_{10}NiP_2$	71.75 4
$C_{36}H_{44}P_2Pt$	71.86	5	$C_{38}H_{28}^{2+}$, $2Cl_6Sb^-$	12.11 3
$C_{36}H_{44}P_2Pt$	71.87	5	$C_{38}H_{30}ClF_3P_2Pt$	72.22 4
$C_{36}H_{44}P_2Pt$	71.88	5	$C_{38}H_{30}ClF_4P_2Rh$	86.84 2
$C_{36}H_{46}N_2$	21.20	3	$C_{38}H_{30}ClIrO_3P_2$, C_6H_6	86.73 3
$C_{36}H_{46}N_4$	49.4	5	$C_{38}H_{30}ClP_2Rh$	72.40 5
$C_{36}H_{47}N_6^{3+}$, $3I^-$	58.118	1	$C_{38}H_{30}Cl_2F_2P_2Pt$	72.23 4
$C_{36}H_{50}N_6Rh^+$, Cl^-, $2H_2O$	49.5	5	$C_{38}H_{30}Cl_4P_2Pt$	72.24 4
$C_{36}H_{51}CrO_6$	77.27	3	$C_{38}H_{30}FeOSn$	72.41 5
$C_{36}H_{51}FO_2$	51.57	4	$C_{38}H_{30}MnNO_3P_2$	86.85 2
$C_{36}H_{52}O_{12}$	38.32	5	$C_{38}H_{30}NO_3OsP_2^+$, ClO_4^-, CH_2Cl_2	86.35 4
$C_{36}H_{54}F_{10}P_2PtS_2$	85.53	3	$C_{38}H_{30}N_4O_5STh$, C_2H_6OS	83.145 3
$C_{36}H_{60}O_{30}$, I_2	61.6	5	$C_{38}H_{31}Cl_2F_2IrP_2$	71.91 5
$C_{36}H_{60}O_{30}$, $2H_2O$	61.7	5	$C_{38}H_{31}IrO_2P_2$	86.74 3
$C_{36}H_{60}O_{30}$, CH_4O	61.8	5	$C_{38}H_{31}N_9Zn$, $0.5C_6H_{15}N$	49.13 4
$C_{36}H_{60}O_{30}$, C_3H_8O	61.9	5	$C_{38}H_{32}Fe_2O_{10}S_2Sn_2$	73.33 4
$C_{36}H_{61}AgO_{11}$, $2H_2O$	50.25	3	$C_{38}H_{33}Cl_3NP_2Re$	86.75 3
$C_{36}H_{62}O_8P_4Ru$	86.55	5	$C_{38}H_{34}Br_2NiP_2$	86.76 3
$C_{36}H_{62}O_{11}$, H_2O	50.11	4	$C_{38}H_{34}Br_2O_2$	20.14 4
$C_{36}H_{63}Br_3O_6$	1.156	1	$C_{38}H_{34}F_6FeOP_2Sn$	73.55 3
$C_{36}H_{66}Cl_2NiP_2$	86.74	2	$C_{38}H_{34}F_6FeOP_2Sn$	73.32 4
$C_{36}H_{66}CuNO_3P_2$	86.34	4	$C_{38}H_{34}NiP_2$	72.77 2
$C_{36}H_{72}AlLiN_4$	68.13	4	$C_{38}H_{34}NiP_2$	72.25 4
$C_{36}H_{74}$	5.42+	1	$C_{38}H_{34}P_2Pt$	72.42 5
$C_{36}H_{84}Pd_6S_{12}$	85.95	2	$C_{38}H_{35}NO_2Si_3$	63.30 2
$C_{37}H_{20}O_9Os_3$	72.38+	5	$C_{38}H_{36}Cl_2CoN_2O_2P_2$	84.28 5
$C_{37}H_{20}O_7Os_3P_2$	71.71	4	$C_{38}H_{38}N_2P_2Pd_2S_2$	86.58 5
$C_{37}H_{27}AsO_2$	65.12	4	$C_{38}H_{41}N_2O_7^+$, Br^-	58.41 4
$C_{37}H_{30}ClIrNO_2P_2^+$, BF_4^-	86.75	2	$C_{38}H_{42}Co_2N_6O_8$	78.26 3
$C_{37}H_{30}ClIrO_3P_2$	86.76	2	$C_{38}H_{44}$	19.20 5
$C_{37}H_{30}ClIrO_3P_2S$	86.77	2	$C_{38}H_{46}BrClO_{13}$, C_2H_6O	54.10 5
$C_{37}H_{30}ClO_3P_2RhS$	86.68	3	$C_{38}H_{46}CuN_4O_8$	83.146 3
$C_{37}H_{30}ClP_2RhS$	86.78	2	$C_{38}H_{46}N_4O_8Pd$	83.147 3
$C_{37}H_{30}CoNO_2P_2$	86.56	5	$C_{38}H_{52}CoP_2$	86.86 2
$C_{37}H_{30}FP_2PtS_2^{2+}$, HF_2^-	85.34	4	$C_{38}H_{53}BrO_5$	56.6 4
$C_{37}H_{30}IIrNO_2P_2^+$, BF_4^-, C_6H_6	86.69	3	$C_{38}H_{56}Cl_2HgO_4$	84.64 2
$C_{37}H_{30}IrO_3P_2$, CH_2Cl_2	86.79	2	$C_{38}H_{67}EuO_7S$	77.19 5
$C_{37}H_{30}IrNO_2P_2$	86.57	5	$C_{38}H_{95}O_{24}Ti_7$	84.15 4
$C_{37}H_{30}O_3P_2Pt$	86.80	2	$C_{39}H_{24}IrO_2P_2S^+$, F_6P^-, C_3H_6O	86.59 5
$C_{37}H_{30}P_2$	64.43	4	$C_{39}H_{27}Fe_2N_2O_5P$	86.36 4
$C_{37}H_{30}P_2PdS_2$	86.81	2	$C_{39}H_{30}ClIrN_2OP_2S$	86.60 5
$C_{37}H_{30}P_2PtS_2$	86.70	2	$C_{39}H_{30}CoGeO_3P$	69.39 3
$C_{37}H_{33}IIrNOP_2$	71.66	3	$C_{39}H_{30}F_4IIrOP_2$	86.87 2
$C_{37}H_{33}IO_2P_2PtS$	71.72	4	$C_{39}H_{30}F_6NiOP_2$	71.92 5
$C_{37}H_{33}IP_2Pt$, O_2S	71.89	5	$C_{39}H_{30}F_{12}NiOP_2$	72.43 5
$C_{37}H_{33}I_2P_2Rh$, C_6H_6	71.55	2	$C_{39}H_{30}O_3OsP_2$	86.88 2
$C_{37}H_{35}MoN_2O_3P_3$	86.71	3	$C_{39}H_{33}Br_2P_3Pt$	86.37 4
$C_{37}H_{37}BrF_2N_2O_5$	51.45	3	$C_{39}H_{34}IP_2Rh$	72.78 2
$C_{37}H_{40}BrNO_7$, C_3H_8O	50.12	4	$C_{39}H_{34}N_8OSi_2$	49.14 4
$C_{37}H_{40}BrNO_7$, $0.5C_3H_8O$	50.42	1	$C_{39}H_{34}P_2Pt$	72.27 4
$C_{37}H_{42}N_2O_6^{2+}$, $2Cl^-$, $5H_2O$	58.46	5	$C_{39}H_{35}Cl_5O_3P_2Re_2$	81.89 2
$C_{37}H_{42}N_4O_7^{2+}$, $2Br^-$	58.65	3	$C_{39}H_{35}IP_2Pd_2$, C_6H_6	72.28 4
$C_{37}H_{43}BrO_8$, $0.5C_4H_8O_2$	51.48	1	$C_{39}H_{36}O_3P_2Pt$	86.89 2
$C_{37}H_{43}FeN_4O_5$	49.19	1	$C_{39}H_{39}BrF_2N_2O_6$, $2C_4H_{10}O$	51.46 3
$C_{37}H_{44}O_5P_4Pt_4$	86.72	3	$C_{39}H_{39}BrO_9S$	59.68 1
$C_{37}H_{51}BrClN_3O_{10}$	58.47	5	$C_{39}H_{39}CuNO_3P_3$	86.61 5
$C_{37}H_{51}IO_{10}S$	59.33	3	$(C_{39}H_{42}N_6Na^+)_n$, nCl^-	67.27+ 4
$C_{37}H_{52}BrNO$, C_2H_6O	56.10	3	$C_{39}H_{44}Br_2N_4O_{10}$	59.27 4
$C_{37}H_{53}BrO_4$, $0.5H_2O$	56.20	1	$C_{39}H_{51}Br_2NO_{11}$, CH_4O	58.120 1
$C_{37}H_{55}BrO_3$	56.5	4	$C_{39}H_{51}Br_2NO_{11}$, C_2H_6O	59.28 4
$C_{37}H_{58}O_8$, C_6H_6	51.49	1	$C_{39}H_{53}BrO_7$	51.50 1
$C_{37}H_{68}NO_{13}^+$, I^-, $2H_2O$	50.43	1	$C_{39}H_{55}IO_5$	51.58 4
$C_{38}H_{18}$	30.33	1	$C_{39}H_{57}BrO_4$	56.12 5

C_{50-99}

$C_{100-149}$

$C_{150-199}$

Ag

Vol. 1 53.10 **Vol. 2** 72.29, 72.53, 74.1, 74.2, 75.2, 75.5, 75.9, 75.16, 75.35, 75.47, 75.67, 75.78, 76.20, 77.19, 79.3, 79.16, 80.41, 81.2, 82.1, 83.16, 83.17, 83.27, 83.44, 83.161, 84.23, 85.2, 85.9, 85.10, 85.11, 85.14, 85.36, 85.64, 85.67, 86.36 **Vol. 3** 50.19, 53.7, 53.15, 53.29. 74.1, 74.6, 74.7, 74.12, 74.13, 74.14, 75.2, 75.4, 75.8, 75.9, 75.37, 76.11, 76.12, 80.21, 81.36, 83.16, 83.61, 86.32 **Vol. 4** 23.4, 23.6, 31.1, 31.13, 74.3, 74.8, 74.9, 75.19, 80.11, 80.20. 81.36, 81.55, 82.1, 82.2, 82.4, 82.5, 82.8, 83.15, 83.16, 83.35, 83.58 **Vol. 5** 71.34, 76.9, 79.14, 81.7, 82.2, 83.22, 83.79, 84.2, 86.81

Am

Vol. 3 77.16

Au

Vol. 2 64.52, 71.16, 71.18, 71.23, 71.52, 73.109, 80.2, 80.8, 80.34, 83.154, 86.1, 86.90 **Vol. 3** 80.11, 86.22, 86.26, 86.99 **Vol. 4** 71.34, 71.43, 71.46, 71.77, 80.11, 80.12, 80.16, 86.71 **Vol. 5** 71.22, 71.60, 71.104, 80.11, 80.15, 80.18, 85.18, 85.19, 85.22, 86.21, 86.27, 86.85, 86.86

Cd

Vol. 2 79.2, 79.5, 79.6, 79.20, 79.21, 79.33, 80.10, 80.11, 80.35, 80.36, 81.61, 82.45, 83.4, 83.5, 83.28, 83.74, 84.2, 84.3, 84.5, 84.8, 85.37, 85.94 **Vol. 3** 61.4, 76.39, 79.4, 79.17, 83.38, 83.39, 84.2, 85.41, 86.58 **Vol. 4** 76.13, 79.4, 81.17, 83.5, 83.7, 83.50, 83.91, 83.92, 83.93, 84.9 **Vol. 5** 61.2, 76.16, 76.33, 76.42, 78.19, 80.2, 80.28, 83.5, 83.36, 83.52, 83.104, 83.106, 84.15

Ce

Vol. 2 77.47, 81.44, 81.67 **Vol. 3** 77.19, 81.11, 84.26 **Vol. 4** 75.31, 81.43 **Vol. 5** 75.26

Co

Vol. 1 3.31 **Vol. 2** 62.33, 71.22, 71.40, 71.44, 71.45, 72.25, 72.35, 72.39, 72.46, 72.52, 72.56, 72.67, 73.9, 73.10, 73.19, 73.31, 73.34, 73.46, 73.58, 73.67, 73.73, 73.108, 73.117, 75.23, 75.55, 75.61, 76.2, 76.4, 76.7, 76.8, 76.9, 76.10, 76.11, 76.12, 76.24, 76.28, 76.34, 76.35, 76.36, 76.37, 76.38, 76.40, 76.49, 76.54, 76.55, 76.56, 76.58, 76.59, 76.75, 76.79, 77.11, 77.20, 77.21, 77.44, 77.51, 77.64, 77.67, 78.15, 78.59, 79.24, 80.5, 80.7, 80.29, 80.30, 81.5, 81.12, 81.34, 81.50, 82.2, 82.24, 82.28, 82.34, 82.37, 82.46, 82.54, 83.61, 83.75, 83.80, 83.89, 83.91, 83.95, 83.103, 83.124, 83.130, 83.131, 83.134, 83.136, 83.146, 83.176, 83.213, 84.12, 84.21, 84.66, 84.68, 85.39, 85.45, 85.55, 85.63, 85.73, 85.74, 85.82, 85.84, 85.86, 86.23, 86.32, 86.44, 86.60, 86.61, 86.73, 86.86, 86.94, 86.112
Vol. 3 62.12, 69.39, 71.23, 71.25, 71.35, 71.38, 71.55, 71.59, 71.60, 71.64, 72.29, 72.39, 73.27, 73.33, 73.37, 76.2, 76.3, 76.6, 76.7, 76.9, 76.10, 76.17, 76.19, 76.23, 76.27, 76.28, 76.30, 76.31, 76.38, 76.40, 76.41, 76.52, 76.54, 76.55, 76.58, 76.61, 76.62, 76.63, 76.64, 76.71, 76.74, 76.75, 76.81, 76.82, 76.85, 77.8, 78.12, 78.14, 78.16, 78.19, 78.26, 78.30, 79.2, 79.5, 79.6, 79.7, 79.21, 81.17, 82.7, 82.12, 82.13, 82.14, 82.15, 82.22, 83.20, 83.26, 83.34, 83.49, 83.51, 83.65, 83.76, 83.88, 83.93, 83.100, 83.115, 83.121, 83.142, 83.148, 83.150, 84.18, 84.24, 85.35, 85.40, 85.45, 86.11, 86.27, 86.32, 86.33, 86.77, 86.93
Vol. 4 60.9, 71.2, 71.40, 71.42, 71.57, 72.29, 73.17, 75.5, 75.20, 75.29, 76.3, 76.10, 76.11, 76.13, 76.14, 76.17, 76.18, 76.19, 76.20, 76.21, 76.22, 76.23, 76.24, 76.25, 76.28, 76.29, 76.30, 76.31, 76.32, 76.39, 76.40, 76.43, 77.9, 77.14, 77.16, 77.20, 78.13, 78.14, 78.15, 79.1, 79.10, 80.1, 81.15, 81.24, 81.30, 81.31, 81.50, 82.10, 83.13, 83.21, 83.24, 83.25, 83.31, 83.32, 83.56, 83.57, 83.67, 83.74, 83.82, 83.94, 83.95, 83.96, 83.97, 83.98, 83.102, 83.120, 83.122, 83.129, 83.130, 84.1, 85.5, 85.22, 86.9, 86.11, 86.13, 86.18, 86.44, 86.48, 86.56 **Vol. 5** 62.9, 67.22, 71.13, 71.15, 71.24, 71.31, 71.42, 71.44, 71.52, 71.57, 71.66, 73.1, 73.2, 73.3, 73.22, 73.23, 73.24, 73.25, 73.28, 73.29, 73.34, 73.35, 73.36, 74.6, 75.10, 75.11, 75.14, 75.27, 75.35, 75.40, 75.41, 75.42, 75.44, 75.45, 75.47, 76.2, 76.4, 76.14, 76.15, 76.17, 76.18, 76.20, 76.21, 76.22, 76.25, 76.26, 76.27, 76.29, 76.30, 76.31, 76.39. 76.40, 76.43, 76.48, 76.49, 76.50, 76.52, 77.3, 77.16, 77.24, 78.5, 78.6, 78.9, 78.13, 78.14, 78.16, 78.20, 79.5, 79.12, 79.23, 79.26, 80.5, 81.6, 81.14, 81.16, 81.36, 81.61, 81.62, 82.10, 82.20, 83.1, 83.6, 83.26, 83.35, 83.41, 83.48, 83.55, 83.60, 83.61, 83.89, 83.91, 83.93, 83.96, 83.105, 83.116, 84.28, 84.30, 86.22, 86.24, 86.25, 86.27, 86.28, 86.32, 86.36, 86.43, 86.50, 86.56, 86.64, 86.72, 86.73, 86.74

Cr

Cu

Dy

Er

Eu

Fe

Gd

Hf

Hg

Ho

Ir

La

Lu

Mn

Vol. 1 3.57 **Vol. 2** 69.14, 69.40. 71.10, 72.51, 73.8, 73.36, 73.41, 73.63, 73.121, 75.34, 75.42, 76.65, 76.77, 77.12, 77.24, 81.9, 81.10, 82.8, 82.56, 83.7, 83.115, 83.153, 84.16, 84.25, 86.25, 86.28, 86.39, 86.41, 86.85, 86.90, 86.91 **Vol. 3** 71.2, 71.14, 75.1, 77.6, 77.31, 81.43, 83.124, 84.10, 86.36 **Vol. 4** 71.50, 71.56, 72.15, 73.6, 76.35, 77.18, 83.81, 83.133, 83.134, 86.3, 86.15, 86.41 **Vol. 5** 71.3, 71.4, 71.5, 71.10, 71.26, 71.39, 71.40, 71.55, 71.61, 71.98, 72.13, 75.43, 78.22, 80.19. 81.15, 81.17, 81.18, 81.60, 83.21, 83.106, 84.21, 84.22, 84.32, 85.14, 85.32, 85.37, 86.8, 86.11, 86.12, 86.14, 86.20

Mo

Vol. 2 71.27, 71.51, 72.66, 73.6, 73.26, 73.27, 73.29, 73.30, 73.32, 73.40, 73.41, 73.66, 73.70, 73.75, 73.96, 74.17, 75.12, 75.20, 75.43, 75.44, 75.69, 75.76, 76.3, 76.61, 77.53, 80.23, 81.1, 81.24, 81.25, 81.59, 82.14, 83.38, 84.18, 85.34, 85.78, 86.17, 86.43, 86.56 **Vol. 3** 64.52, 69.23, 71.6, 71.16, 71.27, 72.12, 72.27, 73.18, 73.19, 73.20, 73.26, 73.28, 73.46, 73.48, 73.51, 73.52, 73.56, 75.25, 75.39, 75.49, 80.5, 80.17, 81.1, 82.16, 82.19, 82.23, 83.17, 83.35, 83.86, 85.17, 86.9, 86.46, 86.71, 86.88 **Vol. 4** 71.10, 71.11, 71.15, 71.53, 71.66, 71.69, 72.9, 72.12, 73.14, 73.19, 73.22, 82.13, 83.46, 83.87, 84.3, 84.10. 85.12, 85.24, 86.1, 86.6, 86.19, 86.23, 86.26, 86.47, 86.52 **Vol. 5** 68.27, 68.29, 71.18, 71.29, 72.27, 73.6, 73.11, 73.13, 73.14, 73.16, 73.17, 73.18, 75.2, 75.24, 75.31, 76.36, 77.2, 80.10. 81.28, 81.50, 82.7, 82.21, 83.53, 84.3, 84.16, 86.5, 86.6, 86.9, 86.33, 86.34, 86.40, 86.53, 86.77

Nb

Vol. 2 72.72, 72.84, 73.11, 83.18, 84.65 **Vol. 3** 72.40, 72.48, 72.53, 83.53, 83.57, 83.129 **Vol. 4** 71.52, 73.29, 80.6, 80.7, 81.3, 81.11, 81.19, 83.47, 83.48 **Vol. 5** 71.48, 73.12, 77.23, 81.10, 83.2

Nd

Vol. 2 77.33, 81.41, 81.43, 81.71, 81.72, 82.52 **Vol. 3** 81.13, 81.14, 81.27, 82.9 **Vol. 4** 77.12, 81.18, 81.51 **Vol. 5** 71.108, 76.37, 84.10

Ni

Vol. 1 3.80 **Vol. 2** 60.155, 65.42, 65.46, 65.47, 72.15, 72.22, 72.27, 72.48, 72.64, 72.77, 73.64,

73.74, 73.82, 73.91, 73.112, 73.113, 73.118, 75.28, 75.53, 75.54, 75.62, 75.63, 76.6, 76.20, 76.21, 76.22, 76.27, 76.29, 76.31, 76.32, 76.39, 76.41, 76.42, 76.43, 76.47, 76.50, 76.69, 76.70, 76.74, 77.13, 77.19, 77.45, 77.46, 77.56, 77.62, 78.9, 78.10, 78.11, 78.22, 78.23, 78.30, 78.32, 78.36, 78.42, 78.43, 78.47, 78.49, 78.52, 78.53, 79.18, 79.23, 79.26, 79.28, 79.31, 79.34, 79.35, 79.36, 80.1, 80.3, 80.17, 80.27, 80.28, 81.11, 81.36, 82.6, 82.19, 82.22, 82.27, 82.30, 82.31, 82.33, 82.48, 83.11, 83.12, 83.13, 83.14, 83.24, 83.25, 83.29, 83.30, 83.31, 83.32, 83.40, 83.43, 83.51, 83.57, 83.83, 83.84, 83.85, 83.86, 83.96, 83.97, 83.99, 83.102, 83.105, 83.106, 83.107, 83.118, 83.120, 83.127, 83.128, 83.138, 83.143, 83.144, 83.165, 83.175, 83.177, 83.184, 83.186, 83.198, 83.199, 83.204, 83.208, 83.215, 83.218, 83.219, 84.7, 84.41, 84.60, 85.1, 85.5, 85.6, 85.7, 85.8, 85.15, 85.16, 85.18, 85.19, 85.24, 85.25, 85.30, 85.32. 85.33, 85.35, 85.38, 85.43, 85.47, 85.49, 85.50, 85.56, 85.57, 85.58, 85.59, 85.60, 85.70, 85.76, 85.81, 85.88, 85.91, 85.93, 85.96, 86.7, 86.9, 86.20, 86.21, 86.26, 86.33, 86.49, 86.51, 86.52, 86.55, 86.65, 86.67, 86.74, 86.97, 86.99 **Vol. 3** 22.2, 60.54, 61.4, 61.8, 71.3, 71.20, 71.43, 72.8, 72.17, 72.43, 73.43, 75.30, 76.11, 76.12, 76.15, 76.21, 76.43, 76.44, 76.49, 76.50, 76.51, 76.56, 76.60, 76.65, 76.67, 76.83, 76.84, 76.86, 76.87, 77.33, 78.6, 78.15, 78.28, 79.13, 79.14, 79.26, 80.8, 80.18, 81.8, 81.16, 81.42, 81.48, 83.3, 83.4, 83.21, 83.30, 83.31, 83.45, 83.48, 83.63, 83.64, 83.71, 83.73, 83.80, 83.84, 83.85, 83.87, 83.90, 83.92, 83.94, 83.95, 83.112, 83.113, 83.120, 83.125, 83.131, 83.133, 83.134, 83.135, 83.136, 83.141, 83.142, 83.143, 84.12, 85.2, 85.3, 85.4, 85.7, 85.12, 85.14, 85.20, 85.23, 85.27, 85.28, 85.29, 85.36, 85.38, 85.43, 85.46, 85.47, 85.50, 85.51, 86.4, 86.14, 86.16, 86.31, 86.39, 86.41, 86.43, 86.44, 86.48, 86.52, 86.53, 86.56, 86.76, 86.80, 86.81, 86.82, 86.85, 86.86 **Vol. 4** 60.11, 71.32, 71.48, 71.65, 71.74, 71.75, 72.13, 72.25, 73.15, 73.29, 75.27, 75.30, 76.7, 76.15, 76.27, 76.42, 76.44, 77.3, 77.4, 77.6, 77.17, 78.3, 78.9, 79.5, 79.12, 80.18, 81.22, 81.33, 81.34, 81.37, 82.3, 82.11, 83.6, 83.10, 83.21, 83.30, 83.33, 83.41, 83.42, 83.44, 83.45, 83.60, 83.73, 83.83, 83.84, 83.99, 83.101, 83.111, 83.113, 83.118, 83.119, 83.125, 83.126, 83.127, 83.131, 83.133, 83.136, 84.7, 85.2, 85.10, 85.15, 85.16, 85.17, 85.20, 85.23, 85.37, 86.17, 86.27, 86.30, 86.46, 86.66 **Vol. 5** 12.11, 61.2, 71.30, 71.41, 71.43, 71.45, 71.49, 71.59, 71.62, 71.73, 71.82, 71.92, 72.30, 72.31, 72.33, 72.34, 72.36, 72.37, 72.43, 72.48, 72.49, 72.50, 72.53, 72.54, 75.13, 75.28, 76.8, 76.12, 76.23, 76.44, 76.46, 76.51, 76.53, 77.12, 78.17, 79.2. 79.3, 79.6, 79.8, 79.9, 79.20, 79.21, 79.27, 80.1. 80.4, 80.7, 80.14, 80.17, 80.20, 80.22, 80.23, 81.44, 81.45, 81.48, 82.15, 82.16, 82.19, 82.26. 83.10, 83.27, 83.28, 83.29, 83.30, 83.31, 83.42, 83.44, 83.45, 83.59, 83.62, 83.67, 83.68, 83.73, 83.85, 83.94, 83.95, 83.102, 83.112, 83.121, 84.13, 84.24, 85.2, 85.6, 85.8, 85.15, 85.24, 85.26, 85.30, 86.19, 86.26, 86.41, 86.42, 86.44, 86.46, 86.49, 86.63